Methods of
Enzymatic Analysis

Methods of Enzymatic Analysis

Third Edition

Editor-in-Chief: Hans Ulrich Bergmeyer
Editors: Jürgen Bergmeyer and Marianne Graßl

Volume I	Fundamentals
Volume II	Samples, Reagents, Assessment of Results
Volume III	Enzymes 1: Oxidoreductases, Transferases
Volume IV	Enzymes 2: Esterases, Glycosidases, Lyases, Ligases
Volume V	Enzymes 3: Peptidases, Proteinases and Their Inhibitors
Volume VI	Metabolites 1: Carbohydrates
Volume VII	Metabolites 2: Tri- and Dicarboxylic Acids, Purine and Pyrimidine Derivatives, Coenzymes, Inorganic Compounds
Volume VIII	Metabolites 3: Lipids, Steroids, Drugs
Volume IX	Metabolites 4: Proteins, Peptides, and Amino Acids
Volume X	Molecular Biologicals (Tentative Title)

Weinheim · Deerfield Beach, Florida · Basel

Methods of Enzymatic Analysis

Third Edition

Editor-in-Chief: Hans Ulrich Bergmeyer
Editors: Jürgen Bergmeyer and Marianne Graßl

Volume IV
Enzymes 2: Esterases, Glycosidases, Lyases, Ligases

Editorial Consultant: Donald W. Moss

Weinheim · Deerfield Beach, Florida · Basel

Editor-in-Chief:
Prof. Dr. rer. nat. Hans Ulrich Bergmeyer
Hauptstraße 88
D-8132 Tutzing
Federal Republic of Germany

Editorial Consultant and Language Editor:
Prof. Donald W. Moss, Ph. D., D. Sc.
Appletrees 42
Greenways Hinchlay Wood
Surrey KT 10 OQD, United Kingdom

Editors:
Dr. rer. nat. Jürgen Bergmeyer
In der Neckarhelle 168
D-6900 Heidelberg
Federal Republic of Germany

Dr. rer. nat. Marianne Graßl
Frauenchiemsee-Straße 20
D-8000 München 80
Federal Republic of Germany

Note

The methods published in this book have not been checked experimentally by the editors. Sole responsibility for the accuracy of the contents of the contributions and the literature cited rests with the authors. Readers are therefore requested to direct all enquiries to the authors (addresses are listed on pp. XVII – XXIII).

Previous editions of "Methods of Enzymatic Analysis":
1st Edition 1963, one volume
2nd printing, revised, 1965
3rd printing, 1968
4th printing, 1971

2nd Edition 1974, four volumes
2nd printing, 1977
3nd printing, 1981

Previous editions of "Methoden der enzymatischen Analyse":
1. Auflage 1962, one volume
2. neubearbeitete und erweiterte Auflage 1970, two volumes
3. neubearbeitete und erweiterte Auflage 1974, two volumes

Deutsche Bibliothek, Cataloguing-in-Publication Data

Methods of enzymatic analysis / ed.-in-chief: Hans Ulrich Bergmeyer. Ed.: Jürgen Bergmeyer and Marianne Graßl. – Weinheim; Deerfield Beach, Florida; Basel: Verlag Chemie
Dt. Ausg. u. d. T.: Methoden der enzymatischen Analyse
NE: Bergmeyer, Hans Ulrich [Hrsg.]
Vol. IV. Enzymes 2: esterases, glycosidases, lyases, ligases / ed. consultant: Donald W. Moss. – 3. ed.
ISBN 3-527-26044-7 (Weinheim, Basel)
ISBN 0-89573-234-3 (Deerfield Beach)
NE: Moss, Donald W. [Hrsg.]

Production manager: Heidi Lenz
Composition: Krebs-Gehlen Druckerei, D-6944 Hemsbach
Printing: Hans Rappold Offsetdruck GmbH, D-6720 Speyer
Bookbinding: Josef Spinner, D-7583 Ottersweier
Printed in the Federal Republic of Germany

Preface to the Series

"Methods of Enzymatic Analysis" appeared for the first time in 1962 as a one-volume treatise in German. Several updated and improved editions in English and German have been published since then. The latest English edition appeared in 1974.

In the meantime, enzymatic analysis has continued to find new applications, refinements and extensions at a pace that justifies – indeed, demands – the preparation of a new and completely revised edition. However, the field has grown so enormously that it can no longer be surveyed adequately by one person. Fortunately, therefore, I am supported in this new enterprise by Dr. M. Graßl, who is highly experienced in biochemical analysis, and Dr. J. Bergmeyer, who represents the younger generation of biochemists.

With the 1974 edition of "Methods of Enzymatic Analysis" as a starting point for our work towards the new edition, it soon became obvious that many chapters had to be eliminated, re-written or added. Moreover, the increased number of analytes that can now be determined enzymatically and of enzymes regularly requiring analysis, especially in the clinical laboratory, together with the emergence of an entirely new field of application through the technique of the enzyme-immunoassay, demanded a new arrangement and subdivision of the contents, if the vast range of material was to be dealt with properly and lucidly.

The result is the plan of the work printed on the page opposite the title page of this volume. Of course, it would be impossible to publish a whole series such as this at one moment and still maintain an equal degree of topicality for all contributions. Therefore, we decided to produce the series at a pace of several volumes per year. The volumes will not necessarily appear in their numerical order, but will be made available as they can be planned and completed.

As before, the purpose of the work is to provide reliable descriptions of well-developed procedures of enzymatic analysis in the broadest sense of the term. Special efforts are being made to arrange every chapter, and to coordinate the contents of all chapters, in such a way that the volumes are useful as laboratory manuals for daily work.

Internationally-agreed enzyme nomenclature as well as quantities and units correlating with the "Système International d'Unités" are used wherever possible in order to make statements and data unambiguous and comparable over time and space.

All contributions are and will be written in English: however, contributors come from all over the world and their manuscripts naturally show various versions of English. These have to be harmonized in style and spelling in order to achieve uniformity throughout the series without, we hope, entirely eliminating each author's personal approach. Professor Donald W. Moss has kindly agreed to undertake this task. We agreed with him to use modern English spelling, but to try to minimize differences between British and American practice. We hope that this will be considered as a fair solution and one which will make the series accessible to as wide a readership as possible.

Thanks are due to the authors in the first place for responding so readily to our invitations, for writing their chapters so diligently within a short time and for communicating their experience and expertise. We are also indebted to all colleagues who gave their advice and to Professor Moss for accepting the task of language editor. Finally I wish to record my gratitude to Verlag Chemie for the fruitful and excellent co-operation during all stages from planning to production.

Tutzing, February 1983 Hans Ulrich Bergmeyer

Preface to Volume IV

This volume continues the detailed description for measuring enzyme activities: esterases, glycosidases, other C – N splitting enzymes, lyases and ligases are considered.

Methods for these enzymes have been enormously improved in recent years. For example, the determination of amylase activity with well-defined pure substrates has been developed for the clinical laboratory, so that results are no longer severely affected by alterations in substrate composition and quality. A convenient routine method is now available for the determination of lipase activity and the same is true for hepatic triglyceride lipase and lipoprotein lipase.

Also biochemists, biochemical pharmacologists, microbiologists, food analysts, plant physiologists and others will all find here the newest methods for the determination of the catalytic activities of glycosidases, the various phosphodiesterases, cyclases and the different ATPases.

Developments in methods for the determination of peptidases and proteinases have been so considerable that these two groups of enzymes could not be adequately treated within the scope of this volume. They have therefore been given a volume of their own (Vol. V).

Again, we have selected – to the best of our knowledge – the most modern procedures and asked highly experienced authors to deal with them. All the chapters are presented in a similar format, and particular care has been taken to ensure uniformity and clarity in nomenclature, symbols and units in the calculations, so that even the less experienced experimenter will be able to work according to the given instructions.

The editors are grateful to the authors for their excellent co-operation, also to the staff of the library in the Biochemical Research Centre Tutzing of Boehringer Mannheim for continuous assistance. Again, thanks are due to Professor Donald W. Moss for his activities as editorial consultant and language editor.

Tutzing, January 1984 Hans Ulrich Bergmeyer

Contents

Contents of Volumes I – X

(Chapter Headings only)

Volume VI
Metabolites 1: Carbohydrates

Volume VII
Metabolites 2: Tri- and Dicarboxylic Acids, Purines, Pyrimidines and Derivatives, Coenzymes, Inorganic Compounds

Volume VIII
Metabolites 3: Lipids, Steroids, Drugs

Volume IX
Metabolites 4: Proteins, Peptides and Amino Acids

Volume X
Molecular Biologicals (tentative)

Contributors

Assmann, Gerd
Zentrallaboratorium der
Medizinischen Einrichtungen der
Westfälischen Wilhelms-Universität
Domagkstraße 3
D-4400 Münster
Federal Republic of Germany

p. 42

Berlin, Alexander
Commission of the European
Comunities, Health and Safety
Directorate, Bâtiment Jean Monnet,
Plateau du Kirchberg
Luxembourg
Luxembourg

p. 363

Böhme, Eycke
Pharmakologisches Institut
der Universität Heidelberg
Im Neuenheimer Feld 366
D-6900 Heidelberg
Federal Republic of Germany

p. 379

Bretaudiere, Jean-Pierre
Dept. of Pathology and
Laboratory Medicine
University of Texas
Medical School
P.O. Box 20708
Houston, Texas 77025
U.S.A.

p. 75, 83, 86

Bruist, Michael F.
Dept. of Biology B-022
University of California
San Diego
La Jolla, CA 92093
U.S.A.

p. 324

Buchholz, Klaus
Institut für Landwirtschaftliche
Technologie und Zuckerindustrie
Langer Kamp 5
D-3300 Braunschweig
Federal Republic of Germany

p. 178

Bush, Karen
The Squibb Institute for
Medical Research
P.O. Box 4000
Princeton, NJ 08540
U.S.A.

p. 280

Cohen, Philip
Biochemistry Department
Dundee University
Medical Sciences Institute
Dundee DD1 4HN
Scotland

p. 120

Colombo, Jean-Pierre
Inselspital Bern
Chemisches Zentrallabor
der Universitätskliniken
CH-3010 Bern
Switzerland

p. 285

Corbishley, Timothy P.
The Liver Unit
King's College Hospital
and Medical School
Denmark Hill
London SE5 9RS
United Kingdom

p. 134

Dahlqvist, Arne
Dept. of Nutrition
University of Lund
Chemical Center
Box 740
S-22007 Lund 7
Sweden

p. 208, 227

Daniels, Lydia B.
Dept. of Biochemistry
University of Pittsburgh
School of Medicine
Pittsburgh, PA 15261
U.S.A.

p. 217

Faber, Christopher N.
Dept. of Biochemistry
University of Pittsburgh
School of Medicine
Pittsburgh, PA 15261
U.S.A.

p. 230

Fishman, William H.
La Jolla Cancer Research Foundation
10901 North Torrey Pines Road
La Jolla, CA 92037
U.S.A.

p. 246

Foo, Ying
Dept. of Chemical Pathology
The Royal Free Hospital &
School of Medicine
London
United Kingdom

p. 167

Galanti, Bruno
Clinical Delle Malattie Infettive
Universitá di Napoli
1[a] Facoltá di Medicina e Chirurgia
Via Ferdinando Russo, 29
I-80123 Napoli
Italy

p. 294, 315

Gerlach, Ulrich
Medizinische Klinik der
Universität Münster
Domagkstraße 3
D-4400 Münster
Federal Republic of Germany

p. 113

Giusti, Giuseppe
Clinica Delle Malattie Infettive
Universitá di Napoli
1[a] Facoltá di Medicina e Chirurgia
Via Ferdinando Russo, 29
I-80123 Napoli
Italy

p. 294, 315

Glew, Robert H.
Dept. of Biochemistry
University of Pittsburgh
School of Medicine
Pittsburgh, PA 15261
U.S.A.

p. 217, 230

Goi, Giancarlo
Istituto di Chimica Biologica
Facoltá di Medicina
e Chirurgica dell' Universitá
di Milano
Via C. Saldini 50
I-20133 Milano
Italy

p. 263

Gröbner, Wolfgang
Medizinische Poliklinik
der Universität München
Pettenkoferstr. 8a
D-8000 München 2
Federal Republic of Germany

p. 328

Haeckel, Rainer
Zentralkrankenhaus
St.-Jürgen-Straße
Institut f. Laboratoriumsmedizin
D-2000 Bremen 1
Federal Republic of Germany

p. 106

Hardie, D. Grahame
Biochemistry Department
Dundee University
Medical Sciences Institute
Dundee DD1 4HN
Scotland

p. 120

Heinz, Fritz
Medizinische Hochschule Hannover
Institut für Klinische Biochemie
und Physiologische Chemie
Arbeitsbereich Enzymologie
Roderbruchstr. 101
D-3000 Hannover 61
Federal Republic of Germany

p. 106, 302, 308

Husen, Norbert van
Medizinische Klinik der
Universität Münster
Domagkstraße 3
D-4400 Münster
Federal Republic of Germany

p. 113

Jabs, Hans-Ulrich
Zentrallaboratorium der
Medizinischen Einrichtungen der
Westfälischen Wilhelms-Universität
Domagkstraße 3
D-4400 Münster
Federal Republic of Germany

p. 42

Jakobs, Karl H.
Pharmakologisches Institut
der Universität Heidelberg
Im Neuenheimer Feld 366
D-6900 Heidelberg
Federal Republic of Germany

p. 369

Johnson, Philip J.
The Liver Unit
King's College Hospital
and Medical School
Denmark Hill
London SE5 9RS
United Kingdom

p. 134

Junge, Wolfgang
Medizinisch-Diagnostisches Labor
Jägersberg 7 – 9
D-2300 Kiel
Federal Republic of Germany

p. 2, 8, 15

Kaspar, Peter
Boehringer Mannheim GmbH
Biochemical Research Center
Bahnhofstr. 9 – 15
D-8132 Tutzing
Federal Republic of Germany

p. 26

Klees, Hubertus
Medizinisch-Diagnostisches Labor
Jägersberg 7 – 9
D-2300 Kiel
Federal Republic of Germany

p. 8

Kinnunen, Paavo K. J.
Dept. of Medical Chemistry
University of Helsinki
Siltavuorenpenger 10
SF-00170 Helsinki 17
Finland

p. 34

Konarska, Liliana
Medical School of Warsaw
Dept. of Clinical Biochemistry
ul. Banacha 1
PL-02-097 Warsaw
Poland

p. 285

Linker, Alfred
Veterans Administration Medical
Center and the Depts. of Biological
Chemistry and Pathology
University of Utah
College of Medicine
Salt Lake City, UT 84148
U.S.A.

p. 256

Lombardo, Adriana
Istituto di Chimica Biologica
Facoltá di Medicina
e Chirurgica dell' Universitá
di Milano
Via C. Saldini 50
I-20133 Milano
Italy

p. 263

Masserini, Massimo
Istituto di Chimica Biologica
Facoltá di Medicina
e Chirurgica dell' Universitá
di Milano
Via C. Saldini 50
I-20133 Milano
Italy

p. 241, 263

Moss, Donald W.
Royal Postgraduate Medical School
University of London
Hammersmith Hospital
London W12
United Kingdom

p. 92

Neumann, Ulrich
Boehringer Mannheim GmbH
Biochemical Research Center
Bahnhofstr. 9 – 15
D-8132 Tutzing
Federal Republic of Germany

p. 26

Nöhle, Ulrich
Biochemisches Institut
Christian-Albrechts-Universität
Olshausenstr. 40 – 60
D-2300 Kiel
Federal Republic of Germany

p. 195

Penefsky, Harvey S.
Public Health Research Institute
455 First Avenue
New York, NY 10016
U.S.A.

p. 324

Pierre, Kenneth J.
Beckman Instruments, Inc.
6200 El Camino Real
Carlsbad, CA 92008
U.S.A.

p. 146

Rauscher, Elli
Boehringer Mannheim GmbH
Biochemical Research Center
Bahnhofstr. 9 – 15
D-8132 Tutzing
Federal Republic of Germany

p. 152, 157

Rapp, Peter
Gesellschaft für
Biotechnologische Forschung mbH
Mascheroder Weg 1
D-3300 Braunschweig
Federal Republic of Germany

p. 178

Reiter, Sebastian
Medizinische Poliklinik
der Universität München
Pettenkofer Straße 8a
D-8000 München 2
Federal Republic of Germany

p. 338

Rosalki, Sidney B.
Dept. of Pathology
The Royal Free Hospital &
School of Medicine
Pond Street, Hamstead
London NW3
United Kingdom

p. 167

Schaller, Karl Heinz
Institut für Arbeits- und
Sozialmedizin und Poliklinik
für Berufskrankheiten
der Universität Erlangen-Nürnberg
Schillerstraße 25 + 29
D-8520 Erlangen
Federal Republic of Germany

p. 363

Schauer, Roland
Biochemisches Institut
Christian-Albrechts-Universität
Olshausenstr. 40 – 60
D-2300 Kiel
Federal Republic of Germany

p. 195

Schultz, Günter
Pharmakologisches Institut
der Universität Heidelberg
Im Neuenheimer Feld 366
D-6900 Heidelberg
Federal Republic of Germany

p. 369, 379

Spillman, Thomas
Dept. of Pathology
and Laboratory Medicine
University of Texas
Medical School
P.O. Box 20708
Houston, TX 77025
U.S.A.

p. 75, 83, 86

Stahl, Philip D.
Dept. of Physiology
and Biophysics
Washington University
School of Medicine
660 South Euclid Avenue
St. Louis, MO 63110
U.S.A.

p. 246

Stirling, John L.
Dept. of Biochemistry
Queen Elizabeth College
Campden Hill
London W8 7AH
United Kingdom

p. 269

Stitt, Mark
Universität Göttingen
Institut Biochemie der Pflanzen
Untere Karspüle 2
D-3400 Göttingen
Federal Republic of Germany

p. 353, 359

Strada, Samuel
Dept. of Pharmakology
University of South Alabama
College of Medicine
Mobile, AL 36688
U.S.A.

p. 127

Sykes, Richard B.
The Squibb Institute
for Medical Research
P.O. Box 4000
Princeton, NJ 08540
U.S.A.

p. 280

Tettamanti, Guido
Istituto di Chimica Biologica
Facoltá di Medicina
e Chirurgica dell' Universitá
di Milano
Via C. Saldini 50
I-20133 Milano
Italy

p. 241, 263

Thompson, W. Joseph
Dept. of Pharmacology
University of South Alabama
College of Medicine
Mobile, AL 36688
U.S.A.

p. 127

Thurén, Tom
Dept. of Medical Chemistry
University of Helsinki
Siltavuorenpenger
SF-00170 Helsinki 17
Finland

p. 34

Tung, Ker-Kong
Beckman Instruments, Inc.
6200 El Camino Real
Carlsbad, CA 92008
U.S.A.

p. 146

Wahlefeld, August Wilhelm
Boehringer Mannheim GmbH
Biochemical Research Center
Bahnhofstr. 9 – 15
D-8132 Tutzing
Federal Republic of Germany

p. 161

Weisner, B.
Universitäts-Krankenhaus
Eppendorf
Martinistraße 52
D-2000 Hamburg 20
Federal Republic of Germany

p. 189

Whittaker, Mary
Dept. of Chemistry
University of Exeter
Devon EX4 4QD
United Kingdom

p. 52

Williams, Roger
The Liver Unit
King's College Hospital
and Medical School
Denmark Hill
London SE5 9RS
United Kingdom

p. 134

Willnow, Peter
Boehringer Mannheim GmbH
Biochemical Research Center
Bahnhofstr. 9–15
D-8132 Tutzing
Federal Republic of Germany

p. 346

Zadražil, František
Institut für Bodenbiologie
Bundesforschungsanstalt für
Landwirtschaft
Bundesallee 50
D-3300 Braunschweig
Federal Republic of Germany

p. 178

Ziegenhorn, Joachim
Boehringer Mannheim GmbH
Biochemical Research Center
Bahnhofstr. 9–15
D-8132 Tutzing
Federal Republic of Germany

p. 26

1 Esterases

1.1 Carboxylesterase

Carboxylic-ester hydrolase, EC 3.1.1.1

Wolfgang Junge

General

Carboxylesterases are a group of enzymes widely distributed in nature. They catalyze the hydrolysis of carboxylic acid esters to the free acid anions and alcohol. With regard to the acyl residue, short chain esters are cleaved at the highest rate, the optimal length being 3 to 6 carbon atoms. Besides aliphatic carboxylesters, aromatic esters and aromatic amides, as well as thioesters, are also substrates of these enzymes. A further reaction catalyzed by many carboxylesterases is the transfer of the ester acyl moiety to nucleophilic acceptors other than water, e.g. alcohols or amino acids. Regarding the physical state of the substrate, carboxylesterases act on esters which are in true or micellar solution as well as on those which are in emulsified form, e.g. tributyrin. This latter property introduces some problems into the differentiation between carboxylesterases and lipases.

It is obvious that, due to the wide and overlapping substrate specificity of carboxylesterases, an unambiguous classification of these enzymes is hardly possible. Therefore esterases have not been differentiated according to their catalytic properties but mainly on the basis of their behaviour towards certain inhibitors. Thirty years ago *Aldridge* [1] introduced a still useful classification, based on the sensitivity towards organophosphates such as diethyl 4-nitrophenylphosphate (Paraoxon, E 600). A-esterases are those enzymes which are not inhibited by organophosphates and even hydrolyze them. Since A-esterases preferentially cleave aromatic esters, e.g. phenyl acetate, they have also been designated arylesterases or aromatic esterases. They are presently classified under EC 3.1.1.2. In contrast to A-esterases, B-esterases are stoichiometrically inhibited by organophosphates through irreversible phosphorylation of the active-site serine. Thus B-esterases, formerly also called aliesterases (ali = aliphatic) or non-specific esterases, are serine hydrolases. On the basis of their sensitivity towards physostigmine, 10^{-5} mol/l [2], B-esterases can be further differentiated into cholinesterases (EC 3.1.1.7 and 3.1.1.8, resp.) and carboxylesterases (EC 3.1.1.1). It is the latter group of enzymes which is dealt with in this chapter.

The most prominent and best-studied carboxylesterases are those occurring in mammalian liver (for reviews cf. [3 – 5]). These enzymes have been purified from many animals and from man. The liver enzyme is attached to the cytoplasmic side of the endoplasmic reticulum membrane [6], probably by hydrophobic binding *via* phospholipids [7, 8]. It comprises about 5 – 7% of all microsomal protein. The esterase preparations from different sources have very similar chemical and structural properties, exhibiting a rather unusual quaternary structure: most of them are trimeric proteins consisting of three identical subunits of molecular weight 60000, each containing one active site. The physiological role of these enzymes is still unknown. They are significant, however, for the metabolism of xenobiotics of the ester or amide type, thus being part of the detoxication system of the body [4, 5, 9].

In man, microsomal carboxylesterase (MCE) is almost exclusively found in liver; only minor amounts were detected in lung, spleen and fat tissue [10]. The gradient liver: serum is >15000. Unlike other enzymes of the endoplasmic reticulum, e.g. glucose-6-phosphatase, MCE remains active after liberation from the membrane. It was detected in high concentration in sera of patients with necrotizing liver diseases by an immunological technique [10].

Application of method: in biochemistry and in clinical chemistry.

Enzyme properties relevant in analysis: in contrast to rat and pig MCE which consist of several isoenzymes with different kinetic properties [11, 12], the human enzyme is rather homogeneous as judged from isoelectric focusing [13]. Only micro-heterogeneity has been observed [14]. In serum, monomeric as well as trimeric MCEs are present (*W. Junge, S. Honegger*), due to the tendency of the enzyme to dissociate in dilute solution [15, 16]. It is not known whether monomeric and trimeric enzyme forms have different catalytic properties.

Methods for determination: for measurement in tissue homogenates, a continuous-monitoring titrimetric technique employing simple aliphatic esters such as methyl or ethyl butyrate is useful [17, 18], because the membrane-bound enzyme can be assayed. For detection of MCE in serum, an immunochemical method was originally developed [9]. Chromogenic esters, e.g. with 4-nitrophenol or naphthol as alkyl residues, cannot be used due to interference by cholinesterase and arylesterase. A specific continuous-monitoring method has not been published so far. Active-site titration with organophosphates is a frequently-used tool in MCE research and of particular importance for control of the purity of enzyme preparations [19]. However, the method is not applicable to tissue extracts or serum because other serine hydrolases may also react, or because the inhibitor is cleaved by arylesterases.

International reference method and standards: neither a standardized method nor a standard reference material exists.

Enzyme effectors: the rate of hydrolysis can be considerably increased by nucleophiles other than water as acceptors for the acyl residue, e.g. methanol [20].

Special activators of MCE are not known. In the method described here, alcohols, except methanol, cannot be used because alcohol dehydrogenase is part of the indicator system.

Assay

Method Design

Principle

(a) Ethyl valerate + H_2O $\xrightarrow{\text{MCE}}$ valerate + ethanol

(b) Ethanol + NAD^+ $\xrightarrow{\text{ADH*}}$ acetaldehyde + NADH + H^+

The amount of ethanol released per unit time is indicated by the increase in absorbance due to reduction of NAD and is the measure of esterase activity.

Optimized conditions for measurements: among the various aliphatic esters which could be used for the proposed methodological principle, valeric acid was chosen because its ester is the substrate hydrolyzed at the highest rate by MCE, the specific activity being 172 U/mg [9]. The *Michaelis* constant was found to be 1.77×10^{-4} mol/l under the assay conditions (*W. Junge, S. Honegger*), which is distinctly lower than values found for the homologous ethyl esters of shorter aliphatic acids [9]. The final concentration of eserine sulphate needed to block interfering cholinesterase activity has no measurable influence on MCE activity. Higher concentrations are, however, inhibitory.

A more sensitive assay in which acetaldehyde is not trapped by semicarbazide but is converted to acetate by aldehyde dehydrogenase (EC 1.2.1.2) to form another molecule of NADH is presently being evaluated in our laboratory. (For trapping enzyme reactions cf. Vol. I, p. 169.)

Temperature conversion factors: not yet established.

Equipment

Spectrophotometer or spectral-line photometer capable of exact measurement at Hg 334, 339 or Hg 365 nm and equipped with a thermostatted cuvette holder; recorder or stopwatch.

Reagents and Solutions

Purity of reagents: there are no special requirements for the purity of the reagents. All reagents needed are commercially available.

---

* Alcohol dehydrogenase from yeast, Alcohol : NAD^+ oxidoreductase, EC 1.1.1.1.

Preparation of solution (for approximately 45 determinations): all solutions to be prepared in re-purified water (cf. Vol. II, chapter 2.1.3.2).

1. Pyrophosphate/glycine buffer (phosphate, 0.15 mol/l, pH 8.0; glycine, 42.6 mmol/l; semicarbazide, 0.15 mmol/l):

   dissolve 66.7 g $Na_4P_2O_7 \cdot 10\ H_2O$, 16.7 g semicarbazide hydrochloride and 3.2 g glycine in about 800 ml water, adjust pH with NaOH to 8.0 and dilute to 1000 ml with water.

2. Eserine sulphate solution (1.1 mmol/l):

   dissolve 71.4 mg eserine sulphate in 100 ml water.

3. NAD solution (75 mmol/l):

   dissolve 250 mg β-NAD (free acid) in 5.0 ml water.

4. Alcohol dehydrogenase solution (ADH, 2400 kU/l):

   dissolve 12 mg of pure yeast ADH (specific activity 400 U/mg corresponding to ca. 20 mg lyophilized material from *Boehringer Mannheim*) in 2 ml water.

5. Working solution (for about 12 determinations):

   dissolve 0.020 ml valeric acid ethyl ester by vortexing in 23 ml buffer (1), then add 0.25 ml eserine sulphate solution (2), 1.25 ml NAD solution (3) and 0.5 ml ADH solution (4).

Stability: on storage at 4°C, solutions (1) and (2) are stable for at least 2 weeks, solutions (3) and (4) for 1 week. The working solution (5) is stable for 8 h at room temperature and should be prepared daily. The eserine sulphate solution must be protected from light and should be stored in a dark bottle.

Procedure

Collection and treatment of sample: serum as well as plasma can be used. The commonly employed anticoagulants such as citrate, oxalate, heparin or EDTA are without effect on MCE. Serum containing fluoride cannot be assayed because the anion is a competitive inhibitor of MCE catalysis [9].

Stability of the enzyme in the sample: MCE is very stable in serum. No loss of activity was observed during 1 week at room temperature or at 4°C, respectively. Frozen samples are stable for several months.

Details for measurement in tissues: as long as the MCE is bound to the endoplasmic reticulum membrane, the UV method described here cannot be employed, but instead a titrimetric technique should be used. However, the enzyme can easily be solubilized by treatment of tissue homogenates with detergents, e.g. Triton X-100, 0.5% final concentration. Bile acids [7, 9] and phospholipase A_2 [7, 8] are also effective. After centrifugation at 100000 *g* for 60 min to sediment the microsomal fraction, MCE activity is quantitatively recovered in the supernatant.

Assay conditions: wavelength 339 (Hg 334, Hg 365) nm; light path 10 mm; final volume 2.2 ml; 25 °C (thermostatted cuvette holder). Measure against air. A blank with isotonic saline instead of sample is run prior to each series of measurement.

Measurement

Pipette into the thermostatted cuvette:			concentration in assay mixture	
working solution	(5)	2.0 ml	phosphate glycine semicarbazide ethyl valerate eserine sulphate NAD ADH	125 mmol/l 35.6 mmol/l 125 mmol/l 4.9 mmol/l 0.01 mmol/l 3.41 mmol/l 44 kU/l
wait for 2–3 min for temperature equilibration				
sample		0.2 ml	volume fraction	0.091
mix, wait for 2 min, then read absorbance and start stopwatch; repeat the reading after 1, 2 and 3 min or monitor the reaction on a recorder.				

Correct readings of sample for blank yielding $\Delta A/\Delta t$. The values $\Delta A/\Delta t$ must be < 0.100/min at Hg 365 nm (< 0.200/min at 339 or Hg 334 nm). Otherwise dilute sample with isotonic saline.

Calculation: use eqn. (k) or (k_1) from "Formulae", Appendix 3. For values of ε cf. Appendix 4. The following relations are valid (Δt in min):

	Hg 334 nm	339 nm	Hg 365 nm	
b =	$1780 \times \Delta A/\Delta t$	$1746 \times \Delta A/\Delta t$	$3235 \times \Delta A/\Delta t$	U/l
b =	$2.96 \times 10^4 \times \Delta A/\Delta t$	$2.90 \times 10^4 \times \Delta A/\Delta t$	$5.40 \times 10^4 \times \Delta A/\Delta t$	nkat/l

Validation of Method

Precision, accuracy, detection limits and sensitivity: for within-run imprecision the relative standard deviation was 8.0% for a serum of 87 U/l and 7.9% for 235 U/l. Data about accuracy are not available due to lack of standard reference material. With a spectral-line photometer (*Eppendorf* photometer) the detection limit is about 1.6 U/l, corresponding to an absorbance change of 0.005/10 min at Hg 365 nm.

Sources of error: the assay is highly sensitive to traces of ethanol present in the sample. This causes an "initial burst" of the reaction before zero order kinetics are obtained. Consumption of ethanol is usually completed within 2 – 3 min. In our experience the detection of serum specimens containing ethanol is not a rare event, even in hospitalized patients. However, it should be mentioned that ethanol contamination can also be brought about by cleaning the skin with ethanolic solutions prior to puncture.

Omission of blank yields falsely elevated results: semicarbazide and NAD react slowly to a compound absorbing light at measuring wavelengths [21].

Specificity: ethyl valerate is not specific for MCE, as already mentioned. Interfering cholinesterase activity in serum is completely blocked by eserine. The following purified enzymes have no measurable activity towards ethyl valerate under the assay conditions: human pancreatic lipase (EC 3.1.1.3), serum arylesterase (EC 3.1.1.2) and non-specific pancreatic carboxylesterase (no EC number). The possibility cannot be excluded that other hydrolases contribute to the activity found in normal serum. However, elevated activity has so far only been observed in liver diseases.

Reference ranges: a logarithmic distribution of MCE activity in human serum was found, the upper normal limit being 18 U/l. In patients suffering from acute liver necrosis, serum activities of up to 6000 U/l were observed [10].

References

[1] *W. N. Aldridge,* Serum Esterases 2. An Enzyme Hydrolyzing Diethyl-p-nitrophenylphosphate (E 600) and its Identity with the A-Esterase of Mammalian Sera, Biochem. J. *53*, 117 – 124 (1953).

[2] *K. B. Augustinsson,* Electrophoretic Separation and Classification of Blood Plasma Esterases, Nature *181*, 1786 – 1789 (1958).

[3] *K. Krisch,* Carboxylesterases, in: *P. D. Boyer* (ed.), The Enzymes, Vol. *V*, Academic Press, New York 1971, pp. 43 – 69.

[4] *W. Junge, K. Krisch,* The Carboxylesterases/Amidases of Mammalian Liver and their Possible Significance, CRC Critical Rev. Toxicol. *3*, 371 – 434 (1975).

[5] *E. Heymann,* Carboxylesterases and Amidases, in: *W. B. Jakoby* (ed.), Enzymatic Basis of Detoxication Vol. II, Academic Press, New York 1980, pp. 291 – 323.

[6] *M. Gratzl, W. Nastainczyk, D. Schwab,* The Spatial Arrangement of Esterases in the Microsomal Membrane, Cytobiology *11*, 123 – 132 (1975).

[7] *T. Akao, T. Omura,* Acetanilide-hydrolyzing Esterase of Rat Liver Microsomes. I. Solubilization, Purification and Intramicrosomal Localization, J. Biochem. *72*, 1245 – 1256 (1972).

[8] *W. Junge, M. Malyusz,* Microsomal Carboxylesterase (EC 3.1.1.1). A Serum Indicator of Endoplasmatic Reticulum Injury in Hypoxemic Liver Disease in: *D. M. Goldberg, M. Werner* (eds.), Progress in Clinical Enzymology, Masson Publishing New York 1980, pp. 187–190.

[9] *W. Junge,* Unspezifische Carboxylesterase vom Menschen, Habilitationsschrift, Universität Kiel, 1981.

[10] *W. Junge,* Human Microsomal Carboxylesterase (EC 3.1.1.1). Distribution in several Tissues and Some Preliminary Observations on Its Appearance in Serum, in: *D. M. Goldberg, J. H. Wilkinson* (eds.), Enzymes in Health and Disease, S. Karger, Basel 1978, pp. 54–58.

[11] *R. Arndt,* Isolierung und Charakterisierung von Carboxylesterasen (EC 3.1.1.1) der Rattenleber. Dissertation, Universität Kiel (1973).

[12] *W. Junge, E. Heymann,* Characterization of the Isoenzymes of Pig Liver Esterase. 2. Kinetic Studies. Eur. J. Biochem. *95*, 519–525 (1979).

[13] *W. Junge, E. Heymann, K. Krisch, H. Hollandt,* Human Liver Carboxylesterase. I. Purification and Molecular Properties, Arch. Biochem. Biophys. *165*, 749–763 (1974).

[14] *N. Atanasov,* Purification and Molecular Weight of Human Liver Carboxylesterase (EC 3.1.1.1), Rev. Roum. Biochim. *13*, 179–184 (1976).

[15] *H. C. Benöhr, K. Krisch,* Carboxylesterase aus Rinderlebermikrosomen. II. Dissoziation und Assoziation, Molekulargewicht, Reaktion mit E 600, Hoppe-Seyler's Z. Physiol. Chem. *348*, 1115–1119 (1967).

[16] *W. Junge, K. Krisch, H. Hollandt,* Further Investigations on the Subunit Structure of Microsomal Carboxylesterases from Pig and Ox Liver, Eur. J. Biochem. *43*, 379–389 (1974).

[17] *K. Krisch,* Eigenschaften und Substratspezifität einer Esterase aus Schweinelebermikrosomen, Biochem. Z. *337*, 546–573 (1963).

[18] *D. J. Horgan, J. K. Stoops, E. C. Webb, B. Zerner,* Carboxylesterases (EC 3.1.1). A Large-scale Purification Procedure of Pig Liver Carboxylesterase, Biochemistry *8*, 2000–2005 (1969).

[19] *K. Krisch,* Reaction of a Microsomal Esterase from Hog Liver with Diethyl-p-nitrophenyl Phosphate, Biochim. Biophys. Acta *122*, 265–280 (1966).

[20] *P. Greenzaid, W. P. Jencks,* Pig Liver Esterase. Reactions with Alcohols, Structure Reactivity Correlations and the Acyl-enzyme Intermediate, Biochemistry *10*, 1210–1222 (1971).

[21] *E. Bernt, H. U. Bergmeyer,* L-Glutamate, UV-Assay with Glutamic Dehydrogenase and NAD, in: *H. U. Bergmeyer* (ed.), Methods of Enzymatic Analysis, Vol. 4, Verlag Chemie Weinheim and Academic Press, New York 1974, pp. 1704–1708.

1.2 Arylesterase

Aryl-ester hydrolase, EC 3.1.1.2

Wolfgang Junge and Hubertus Klees

General

Arylesterases (ARE), formerly designated A-esterases [1] or aromatic esterases, occur mainly in mammalian plasma. *Augustinsson* [2] was the first to obtain a clear differentiation of ARE activities in various animals and in man by means of preparative column electrophoresis on cellulose. At first sight the term arylesterase seems to be a poor definition, since arylesters are split by numerous hydrolases. Besides various esterases, carbonic anhydrase and proteinases are also active on such compounds. However, the AREs designated by the EC 3.1.1.2 number represent a rather well-defined group of esterases. The enzymes from serum are characterized by their resistance to organophosphates and eserine, and by their sensitivity to both sulphydryl reagents, e.g. 4-hydroxymercuribenzoate, and heavy metal ions such as Hg^{2+}, Cu^{2+} or La^{3+} [3]. With respect to substrate specificity, AREs preferentially cleave aromatic esters of acetic acid, the best substrate being phenyl acetate. Except for vinyl esters, aliphatic esters are not split [4]. As AREs are not inhibited by organophosphates but sometimes hydrolyze these compounds, e.g. diethyl 4-nitrophenylphosphate, it would be justifiable also to classify these enzymes as phosphatases (EC 3.1.3.1).

The site of synthesis of serum ARE is not known, nor is the physiological role of the enzyme. It was suggested that ARE originates from the kidney [5], but indirect evidence indicates that also the liver might be the source: in patients with impaired liver function (cirrhosis, cachexia) reduced ARE activity was found [6, 7]. Employing 2-naphthyl acetate as substrate for ARE, *Burlina et al.* [8] found high activity in liver, where it is mainly located in the microsomal fraction. However, their conclusion that serum ARE originates from liver is not justified as the microsomal carboxylesterase (EC 3.1.1.1) has high activity towards naphthol esters and it is likely that their method measured that enzyme.

There is little information about the molecular properties of serum ARE. So far, the enzyme has only been isolated from bovine serum, with a molecular weight of 440000 [9]. We obtained a value of approximately 240000 for human ARE by gel filtration. A highly interesting finding is that ARE is associated with serum lipoproteins; very probably with those belonging to the high density lipoprotein fraction [10]. Purification of ARE leads to a lipid-rich material containing triglycerides, cholesterol and β-carotene [9]. The significance of this observation is unknown at present, but it suggests an involvement of ARE in lipid metabolism. This possibility was already discussed by *Pilz* [11] who studied transalkylation reactions catalyzed by ARE.

Application of method: in biochemistry and in clinical chemistry.

Enzyme properties relevant in analysis: the existence of ARE isoenzymes in human as well as in animal serum has been repeatedly reported [7, 8, 11 – 15]. However, in almost all studies dealing with this subject, esters of naphthol have been used for detection of ARE activity, e.g. for staining of zymograms. This technique is of limited value, since these compounds are hydrolyzed by other serum esterases as well, particularly by the cholinesterase isoenzymes. Isoenzymes of ARE have not unequivocally been demonstrated in human serum so far. *Augustinsson* showed [2] that more than 95% of the phenyl acetate-hydrolyzing activity of human serum is found in a single, symmetrical peak after preparative electrophoresis. The remaining activity is due to cholinesterase [2]. We recently confirmed this result by gel filtration and ion-exchange chromatography. Furthermore, as expected from its association with high-density lipoproteins, which represent a rather heterogeneous material with regard to size and charge, ARE activity is found in a broad zone after disc electrophoresis. The same is true for the purified enzyme after staining with 2-naphthyl acetate (*W. Junge, H. Klees*).

Methods of determination: methods based on directly chromogenic substrates, e.g. 4-nitrophenyl acetate [16, 17] or on the azo dye-forming naphthol esters [11, 18] are not suitable due to their lack of specificity. 1-Naphthyl acetate is not cleaved by ARE but by cholinesterase [10, 19], 2-naphthyl acetate by both ARE and cholinesterase [7, 10]. Diethyl 4-nitrophenyl phosphate (Paraoxon) was suggested by *Krisch* [20] for specific ARE determination. This substrate brings about an interesting problem which cannot be explained yet: whereas with phenyl acetate as substrate ARE activity exhibits a *Gaussian* distribution in healthy individuals, a binominal distribution is found with Paraoxon [20]. It is surprising that both activities cannot be separated by conventional preparative techniques, e.g. gel chromatography or isoelectric focusing (*W. Junge, H. Klees*).

It is agreed that phenyl acetate is the substrate of choice for measurement of ARE. Besides the continuous-monitoring UV-test to be described, hydrolysis of phenyl acetate can alternatively be determined by oxidative coupling of phenol with 4-aminoantipyrine [21], but this has the drawbacks of a 2-point assay. In the past, release of acetic acid from the substrate was followed by the *Warburg* technique in bicarbonate/CO_2 buffer [5].

International reference methods and standards: neither a standardized method nor standard reference material exists.

Enzyme effectors: calcium ions are necessary for full activity of serum ARE, their role being unknown. Other special activators are not known.

Assay

Method Design

Principle

$$\text{Phenyl acetate} + H_2O \xrightarrow{\text{ARE}} \text{phenol} + \text{acetic acid}.$$

The release of phenol is continuously monitored at 270 nm. Equilibrium is far in favour of the cleavage products. Due to slow spontaneous hydrolysis, a blank should be performed prior to each series of measurement.

Optimized conditions for measurement: although ARE activity is slightly higher at pH 8 [7, 14], pH 7.4 is preferred to suppress spontaneous hydrolysis. The enzyme is distinctly more active in Tris than in phosphate buffer [22]. The K_m for human serum ARE is about 1 mmol/l [7, 10, 12] with phenyl acetate. Although cholinesterase is reported to be active on phenyl acetate as well, inclusion of eserine, 0.01 mmol/l, into the assay mixture does not cause measurable reduction of serum activity. The final concentration of Ca^{2+} ensures full activity of ARE [20, 22].

Temperature conversion factors: according to [7] a factor of 1.16 can be used for conversion from 25 °C to 30 °C.

Equipment

Spectrophotometer capable of exact measurement at 270 nm, equipped with thermostatted cuvette holder, water-bath; recorder or stopwatch.

Reagents and Solutions

Purity of reagents: phenyl acetate should be as pure as possible.

Preparation of solutions (for about 100 determinations): all solutions in re-purified water (cf. Vol. II, chapter 2.1.3.2).

1. Tris/acetate buffer (Tris, 100 mmol/l, pH 7.4; calcium chloride, 10 mmol/l):

   dissolve 12.1 g Tris and 1.47 g $CaCl_2 \cdot 2H_2O$ in about 800 ml water, adjust pH to 7.4 with acetic acid and dilute to 1000 ml.

2. Phenyl acetate solution (8.08 mmol/l):

   dissolve 0.1 ml phenyl acetate in 97.5 ml water.

3. Dilution buffer (Tris, 50 mmol/l; calcium chloride, 5 mmol/l):

   dilute Tris buffer (1) 1 : 2 with water.

Stability of solutions: solutions (1) and (3) are stable for several months, solution (2) for 5 days at 4 °C.

Procedure

Collection and treatment of sample: plasma or serum can be used, but the samples should not contain Ca-binding additives such as citrate, oxalate or EDTA. Heparin is without effect. Serum must be diluted 1 + 20 with dilution buffer (3).

Stability of the enzyme in serum: according to [7] ARE is stable for 5 days at 20 °C, for 6 weeks at 4 °C and for several months when stored frozen.

Details for measurement in tissue: like cholinesterase, ARE is probably a secretory enzyme; however, its site of synthesis is unknown. If arylesterase activity is to be measured in tissue homogenates or tissue extracts, respectively, interference by other esterases has to be considered. Serine esterases, e.g. microsomal carboxylesterase or cholinesterases, can be completely blocked by organophosphates. For example, a Paraoxon concentration of 0.01 mmol/l will inhibit both enzymes within minutes.

Assay conditions: wavelength 270 nm; light path 10 mm; final volume 2.02 ml; 25 °C (thermostatted cuvette holder). Measure against air. Run a reagent blank with isotonic saline instead of sample prior to each series of measurement.

Measurement

Pipette successively into cuvettes:			concentration in assay mixture	
Tris buffer	(1)	1.0 ml	Tris	49.5 mmol/l
phenyl acetate solution	(2)	1.0 ml	phenyl acetate	4 mmol/l
mix thoroughly with a plastic spatula; record blank for 2 – 3 min				
sample		0.02 ml	volume fraction	0.099
mix, read absorbance and start stopwatch; repeat the reading after 1, 2 and 3 min or monitor the reaction on a recorder.				

Calculation: use eqn. (k) or (k_1), cf. "Formulae", Appendix 3. The molar absorption coefficient is 148 $l \times mol^{-1} \times mm^{-1}$ at 270 nm [23]. The catalytic activity concentration of a serum sample is then calculated from

$$b = \frac{2.02 \times 1000}{0.148 \times 10 \times 0.02} \times F \times \Delta A/\Delta t = 68243 \times F \times \Delta A/\Delta t \quad U/l$$

$$b = 1.137 \times 10^{6} \times F \times \Delta A/\Delta t \quad nkat/l$$

F is dilution factor of serum (e.g. F = 21).

It is more suitable to use kU/l instead of U/l or mkat/l instead of nkat/l.

Validation of Method

Precision, accuracy, detection limits and sensitivity: for a catalytic activity in serum of 90.8 kU/l an imprecision of ± 6.1 kU/l has been obtained, corresponding to a relative standard deviation of 6.8%. Data about accuracy are not available as no standard reference material exists. The lowest possible value which can be measured is about 70 U/l if serum is not diluted. Sensitivity is dependent on the quality of the spectrophotometer used.

Sources of error: the influence of 36 drugs on human ARE was studied by *Marton & Kalow* [24]. The strongest inhibitors were chlorpromazine (70% inhibition), naphazoline (50%) and atropine (36%), all at a final test concentration of 0.1 mmol/l. The significance of these findings for the measurement of human serum ARE is, however, low, due to the much lower *in vivo* concentrations of these drugs and due to the small volume fraction of serum in the assay, i.e. 0.00047. Ca^{2+}-binding agents such as EDTA may interfere as ARE activity is dependent on calcium ions.

Specificity: since phenyl acetate is not specific for ARE, interference is theoretically possible by other esterases. Microsomal carboxylesterase is found in serum in necrotizing liver diseases. However, the highest level found (approximately 6 kU/l with ethyl valerate as substrate) would contribute 1.5 kU/l only, which is less than 2% of the activity measured for ARE activity in healthy subjects. A similar calculation is possibly valid for the interference by other esterases.

Reference ranges: in 117 healthy individuals (adults only) a normal range of 48 – 188 kU/l was calculated (mean ± 2 s). These values are identical to those found by *Lorentz et al.* [7], obtained by the oxidative coupling technique. In newborns the activities are distinctly lower [5], ranging from 2 – 52 kU/l, but are fully expressed at the age of two years [25].

References

[1] *W. N. Aldridge,* Serum esterases 2. An Enzyme Hydrolyzing Diethyl-p-nitrophenylphosphate (E 600) and its Identity with the A-esterase of Mammalian Sera, Biochem. J. *53,* 117 – 124 (1953).
[2] *K.-B. Augustinsson,* Electrophoretic Studies on Blood Plasma Esterases, Acta Chem. Scand. *13,* 571 – 592 (1959).
[3] *K.-B. Augustinsson,* Multiple Forms of Esterase in Vertebrate Blood Plasma, Ann. N. Y. Acad. Sci. *94,* 844 – 860 (1961).
[4] *K.-B. Augustinsson,* Arylesterases, J. Histochem. Cytochem. *12,* 744 – 747 (1964).
[5] *K.-B. Augustinsson, S. Brody,* Plasma Arylesterase Activity in Adults and Newborn Infants, Clin. Chim. Acta *7,* 560 – 565 (1962).
[6] *A. Marton, W. Kalow,* Studies on Aromatic Esterase and Cholinesterase of Human Serum, Cand. J. Biochem. *37,* 1367 – 1373 (1959).
[7] *K. Lorentz, B. Flatter, E. Augustin,* Arylesterase in Serum: Elaboration and Clinical Application of a Fixed-Incubation Method, Clin. Chem. *25,* 1714 – 1720 (1979).

[8] *A. Burlina, E. Michielin, L. Galzigna,* Characteristics and Behaviour of Arylesterase in Human Serum and Liver, Eur. J. Clin. Invest. *7*, 7 – 20 (1977).

[9] *B. J. Kitchen, C. J. Masters, D. J. Winzor,* Effects of Lipid Removal on the Molecular Size and Kinetic Properties of Bovine Plasma Arylesterase, Biochem. J. *135*, 93 – 99 (1973).

[10] *C. H. Brogren, T. C. Bog-Hansen,* Enzyme Characterization in Quantitative Immunelectrophoresis. An Enzymological Study of Human Serum Esterases, Scand. J. Immunol. *4*, Suppl. 2, 37 – 51 (1975).

[11] *W. Pilz,* Arylesterases, in: *H. U. Bergmeyer* (ed.), Methods of Enzymatic Analysis, Vol. 2, Verlag Chemie, Weinheim and Academic Press, New York 1974, pp. 806 – 813.

[12] *N. Kingsburg, C. J. Masters,* Molecular Weight Interrelationsship in the Vertebrate Esterases, Biochim. Biophys. Acta *200*, 58 – 69 (1970).

[13] *A. Burlina, L. Galzigna,* Serum Arylesterase Isoenzymes in Chronic Hepatitis, Clin. Biochem. *7*, 202 – 205 (1974).

[14] *C. E. Wilde, R. G. O. Keckwick,* The Arylesterases of Human Serum, Biochem. J. *91*, 297 – 307 (1964).

[15] *R. S. Holmes, C. J. Masters,* The Developmental Multiplicity and Isoenzyme Status of Cavian Esterases, Biochim. Biophys. Acta *132*, 379 – 399 (1967).

[16] *W. N. Aldridge,* Serum Esterases 1. Two Types of Esterase (A and B) Hydrolyzing p-Nitrophenyl Acetate, Propionate and Butyrate and a Method for Their Determination, Biochem. J. *53*, 110 – 117 (1953).

[17] *C. D. Koch, H. Block, J. Molz,* Zur Bestimmung der Arylesteraseaktivität im menschlichen Serum mit 4-Nitrophenylazetat, Z. Gastroenterologie *13*, 695 – 703 (1975).

[18] *A. Burlina, L. Galzigna,* A New and Simple Procedure for Serum Arylesterase, Clin. Chim. Acta *39*, 255 – 257 (1972).

[19] *P. W. Zapf, C. M. Coghan,* A Kinetic Method for the Estimation of Pseudocholinesterase Using Naphthyl Acetate Substrate, Clin. Chim. Acta *43*, 237 – 242 (1974).

[20] *K. Krisch,* Enzymatische Hydrolyse von Diäthyl-p-nitrophenylphosphat (E 600) durch menschliches Serum, J. Clin. Chem. Clin. Biochem. *6*, 43 – 45 (1968).

[21] *N. Aldridge,* Some Esterases of the Rat, Biochem. J. *57*, 692 – 702 (1954).

[22] *E. G. Erdös, C. R. Debay, M. P. Westerman,* Arylesterases in Blood: Effect of Calcium and Inhibitors. Biochem. Pharmacol. *5*, 173 – 186 (1960).

[23] *P. Greenzaid, W. P. Jencks,* Pig Liver Esterase. Reactions with Alcohols, Structure-activity Correlations, and the Acyl-enzyme Intermediate, Biochemistry *10*, 1210 – 1222 (1971).

[24] *A. V. Marton, W. Kalow,* Inhibitors of Aromatic Esterase of Human Serum, Biochem. Pharmacol. *3*, 149 – 154 (1960).

[25] *K.-B. Augustinsson, M. Barr,* Age Variation in Plasma Arylesterase Activity in Children, Clin. Chim. Acta *8*, 568 – 573 (1963).

1.3 Lipases

Triacylglycerol acylhydrolase, EC 3.1.1.3

1.3.1 Pancreatic Lipase

1.3.1.1 Titrimetric Method

Wolfgang Junge

General

The problem of characterizing and classifying the enzymes belonging to the group of carboxylester hydrolases (EC 3.1.1) becomes particularly apparent in the case of lipases. The definition of these enzymes is not based on the chemical structure of a substrate but rather on its physical state. However, the definition given by the Enzyme Nomenclature Committee that pancreatic lipase "... acts only on an ester-water interface ..." [1] is, for two main reasons, neither satisfying nor unambiguous. First, lipase splits not only emulsified substrates but also truly dissolved ones (e.g. 4-nitrophenyl acetate [2] or substrates in micellar solution [3, 4]. Secondly, the definition applies not only to lipases but also to esterases; e.g., microsomal carboxylesterase which shows high activity towards short-chain emulsified triglycerides such as tributyrin [5]. A practical approach to the problem of distinguishing both groups of enzymes is to regard as a lipase any enzyme which is capable of splitting emulsified long chain triglycerides, irrespective of whether or not it is also active on typical esterase substrates.

Lipases have been found in various tissues of many animals as well as in microorganisms and in plants. Among these enzymes, pancreatic lipase, which is present in the acinar cells of pancreatic tissue in high concentration, has attracted most interest, because it is easily available in large quantities, and because studies on its biochemical properties are directly relevant to the understanding of its physiological role in man. Furthermore, the determination of serum lipase activity is of outstanding clinical value in the diagnosis of acute pancreatic disorders.

Pancreatic lipase has been isolated from various mammals, including man. The enzymes are structurally closely related as judged from amino acid analyses, and from molecular and immunological properties. The diagnostic value of the enzyme is due to its absolute organ-specificity. The gradient of activity between pancreatic tissue and serum is approximately 20000. Due to its low molecular weight, lipase, like amylase, is rapidly removed from the circulation by glomerular filtration [6] but, in contrast to amylase, it is completely re-absorbed in the proximal tubules [7]. That is why lipase activity is not found in urine of pancreatitis patients, provided they have no renal dysfunction.

Knowledge of the mechanism of catalysis by lipase has been enormously increased during the last decade since the detection and purification of a physiologically important specific cofactor of lipase, so-called colipase, a small protein of molecular weight 10000 (for reviews cf. [8, 9]). In short, the function of colipase is to anchor the lipase molecule on the substrate surface in the presence of amphipaths, e.g. bile acids [10], proteins [11, 12], phospholipids [13] or free fatty acids [13], all of them occurring in the natural environment of lipase. If colipase is absent, access of lipase to the oil-water interface may be blocked in a competitive manner by the compounds [11, 14]. As with most effectors of lipase activity, their action is directed to the interface rather than to the enzyme itself, and therefore lipase is also active without its cofactor provided the concentration of amphipaths is not too high. Recent results indicate that colipase is an evolutionarily old protein, probably emerging simultaneously with the bile acids in vertebrates [15]. In contrast to lipase, colipase is synthesized as a precursor – procolipase – in the pancreas and is activated by trypsin [16]. Colipases purified from various animals and from man exhibit very similar properties: their similarity is illustrated particularly by the fact that they usually "co-operate" also with heterologous lipases.

Application of method: in biochemistry, clinical and pharmaceutical chemistry.

Enzyme properties relevant in analysis: in the presence of surface-active agents, e.g. bile acids or proteins, lipase activity is influenced by the colipase content of the sample. Sera from pancreatitis patients have been shown to contain variable amounts of the cofactor [17]; the same is true for duodenal juice [18]. There is no correlation with the lipase concentration of these fluids and, since the colipase concentration is usually well below that neccessary to ensure full lipase activity under assay conditions, an excess of the cofactor has to be added to the reagent mixture to exclude any influence of the colipase content of the sample on the result. At the oil-water interface, lipase undergoes progressive irreversible inactivation due to unfolding of the molecule. This process can largely be overcome by bile acids [19], which must therefore be included in the assay. Isoenzymes of lipase have not been demonstrated so far in serum. The activity measured in human serum is not due to lipase alone, but also to another esterolytic enzyme of pancreatic origin [20].

Methods of determination: the numerous methods described for measurement of lipase activity can be divided into three types: either the liberated alkyl or acyl moiety is determined, or the disappearance of substrate is followed provided it is present as an emulsion.

Assays based on measuring alkyl-release have hardly found application due to lack of specificity: almost all n-monoacyl esters of directly chromogenic alcohols (4-nitrophenyl, fluorescein), or of those forming azo-dyes (phenol, naphthol) are cleaved by carboxylesterases. Based on the observation that ω-phenyloctanoate vinylester is rapidly hydrolyzed by pig pancreatic lipase [21], *Myrick* [3] developed an UV-method for serum lipase. However, in contrast to pig lipase, the human enzyme has only low activity towards this substrate, but approximately 90% of the activity found in human pancreatic tissue is due to another enzyme, provisionally called non-specific pancreatic carboxylesterase [20]. Furthermore, the vinyl ester is also split by human liver microsomal carboxylesterase [4] and by arylesterase (*G. Bürkle, W. Junge*).

In another new method, 2,3-dimercaptopropan-1-ol tributyrate is employed as substrate [22, 24], the SH-groups formed on hydrolysis being followed by reaction with DTNB. Interfering carboxylesterases are blocked with phenyl methane sulphonyl fluoride, but the efficiency of the inhibition step has not been proved so far. Phenyl methane sulphonyl fluoride is extremely unstable at pH values above 7 [25] and, in addition, arylesterase is not inhibited by this compound [26].

In turbidimetric or nephelometric methods the clearing of a bile acid-stabilized substrate emulsion, usually olive oil or triolein, is followed photometrically. These assays are easy to perform and highly specific, and therefore have found wide application in the routine clinical laboratory. *Guth* [27] was the first to introduce this technique for diagnostic purposes and modifications of his method were described by a number of authors. A radial diffusion method also belongs to this type of assay [28].

It was recently shown, however, that pure human lipase is almost completely inactive in turbidimetric assays [14]. The inhibition is due to the high concentration of deoxycholate which can only be overcome by colipase [14]. Sera of patients with pancreatitis exhibit elevated lipase activity only because they contain colipase [17]. The activity as measured by turbidimetric tests is therefore not a function of the lipase concentration alone, but of the molar ratio of colipase to lipase which shows large variation [17]. Consequently, these assays have to be carried out in the presence of a sufficient colipase concentration (>250 molar excess) to exclude any influence of the endogenous colipase content of the sample.

A turbidimetric method which includes an excess of colipase has been elaborated [29] and is described in the following chapter 1.3.1.2, p. 26. A minor drawback of turbidimetric methods is that the photometric signal cannot be converted directly to U/l, but a standard of known lipase activity has to be used [29]. The clearing reaction can also be related to the formation of fatty acids [30] measured by the copper-salt method.

It is generally accepted that titrimetric methods employing long-chain triglycerides such as triolein (or olive oil as its natural substitute) are the most reliable ones. Short- and medium-chain triglycerides are split at a higher rate than triolein but these substrates are not specific for lipase. However, emulsified tributyrin is a widely used model substrate for studies on the catalytic properties of purified pancreatic lipase, due to its favourable physicochemical properties.

The classical discontinuous technique as first described by *Cherry & Crandall* [31] and modified by several authors has the known drawbacks of non-linearity with time, so that initial velocity is not measured. The shortcomings of discontinuous techniques were overcome by introducing a continuous-monitoring titrimetric assay using a pH-stat equipment. The original procedure [32] was adapted to serum lipase determination [33, 34] and is also the basis of the method described here. These continuous assays are regarded as reference methods for serum lipase. However, for reasons given below, this seems to be no longer justified.

Formation of fatty acids can be measured by the copper-salt method [35] or by means of an acid-base indicator present in the assay [36, 37] instead of by titration. Another approach was developed by *Proelss & Wright* [38] employing trilinolein as substrate. The liberated fatty acid is reacted with O_2 by lipoxygenase catalysis fo form a hydroperoxide which itself oxidizes Fe^{2+} to Fe^{3+}, the latter forming a coloured product with thiocyanate. It was recently shown that this assay does not give satisfactory results [39].

No method for active-site titration is known. Interestingly, the serine which is phosphorylated by diethyl 4-nitrophenyl phosphate in a stoichiometric reaction is not part of the active site of lipase but of its binding centre [40].

An immunochemical method for lipase was recently reported by *Grenner et al.* [41].

International reference methods and standards: the assay used in clinical chemistry has not been standardized at the international level nor is the existence of an internationally adopted standard reference material known. It will be difficult to establish a reference method due to the specificity problem associated with lipase substrates*.

Enzyme effectors

Calcium, sodium chloride, gum acacia: calcium ions stabilize lipase activity by formation of the respective salts with the free fatty acids released at the interface [42]. It was shown [13] recently that inhibition by oleic acid of lipase-catalyzed hydrolysis of tributyrin is stronger in the absence than in the presence of calcium. The activating effect of sodium chloride on lipase activity is not due to a stimulation of the enzyme but is explained by lowering the oleic acid pK value. Hence oleic acid is titrated more quantitatively [42]. In a triolein emulsion stabilized by gum *acacia*, more than 90% of the released fatty acids are titrated at pH 8.3 [43].

Bile acids: an enormous amount of literature exists on the effects of bile acids on lipase activity. Depending on the assay system used and on their final concentration in the reaction mixture, these detergents have been reported to be activators as well as inhibitors of the enzyme [19]. However, most studies carried out before the discovery of colipase in 1972 are difficult to comment on because the colipase content of the

* A standardized method was published by the Fédération International Pharmaceutique (FIP): Description of the Assay Method for Microbial Lipases, in: *R. Ruyssen, A. Lauwers,* Pharmaceutical Enzymes, Properties, Assay Methods, E. Story-Scientia, Ghent 1978, pp. 210–213.

lipase preparations used is unknown. The role of bile acids in lipase activity is presently understood as follows: a) at inframicellar concentration, bile acids stabilize lipase against unfolding at the interface (lipase is not activated); b) by increasing the concentration, lipase activity is progressively inhibited due to blockage of the interface (or by displacing the enzyme from the interface). This inhibition can be reversed by colipase; c) comparable to their effect on lipase, bile acids clear the interface from various other amphipaths such as protein or phospholipids, hence enabling the lipase-colipase complex to bind to the substrate; d) bile acids are important for the binding between lipase and colipase by forming a ternary complex with both proteins [8, 9].

Proteins: proteins exhibit a similar effect to bile acids on lipase activity. It was found that serum albumin at low concentration ($<10^{-7}$ mol/l) protects the lipase molecule from unfolding at the interface [19], but at higher concentration sterically hinders adsorption of the enzyme. A *Lineweaver-Burk* plot showed the inhibition of lipase by albumin [11], and also by whole serum [14], to be competitive. Colipase alone cannot restore lipase activity caused by protein inhibition, but only in the presence of bile acids [11]. A high protein concentration usually exists in continuous titrimetric assays due to the relatively large volumes of serum used. In the method of [33] its volume fraction may not exceed 0.067, otherwise no linearity with amount of sample is obtained. The decreased activity seen in many lipase assays when too great a sample volume is assayed is not due to the existence of serum lipase inhibitors (as discussed by several investigators [28, 33, 44, 45]), but simply to protein inhibition. In the assay to be described the sample volume comprises 33% of the final volume. However, inhibiting effects are overcome by the presence of bile acid and colipase.

Colipase: the various effects of colipase have already been mentioned. Although assay conditions may be experimentally established under which the cofactor effects no increase in activity, its exclusion from an assay for lipase is not useful, because in that case optimal conditions have to be determined for each individual sample.

Assay

Method Design

Principle

$$\text{Glyceryl trioleate} + H_2O \xrightarrow{\text{lipase}} \text{glyceryl dioleate} + \text{oleic anion} + H^+$$

The protons formed are continuously titrated with NaOH, 10 mmol/l, at pH 8.8.

Optimized conditions for measurement: with regard to optimal concentrations of substrate and gum *acacia* the method of [33] is followed. Colipase is added at a final concentration of 13 mg/l to ensure full activity of lipase. Without the cofactor a narrow concentration optimum for glycocholate exists which is dependent on the volume, as well as on the colipase content, of the sample (Fig. 1). Thus an optimal bile acid concentration can only be given for an individual sample. *Hockeborn & Rick* [46] reported concentration optima differing by 100% for glycocholate lots available before and after 1980, assuming differing purity of the batches. However, gas-chromatographic analysis showed identical purity of the respective lots (*S. Führ, W. Junge*). It appears more likely that the results were obtained with samples of different colipase content. Furthermore, the non-linear response of activity to the amount of serum added above a volume fraction of 0.067 [33] is abolished by colipase (Fig. 2).

pH 8.8 is optimal for lipase which exhibits a broad optimum between pH 8 and 9 [19].

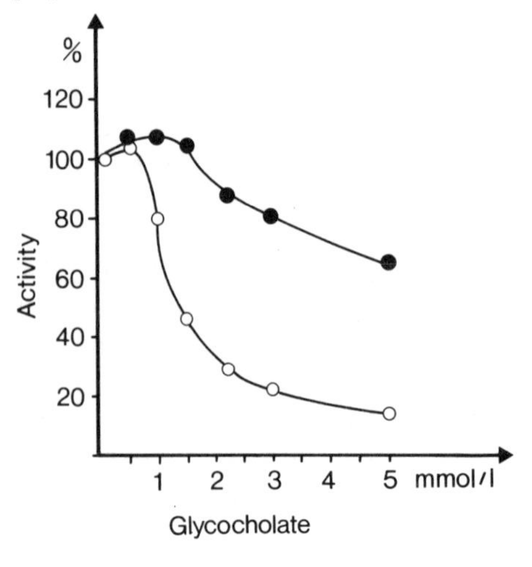

Fig. 1. Influence of glycocholate on the activity of two pancreatitis sera with different colipase contents. A: lipase 1125 U/l, colipase/lipase molar ratio 1.2; B: lipase 2813 U/l, ratio 0.21.

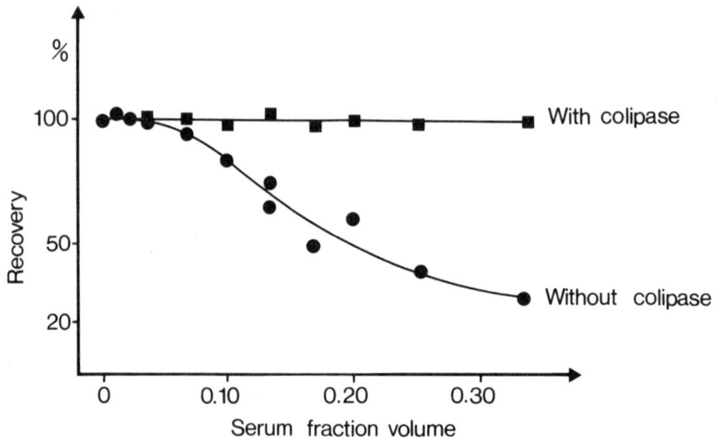

Fig. 2. Influence of pooled serum on the recovery of pure human pancreatic lipase in the presence and absence of colipase (final concentration 2.4 mg/l). Titrimetric test with glycocholate, 0.5 mmol/l.

Due to the presence of atmospheric CO_2 which will be absorbed by the assay mixture, a blank must be measured prior to each series of measurement. Alternatively, a stream of nitrogen may be blown over the reaction mixture to prevent CO_2-absorption [33, 34].

Temperature conversion factors: temperature conversion factors are not yet established for lipase. In assays without colipase, lipase undergoes temperature dependent inactivation at the interface [47, 48].

Equipment

pH Meter connected with titrator, automatic burette and recorder (e.g. equipment as available from *Radiometer*, Copenhagen, Denmark, consisting of pH-meter PHM 61 or pH-meter 26; titrators TTT 60, Titrator 11 or TTT 1c; autoburette ABU 12 or ABU 13 with burette B 230; recorder Servograph REC 61 with mechanical drive or interface REA 200, thermostatted reaction vessel for 3 – 5 ml, micro-electrodes G 2222 C (glass) and K 4112 (calomel). The titrant should be protected from atmospheric CO_2 by $Ca(OH)_2$. *Waring* blender. Water-bath.

Reagents and Solutions

Purity of reagents: the method described here was evaluated with a triolein lot of >95% purity (*Serva-Chemie,* Heidelberg). Glycocholate should be of analytical grade. Colipase must be free from lipase.

Preparation of solutions (for about 50 determinations): all solutions in re-purified water (cf. Vol. II, chapter 2.1.3.2).

1. Gum *acacia* solution (100 g/l):

   dissolve 10 g gum *acacia* in about 80 ml water and dilute to 100 ml with water.

2. Triolein emulsion (206 mmol/l):

   add 20 ml triolein to 80 ml gum *acacia* solution (1), emulsify for at least 5 min in a *Waring* blender. Adjust pH to 8.8 with NaOH, 1 mol/l, mix again for 2 – 3 min; control pH.

3. Sodium glycocholate (155 mmol/l):

   dissolve 378 mg glycocholate, sodium salt, in 5 ml water.

4. Sodium hydroxide (10 mmol/l):

   dilute 50 ml of a standard NaOH solution, 0.1 mol/l, with 450 ml water.

5. Colipase solution (2 mg/ml):

   dissolve 5 mg pig colipase in 2.5 ml water.

Stability of solutions: solutions (1) – (5) are stable for several weeks on storage at 4 °C, provided contamination by micro-organisms is excluded. The substrate should be re-emulsified for 2 – 3 min after longer storage.

Procedure

Collection and treatment of sample: serum or plasma can be used. Samples containing Ca^{2+}-binding additives (e.g. EDTA, citrate, oxalate) should not be used.

Stability of the enzyme in the sample: lipase has excellent stability in serum. At room temperature and at 4 °C the enzyme is stable for at least 3 weeks. Kept frozen at – 20 °C, lipase was reported to be stable for more than 10 years [48].

Assay conditions: pH 8.8; final volume 3.06 ml; 25 °C (thermostatted reaction vessel). Before starting adjust temperature to 25 °C. Run a blank (cf. p. 21) with isotonic saline instead of sample.

Measurement

Pipette successively into the reaction vessel:			concentration in assay mixture	
triolein emulsion	(2)	2.0 ml	triolein gum *acacia*	135 mmol/l 52.3 g/l
glycocholate solution	(3)	0.04 ml	glycocholate	2.0 mmol/l
colipase solution	(4)	0.02 ml	colipase	13 mg/l
sample or blank		1.0 ml	volume fraction	0.33
start stirrer and adjust pH slightly above 8.8 by addition of small volumes of NaOH, 0.1 mol/l; start recorder and monitor the reaction for up to 10 min, depending on the activity; for blank determination bring pH slightly below 8.8 until the burette valve is open (observe valve control light on titrator); start titration at lowest speed of the burette motor. If a sample kept at 4 °C is added, temperature equilibrium will be reached in less than 2 min.				

If NaOH consumption exceeds 75 μl/min, corresponding to 30% of the burette volume, dilute sample with isotonic saline.

Calculation: 0.1 ml of titrant corresponds to 1 micromole NaOH. Accordingly the catalytic activity concentration of a serum sample is

$$b = 10 \times V/\Delta t \quad U/l$$

$$b = 167 \times V/\Delta t \quad nkat/l$$

where V is the volume of titrant in μl.

Validation of Method

Precision, accuracy, detection limits and sensitivity: the imprecision was found to be ± 44 U/l with a serum of catalytic activity 745 U/l, the standard deviation for within-run imprecision being ± 30 U/l. Due to lack of standard reference material data about accuracy are not yet available. Detection limit is approximately 10 – 15 U/l. The sensitivity depends largely on the quality and age of the electrodes.

Sources of error: if equipment of proper quality and sensitivity is used, disturbances are mainly caused by contamination of the electrodes with protein and/or fat. The reaction vessel, the micro-tip delivering the titrant, and the electrodes should be carefully cleaned after each determination. Other sources of error are technical defects such as improper mixing, air-bubbles in the titrant-tube system or leakage of three-way cock. Haemoglobin, even in high concentration, does not interfere [46]. The influence of drugs is unknown.

Specificity: triolein is not a specific substrate for pancreatic lipase. Patients under heparin therapy will exhibit slightly elevated serum triolein-hydrolyzing activity due to lipoprotein lipase (cf. chapter 1.3.3). A further enzyme which was found to contribute up to 12% of serum lipase catalytic activity in the titrimetric test is non-specific pancreatic carboxylesterase [20]. This enzyme is pancreas-specific and is released into the serum together with lipase in pancreatic disorders: thus it does not impair the diagnostic significance of the assay. However, the continuous titrimetric assay can no longer be considered as a standard or reference method for serum lipase, and it is questionable whether the establishment of such a test is at all possible.

Reference ranges: in 29 sera from healthy individuals we found catalytic concentrations between 15 and 80 U/l. These preliminary figures indicate that the normal range for the method described here is probably similar to that given by *Rick* [33] for his method, i.e. 11 – 138 U/l.

References

[1] Enzyme Nomenclature 1978. Recommendations of the Nomenclature Committee of the International Union of Biochemistry on the Nomenclature and Classification of Enzymes, Academic Press, New York 1979.

[2] *M. Semeriva, C. Chapus, C. Bovier-Lapierre, P. Desnuelle,* On the Transient Formation of an Acetyl Enzyme Intermediate During the Hydrolysis of p-Nitrophenyl Acetate by Pancreatic Lipase, Biochem. Biophys. Res. Commun. *58*, 808 – 813 (1974).

[3] *J. E. Myrick,* A Spectrophotometric Method for the Kinetic Measurement of Lipase Activity, A Dissertation, Birmingham, Alabama 1976.

[4] *W. Junge, K. Leybold,* Hydrolyse des neu eingeführten Lipasesubstrats 8-Phenyloctansäurevinylester durch Carboxylesterase aus Menschenleber, Laboratoriumsmedizin *1*, 149 – 150 (1977).

[5] *K. Krisch,* Isolierung einer Esterase aus Schweinelebermikrosomen, Biochem. Z. *337*, 546 – 573 (1963).

[6] *W. Junge, M. Malyusz,* The Role of the Kidney in the Elimination of Pancreatic Lipase from Serum. V. Eur. Congr. Clin. Chem., Budapest 1983, p. 298.

[7] *M. Malyusz, H. J. Ehrens, W. Junge,* Degradation and Elimination of ^{125}I-lipase by the Kidney. V. Eur. Congr. Clin. Chem., Budapest 1983, p. 352.

[8] *M. Semeriva, P. Desnuelle,* Pancreatic Lipase and Colipase. An Example of Heterogenous Biocatalysis, in: *A. Meister* (ed.), Adv. in Enzymology, Vol. 48, J. Wiley & Sons, New York, Chichester, Brisbane, Toronto 1979, pp. 319 – 370.

[9] *B. Borgström, C. Erlanson-Albertsson, T. Wieloch,* Pancreatic Colipase: Chemistry and Physiology, J. Lipid Res. *20*, 805 – 816 (1979).

[10] *B. Borgström, C. Erlanson,* Pancreatic Lipase and Colipase, Eur. J. Biochem. *37*, 60 – 68 (1973).

[11] *B. Borgström, C. Erlanson,* Interactions of Serum Albumin and Other Proteins with Porcine Pancreatic Lipase, Gastroenterology *75*, 382 – 386 (1978).

[12] *L. Bläckberg, O. Hernell, G. Bengtsson, T. Olivecrona,* Colipase Enhances Hydrolysis of Dietary Triglycerides in the Absence of Bile Salts, J. Clin. Invest. *64*, 1303 – 1308 (1979).

[13] *A. Larsson, C. Erlanson-Albertsson,* The Importance of Bile Salt for the Reactivation of Pancreatic Lipase by Colipase, Biochim. Biophys. Acta *750*, 171 – 177 (1983).

[14] *W. Junge, K. Leybold, B. Kraack,* Influence of Colipase on the Turbidimetric Determination of Pancreatic Lipase Catalytic Activity, J. Clin. Chem. Clin. Biochem. *21*, 445 – 451 (1983).

[15] *B. Sternby, A. Larsson and B. Bergström,* Evolutionary Studies on Pancreatic Colipase, Biochim. Biophys. Acta *750*, 340 – 345 (1983).

[16] *T. Wieloch, B. Borgström, G. Pieroni, F. Pattus, R. Verger,* Porcine Pancreatic Procolipase and its Trypsin-activated Form, FEBS Letters *128*, 217 – 220 (1981).

[17] *W. Junge, K. Leybold, B. Kraack,* Detection of Colipase in Serum and Urine of Pancreatitis Patients, Clin. Chim. Acta *123*, 293 – 302 (1982).

[18] *K. J. Gaskin, P. R. Durie, R. E. Hill, L. M. Lee, G. G. Forstner,* Colipase and Maximally Activated Pancreatic Lipase in Normal Subjects and Patients with Steatorrhea, J. Clin. Invest. *69*, 427 – 434 (1982).

[19] *H. Brockerhoff, R. G. Jensen,* Lipolytic Enzymes, Academic Press, New York 1974.

[20] *W. Junge, K. Leybold, B. Philipp,* Identification of a Non-specific Carboxylesterase in Human Pancreas using Vinyl 8-Phenyloctanoate as a Substrate, Clin. Chim. Acta *94*, 109 – 114 (1974).

[21] *H. Brockerhoff,* Substrate Specificity of Pancreatic Lipase. Influence of the Structure of Fatty Acids on the Reactivity of Esters, Biochim. Biophys. Acta *212*, 92 – 101 (1970).

[22] *S. Kurooka, M. Hashimoto, M. Tomita, A. Maki, Y. Yoshimura,* Relationship between the Structures of S-Acyl Thiol Compounds and Their Rates of Hydrolysis by Pancreatic Lipase and Hepatic Carboxylic Esterase, J. Biochem. *79*, 533 – 541 (1976).

[23] *S. Kurooka, S. Okamoto, M. Hashimoto,* A Novel and Simple Colorimetric Assay for Human Serum Lipase, J. Biochem. (Tokyo) *81*, 361 – 369 (1977).

[24] *I. Furukawa, S. Kuroowa, K. Aisue, K. Kohda, C. Hayashi,* Assays of Serum Lipase by the "BALB-DTNB Method" Mechanized for Use with Discrete and Continuous-Flow-Analyzers, Clin. Chem. *28*, 110 – 113 (1982).

[25] *E. Heymann, W. Junge, K. Krisch,* Carboxylesterase aus Schweinelebermikrosomen, Reaktion mit Phenylmethansulfonylfluorid und Nachweis von Isoenzymen, Hoppe-Seyler's Z. Physiol. Chem. *353*, 576 – 588 (1972).
[26] *K. Lorentz, B. Flatter, E. Augustin,* Arylesterase in Serum: Elaboration and Clinical Application of a Fixed-Incubation Method, Clin. Chem. *25*, 1714 – 1720 (1979).
[27] *P. H. Guth,* Evaluation of Phototurbidimetric Technics for the Determination of Serum Amylase, Lipase and Esterase, Am. J. Gastroenterol. *33*, 319 – 334 (1960).
[28] *J. M. Goldberg, P. Pagast,* Evaluation of Lipase Activity in Serum by Radial Enzyme Diffusion, Clin. Chem. *22*, 633 – 637 (1976).
[29] *J. Ziegenhorn, U. Neumann, K. W. Knitsch, W. Zwez, A. Roeder, H. Lenz,* Lipase – eine Testcharakteristik, Medica *1*, 919 – 925 (1980).
[30] *P. A. Verduin, J. M. H. M. Punt, H. H. Kreutzer,* Studies on the Determination of Lipase Activity, Clin. Chim. Acta *46*, 11 – 19 (1973).
[31] *I. S. Cherry, L. A. Crandall,* The Specificity of Pancreatic Lipase: Its Appearance in the Blood after Pancreatic Injury, Am. J. Physiol. *100*, 266 – 273 (1932).
[32] *G. Marchis-Mouren, L. Sarda, P. Desnuelle,* Purification of Hog Pancreatic Lipase, Arch. Biochem. Biophys. *83*, 309 – 319 (1959).
[33] *W. Rick,* Kinetischer Test zur Bestimmung der Serumlipaseaktivität, Z. klin. Chem. klin. Biochem. *7*, 530 – 539 (1969).
[34] *N. W. Tietz, E. V. Repique,* Proposed Standard Method for Measuring Lipase Activity in Serum by Continuous Sampling Technique, Clin. Chem. *19*, 1268 – 1275 (1973).
[35] *R. Fried, J. Hoeflmayr,* Die Bestimmung der Lipaseaktivität unter Verwendung von Olivenöl in neuer Form, J. Clin. Chem. Clin. Biochem. *11*, 189 – 192 (1973).
[36] *A. Härtel, D. Banauch, R. Helger,* Ein Suchtest für Lipase in Serum, J. Clin. Chem. Clin. Biochem. *9*, 396 – 397 (1971).
[37] *G. Hillmann, G. Weidemann,* Continuous Photometric Measurement of Lipase Activity with the Substrate Triolein, J. Clin. Chem. Clin. Biochem. *11*, 257 – 258 (1973).
[38] *H. F. Proelss, B. W. Wright,* Lipoxygenic Micromethod for Specific Determination of Lipase Activity in Serum and Duodenal Fluid, Clin. Chem. *23*, 522 – 531 (1977).
[39] *W. Rick, M. Hockeborn,* Zur Bestimmung der katalytischen Aktivität der Lipase mit Trilinolein als Substrat, J. Clin. Chem. Clin. Biochem. *20*, 745 – 752 (1982).
[40] *A. Guidoni, F. Benkonka, J. C. de Caro, M. Rovery,* Characterization of the Serine Reacting with Diethyl p-Nitrophenyl Phosphate in Porcine Pancreatic Lipase, Biochim. Biophys. Acta *660*, 148 – 150 (1981).
[41] *G. Grenner, G. Deutsch, R. Schmidtberger, F. Dati,* Hochempfindlicher Enzymimmunoassay zur Bestimmung der Human-Pankreas-Lipase, J. Clin. Chem. Clin. Biochem. *20*, 515 – 519 (1982).
[42] *G. Benzonana, P. Desnuelle,* Action of some Effectors on the Hydrolysis of Long-chain Triglycerides by Pancreatic Lipase, Biochim. Biophys. Acta *164*, 47 – 58 (1968).
[43] *M. W. Ringler, J. S. Patton,* The Production of Liquid Crystalline Product Phases by Pancreatic Lipase in the Absence of Bile Salts, Biochim. Biophys. Acta *751*, 444 – 454 (1983).
[44] *R. W. Forsman, H. Steige, J. D. Jones, M. Sass, J. D. Storey,* An Evaluation of a Lipase Method for the *Du Pont* ACA, *Du Pont*, Wilmington 1977.
[45] *J. Prokopowicz, M. Rublewska,* Pancreatic Lipase and its Inhibitors, in: *D. M. Goldberg, J. H. Wilkinson* (eds.), Enzymes in Health and Disease, S. Karger, Basel 1978, pp. 123 – 126.
[46] *M. Hockeborn, W. Rick,* Zur Bestimmung der katalytischen Aktivität der Lipase mit dem kontinuierlichen titrimetrischen Test, J. Clin. Chem. Clin. Biochem. *20*, 773 – 785 (1982).
[47] *S. Granon, M. Semeriva,* Effect of Taurodeoxycholate, Colipase and Temperature on the Interfacial Inactivation of Porcine Pancreatic Lipase, Eur. J. Biochem. *111*, 117 – 124 (1980).
[48] *B. Borgström,* The Temperature-dependent Interfacial Inactivation of Porcine Pancreatic Lipase. Effect of Colipase and Bile Salt, Biochim. Biophys. Acta *712*, 490 – 497 (1982).

1.3.1.2 Turbidimetric Method

Ulrich Neumann, Peter Kaspar and Joachim Ziegenhorn

General

Lipases are characterized by their ability to liberate fatty acids from emulsified triglycerides and other emulsified esters. They can be distinguished from other esterases by their preferential action at the oil-water interface whereas the latter act in solution.

Although the existence of lipase in pancreas had already been demonstrated in 1856 by *Bernard* [1], successful purification was not achieved until 1957 by *Desnuelle* and co-workers [2]. Lipases were isolated from a variety of vertebrates, non-vertebrates, plants and micro-organisms [3, 4]. The lipases from vertebrates can be classified into lipases from the gastro-intestinal tract, milk lipases, and tissue lipases. The lipoprotein-lipase was considered for a long time a member of the latter group but nowadays it has been assigned its own EC number, 3.1.1.34.

Pancreatic lipase, which is responsible for the digestion of dietary fats, has been investigated with particular thoroughness [5]. Human pancreatic lipase is a glycoprotein with a molecular weight of approx. 46000 – 48000 [6 – 7]. It acts under physiological conditions together with colipase. Human colipase is described as a glycoprotein with a molecular weight of approx. 10000 [5, 8, 9]. This co-factor is secreted by the pancreas as an inactive precursor form. After cleavage of an N-terminal oligopeptide by trypsin, an activated molecule is obtained which can form a complex with lipase [10, 11]. According to [12], the true lipolytic entity *in vivo* has to be considered as a complex of lipase, colipase, and bile lipoproteins.

Application of method: its application in clinical chemistry is in the diagnosis of acute pancreatitis. Furthermore, it can be used for testing pancreatin-containing preparations in pharmacy. The same system is also suitable for the determination of colipase samples, if one replaces colipase by lipase in the assay mixture.

Enzyme properties relevant in analysis: lipase is a rather non-specific esterase which splits not only triglycerides but also emulsified esters of n-alcohols and fatty acids without branch-points near the ester bond [14]. The water-insolubility of a substrate is of utmost importance: e.g. esters of limited water-solubility such as triacetin are poor substrates for lipase but good substrates for esterases in concentrations below their limit of solubility, whereas at supersaturating substrate concentrations lipase activity against the same substrates is strongly enhanced [15].

As the lipase acts only at the oil-water interface the reaction velocity is not determined by the concentration of the emulsified ester in the aqueous phase, but rather by the surface area of the emulsified particles per unit volume [16]. Consequently, the particle-size distribution of the emulsion and its stability strongly influences the observed enzyme activity.

Bile acids – physiological emulsifiers in the intestine – exert a slightly activating effect at low concentrations *in vitro* while higher concentrations strongly inhibit lipase activity [4, 14]. This inhibition can be overcome by addition of colipase [13, 17]. Because this effect is typical of pancreatic lipase, it can be used to enhance the specificity of the assay for the pancreatic enzyme [13, 25]. As a result of the high degree of similarity between colipases of different species, activation of human lipase can also be achieved by porcine or bovine colipase [1, 13, 18].

The occurrence and duration of a lag phase depend on the reaction conditions (kind of substrate, pH, concentration of bile acids and free fatty acids) [13, 17].

Methods of determination: for determination of lipase activity, titrimetric, turbidimetric, nephelometric, fluorimetric, photometric, polarographic and immunological methods are described [13, 19–22]. Among the continuous monitoring methods, only the titrimetric methods offer the advantage of providing results directly in international units [23, 24]. However, they are time-consuming and require special equipment. In contrast, measurements by the turbidimetric lipase method described here can easily and reproducibly be carried out with photometers and analyzers commonly available in the routine clinical laboratory [13, 20, 25]. In addition, this method provides improved specificity in comparison with the lipase assays so far reported because it discriminates between the pancreatic enzyme and serum esterases, or lipases of other origin [13, 20, 25]. However, the turbidimetric procedure requires an enzyme standard which has been calibrated by a titrimetric method (cf. chapter 1.3.1.1).

International reference method and standards: so far, neither an international reference method nor an international enzyme standard exists for the lipase assay. For the turbidimetric method described here, calibration of an enzyme standard by a titrimetric method is recommended.

Enzyme effectors: the influence of bile acids and colipase has already been described. NaCl can activate lipase within certain concentration ranges which depend on the whole assay conditions [4, 10, 26]. Calcium ions improve the thermal stability of the enzyme and can enhance lipase activity (presumably by binding liberated fatty acids in the form of calcium soaps). In the basic pH range, calcium ions are essential for the fixation of lipase at the oil-water interface [4, 14]. The presence of bile phospholipids may improve the adsorption of the lipase-colipase complex onto the surface of the oil droplets [12].

Assay

Method Design

Principle: the fatty acid residues of a triacylglycerol molecule are hydrolyzed stepwise by the action of lipase.

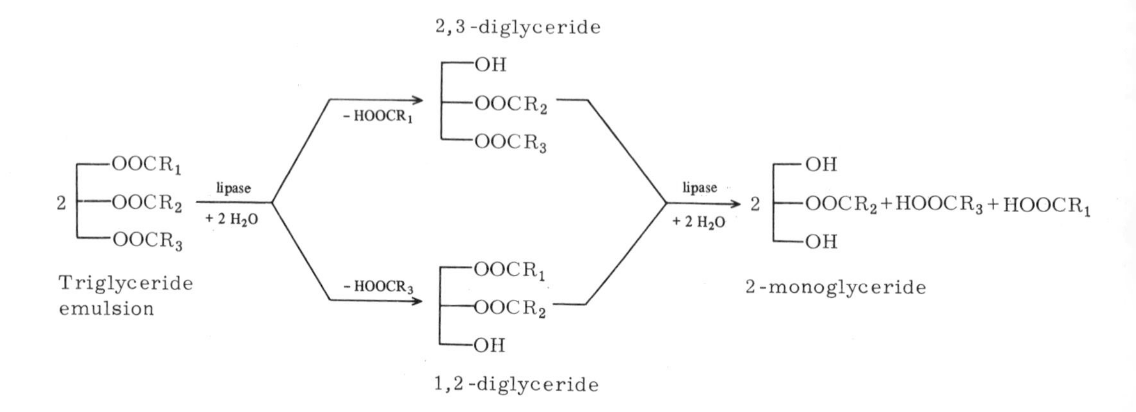

Human pancreatic lipase hydrolyzes triglycerides through a fast step to diglycerides, followed by slower cleavage to monoglycerides. The final hydrolysis to glycerol is the slowest reaction. Lipase always hydrolyzes the α-ester bond. An isomerization of a β-fatty acid residue to the α-position is described [14].

By the action of lipase on the oil emulsion, the diameters of the oil droplets are continuously reduced, while the free fatty acids dissolve in the alkaline buffer solution. The resulting decrease of turbidity of the emulsion in a given time is proportional to the activity of lipase. The best sensitivity is obtained by measuring at shorter wavelengths.

Under the assay conditions described, linear kinetics are obtained between the 4th to the 10th minute. A lag phase may occur at the beginning of the reaction [13].

As turbidimetric signals are strongly dependent on the construction of the optical system of the photometer used, they cannot be converted into enzyme activities by using a constant factor. The factor has to be established for each photometer by the use of an enzyme standard. The activity of this standard has to be determined by a method which measures directly the liberation of fatty acids in unit time.

Optimized conditions for measurement [13]

pH Range: the optimum pH for human pancreatic lipase activity lies between 8.5 – 9.4 [27] using sodium deoxycholate as emulsifier. The pH optimum may be shifted to lower pH values if other bile salts are employed.

Deoxycholate: high deoxycholate concentrations in the presence of colipase inhibit serum lipases and esterases which do not originate from pancreatic tissue while pancreatic lipase activity is retained. Furthermore, they minimize lag phases and ensure that the reaction follows zero order kinetics over a wide range of lipase activity concentration. In the present method, the optimum concentration of deoxycholate was established to be 18 mmol/l. Deoxycholate is available in high purity.

Colipase: in order to keep pancreatic lipase active at a deoxycholate concentration of 18 mmol/l, colipase concentrations of 7×10^4 to 2×10^6 U/l are needed.

Triolein: in contrast to olive oil which has been widely employed as lipase substrate in the past [23], triolein can be obtained commercially in high purity and consistent quality. Therefore, this substance is used as substrate in the following method. The optimum triolein concentration is about 0.3 mmol/l.

Chloride: lipase activity cannot be detected in the complete absence of salts at pH 9.0. Maximum lipase activity can be obtained at a NaCl concentration of about 140 mmol/l [26]; however, to obtain a stable emulsion, concentrations of 30 – 40 mmol/l should not be exceeded.

Calcium: the presence of calcium ions is necessary to obtain linear kinetics. 1 mmol/l is used. At concentrations higher than 2 mmol/l precipitation of insoluble calcium deoxycholate is possible.

Equipment

A spectrophotometer or spectral-line photometer capable of measurement at 339 nm (Hg 334 or Hg 365 nm) is preferred. It is also possible to measure at Hg 405 nm but with loss of detectability. A thermostatted cuvette holder, and a recorder or a stop-watch, are necessary. The assay can also be carried out with automated instruments [13, 20].

Reagents and Solutions

Purity of reagents: the reagents must be free from heavy metal ions, detergents interacting with surface of the emulsion droplets, and substances capable of inhibiting serine enzymes.

Preparation of solutions (for about 25 determinations): all solutions in re-purified water (cf. Vol. II, chapter 2.1.3.2).

1. Tris buffer (0.125 mol/l; pH 9.2):

   dissolve 1.535 g Tris in about 80 ml water, adjust to pH 9.2 with HCl, 1 mol/l, and dilute to 100 ml with water.

2. Calcium solution (100 mol/l):

   dissolve 1.47 g $CaCl_2 \cdot 2\,H_2O$ in 100 ml water.

3. Colipase solution (3000 U/l, 25 °C):

   dissolve 3 mg lyophilized colipase (ca. 70000 units*/mg, *Boehringer Mannheim*) in 1.0 ml water.

4. Buffer/reagent solution:

   add to 10 ml Tris buffer (1) 0.05 ml calcium solution (2), 0.1 ml colipase solution (3), about 30 ml water and dissolve therein 373 mg sodium deoxycholate and 106 mg sodium chloride and make up to 50 ml with water.

5. Triolein solution (11.3 mmol/l):

   dissolve 200 mg triolein (ca. 95%) in 20 ml *n*-propanol.

6. Working emulsion:

   pipette 3 ml of buffer/reagent solution (4) into a bottle of approximately 10 ml capacity. Inject therein rapidly 2 ml solution (5) by means of a pipette, keeping the tip beneath the liquid surface. Close the bottle and agitate vigorously to achieve a good emulsification.
   The working emulsion can likewise be obtained by dissolving a dry emulsion of the components of the assay with water [13, 28]. The dry emulsion is commercially available (*Boehringer Mannheim*). Due to its composition, this reagent reproducibly provides the same distribution of the particle size of triolein after reconstitution [13, 28].

7. Lipase standard (about 500 U/l):

   porcine pancreatic lipase is isolated according to [33] and dissolved in an aqueous bovine serum albumin solution (5%) to give a catalytic activity concentration of about 500 U/l; titration procedure according to chapter 1.1.1. Alternatively, commercially available standard preparations can be employed (*Boehringer Mannheim*) which are calibrated by the titrimetric lipase method according to [23], cf. also p. 15.

Stability of solutions: store all solutions at 4 °C. Solutions (1) – (5) are stable if no microbial contamination occurs. The emulsion (6) is stable for about 3 weeks at 4 °C or 1 week at 25 °C. Solution (7) is stable for at least 3 days at 4 °C or 1 day at 25 °C.

Procedure

Collection and treatment of specimen: serum and heparin-plasma can be used. Addition of calcium complexing anticoagulants (e.g. EDTA, citrate) should be avoided. Blood should be centrifuged at 3000 *g* for 10 min.

---

* One colipase unit corresponds to the liberation of one μmole butyric acid per minute under the conditions described by *C. Erlander* (Biochim. Biophys. Acta, *310,* 437 – 445 (1973).

Stability of the enzyme in the sample: no loss of lipase activity in serum has been observed during 2 weeks storage at 2–6 °C. Serum samples frozen at –20 °C showed no decrease of activity after a storage period of one year.

Assay conditions: because of better detectability, measurement at wavelengths 339, Hg 334 or Hg 365 nm is preferred, but Hg 405 nm can be used. Light path 10 mm, final volume 2.80 ml; temperature: 25 °C, 30 °C or 37 °C (thermostatted cuvette holder). Measurement against air.

For each series of measurements, a standard lipase solution (7) of known catalytic activity concentration, assigned by the titrimetric procedure (cf. p. 15), is analyzed with the unknown samples.

Measurement

Pipette successively into the cuvette:			concentration in assay mixture	
buffer/reagent	(4)	2.50 ml	Tris	24 mmol/l
working emulsion	(6)	0.200 ml	NaCl	35 mmol/l
			$CaCl_2$	1.04 mmol/l
			colipase	3.9×10^5 U/l
			deoxycholate	17.5 mmol/l
			triolein	0.28 mmol/l
mix, allow temperature equilibrium to be reached				
sample (or standard 7)		0.100 ml		
mix, read absorbance after 4 min, start stop-watch, second reading 5 min later, or monitor the reaction on a recorder.				

Calculation: use calculation formula (m_1) (cf. "Formulae", Appendix 3).

$$b = \frac{\Delta A_{\text{sample}}}{\Delta A_{\text{standard}}} \times b_{\text{standard}} \qquad \text{U/l}$$

Validation of Method

Precision, detection limits, sensitivity and accuracy: in the clinical decision range (200 U/l) the within-run imprecision is about 3%, and the day-to-day imprecision about 5% using an automated analyzer [20]. At low activities (≦100 U/l), the standard deviation is 10 U/l. The detection limit is 20 U/l. Sensitivity is found to be $\Delta A/\Delta t = 0.01\ \text{min}^{-1}$ (*Eppendorf* photometer, Hg 365 nm) [20]. The linear range for determination of lipase activity in human sera at 25 °C extends at least up to 700 U/l using a reconstituted dry emulsion as substrate [13, 20, 25, 28].

Accuracy can only be judged in comparison to the titrimetric reference method [23]. A method-comparison between the titrimetric and the turbidimetric methods with human sera is shown in Fig. 1. Human lipase added to human sera with endogeneous lipase was recovered to 100 ± 10% [13].

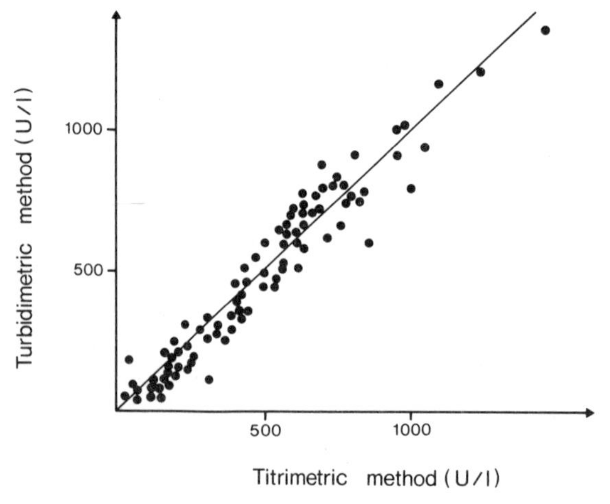

Fig. 1. Comparison of the turbidimetric method (y) with the titrimetric method (x) for lipase according to *Rick* [23].
Samples: human sera. $n = 115$; $y = 0.99 \times - 19.7$; $r = 0.965$.

Sources of error: there is no interference by lipaemic sera, bilirubin up to 340 μmol/l (20 mg/100 ml) and haemoglobin up to 125 μmol/l (200 mg/100 ml). The following drugs do not influence the assay at twice their toxic concentrations: acetylsalicylic acid, ampicillin, amidotrizoic acid (methylglucamine salt), ascorbic acid, adipiodone (methylglucamine salt), glibenclamid, carbochromen, quinidine (bisulphate), chloramphenicol, chlordiazepoxide, bezafibrat, caffeine, dextran, dipyramidole, ethaverine, furosemide, gelatin, indomethacin, methaqualone, methyldopa, nicotinic acid, nitrofurantoin, noramidopyrine, oxazepam, oxyphenbutazone, oxytetracycline, paracetamol, phenazopyridine, phenobarbital, phenprocoumon, phenytoin, probenecid, procaine, pyrithioxin, sulfamethoxazole, theophylline, trimethoprim (*R. Portenhauser,* unpublished results).

Emulsified inhibitors of serine enzymes disturb lipase activity by preventing the enzyme fixation at the oil water interface [29].

Sera containing normal or subnormal lipase concentrations may rarely exhibit nonlinear or increasing absorbance changes between the 4th and the 9th minute of measurement. In these cases a recorder should be employed and the linear decreasing part of the reaction curve should be used for calculation.

Specificity: post-heparin lipase and other esterases do not interfere with the assay [13, 30]. Beside human pancreatic lipase, lipases from other species can also be measured if they are able to utilize the colipase used in the assay.

Reference ranges: the upper limit of the normal range for human sera is reported to be 190 U/l [31, 32].

References

[1] *C. Bernard,* Mémoire sur le pancréas et sur la rôle du suc pancréatique dans les phénomènes digestives, particulièrement dans la digestion des matières grasses neutres, Baillière et Fils, Paris 1856.

[2] *L. Sarda, G. Marchis-Mouren, M. J. Constantin, P. Desnuelle,* Sur quelques essaies de purification de la lipase pancréatique, Biochim. Biophys. Acta *23*, 264–274 (1957).

[3] *E. D. Wills,* Lipases, in: *R. Paoletti, D. Kritchevsky* (eds.), Adv. in Lipid Res., Vol. *3*, Academic Press, New York 1965, pp. 197–240.

[4] *P. Desnuelle,* The Lipases, in: *P. D. Boyer* (ed.), The Enzymes, 3rd. edit., Vol. *VII*, Academic Press, New York 1972, pp. 575–616.

[5] *M. Sémériva, P. Desnuelle,* Pancreatic Lipase and Colipase. An Example of Heterogenous Biocatalysis, in: *A. Meister* (ed.), Adv. in Enzymology, Vol. 48, J. Wiley & Sons, New York 1979, pp. 319–370.

[6] *A. Vandermeers, M. C. Vandermeers-Piret, J. Rathé, J. Christophe,* On Human Pancreatic Triglycerol Lipase: Isolation and Some Properties, Biochim. Biophys. Acta *370*, 257–268 (1974).

[7] *A. de Caro, C. Figorella, J. Amic, R. Michel, O. Guy,* Human Pancreatic Lipase: A Glycoprotein, Biochim. Biophys. Acta *490*, 411–419 (1977).

[8] *B. Sternby, B. Borgström,* Purification and Characterization of Human Pancreatic Colipase, Biochim. Biophys. Acta *572*, 235–243 (1979).

[9] *B. Borgström, C. Erlanson-Albertsson, T. Wieloch,* Pancreatic Colipase: Chemistry and Physiology, J. Lipid Res. *20*, 805–816 (1979).

[10] *T. Wieloch, B. Borgström, G. Piéroni, F. Pattus, R. Verger,* Porcine Pancreatic Procolipase and its Trypsin-activated Form, FEBS Letters *128*, 217–220 (1981).

[11] *C. Chapus, H. Sari, M. Sémériva, P. Desnuelle,* Role of Colipase in the Interfacial Adsorption of Pancreatic Lipase at Hydrophilic Interfaces, FEBS Letters *58*, 155–158 (1975).

[12] *D. Lairon, G. Nalbone, H. Lafont, J. Leonardi, J.-L. Vigne, C. Chabert, J. C. Hauton, R. Verger,* Effects of Bile Lipids on the Adsorption and Activity of Pancreatic Lipase on Triacylglycerol Emulsions, Biochim. Biophys. Acta *618*, 119–128 (1980).

[13] *J. Ziegenhorn, U. Neumann, K. W. Knitsch, W. Zwez, A. Roeder, H. Lenz,* Lipase – eine Testcharakteristik, Medica *1*, 919–925 (1980).

[14] *H. Brockerhoff, R. G. Jensen,* Lipolytic Enzymes, Academic Press, New York 1974, pp. 49–73.

[15] *L. Sarda, P. Desnuelle,* Actión de la lipase pancréatique sur les esters en émulsion, Biochim. Biophys. Acta *30*, 513–521 (1958).

[16] *G. Benzonana, P. Desnuelle,* Etude cinétique de l'actión de la lipase pancréatique sur les triglycerides en émulsion. Essai d'une enzymologie en milieu heterogène, Biochim. Biophys. Acta *105*, 121–136 (1965).

[17] *B. Borgström,* The Action of Bile Salts and Other Detergents on Pancreatic Lipase and the Interaction with Colipase, Biochim. Biophys. Acta *488*, 381–391 (1977).

[18] *K. J. Gaskin, P. R. Durie, R. E. Hill, L. M. Lee, G. G. Forstner,* Colipase and Maximally Activated Pancreatic Lipase in Normal Subjects and Patients with Steatorrhea, J. Clin. Invest. *69*, 427–434 (1982).

[19] *K. Lorentz, T. Weiß,* Pankreaslipase – eine Übersicht, Med. Lab. *34*, 272–277 (1981).

[20] *U. Neumann, J. Ziegenhorn,* Determination of Serum Lipase with Automated Systems, in: *G. Siest, M. M. Galteau* (eds.), Comptes Rendus du 4e colloque de Pont-à-Mousson – Biologie prospective, Masson, Paris 1979, pp. 627–634.

[21] *P. K. J. Kinnunen, T. M. Schroeder, J. A. Virtanen,* Method of Fluorometrically Measuring the Activity of Fat-degrading Enzymes and Means for Carrying out the Method, European Patent Application 0037583.
[22] *G. Grenner, G. Deutsch, R. Schmidtberger, F. Dati,* Hochempfindlicher Enzymimmunoassay zur Bestimmung der Human-Pankreas-Lipase, J. Clin. Chem. Clin. Biochem. *20,* 515 – 519 (1982).
[23] *W. Rick,* Kinetischer Test zur Bestimmung der Serumlipaseaktivität, Z. Klin. Chem. Klin. Biochem. *7,* 530 – 539 (1969).
[24] *N. W. Tietz, E. V. Repique,* Proposed Standard Method for Measuring Lipase Activity in Serum by a Continuous Sampling Technique, Clin. Chem. *19,* 1268 – 1276 (1973).
[25] *J. Ziegenhorn, U. Neumann, K. W. Knitsch, W. Zwez,* Determination of Serum Lipase, Clin. Chem. *25,* 1067 (1979).
[26] *P. C. Arzoglou, G. Férard, K. Khalfa, P. Métais,* Conditions for the Assay of Pancreatic Lipase from Human Plasma Using a Nephelometric Technique, Clin. Chim. Acta *119,* 329 – 335 (1982).
[27] *B. Borgström, C. Erlanson,* Pancreatic Lipase and Co-Lipase, Eur. J. Biochem. *37,* 60 – 68 (1973).
[28] *U. Neumann, K. W. Knitsch, J. Ziegenhorn, A. Röder, W. Zwez, W. Krämer,* Reagenz zur Lipasebestimmung und Verfahren zu seiner Herstellung, German Patent 2904305, US Patent 4343897.
[29] *A. Guidoni, F. Benkonka, J. C. de Caro, M. Rovery,* Characterization of the Serine Reacting with Diethyl p-Nitrophenol Phosphate in Porcine Pancreatic Lipase, Biochim. Biophys. Acta *660,* 148 – 150 (1981).
[30] *G. E. Hoffmann, L. Weiss,* Specific Serum Pancreatic Lipase Determination, with Use of Purified Colipase, Clin. Chem. *26,* 1732 – 1733 (1980).
[31] *H. D. Weißhaar, H. Sudhoff, P. U. Koller, W. Bablok,* Pankreaslipase: Referenzwerte für 25 °C, Dtsch. med. Wochenschr. *106,* 239 – 241 (1981).
[32] *R. Allner, E. Gaschler, B. Herrmann,* Referenzwerte für die Lipaseaktivität im Serum, Med. Welt *32,* 1533 – 1537 (1981).
[33] *C. W. Garner, L. C. Smith,* Porcine Pancreatic Lipase, J. Biol. Chem. *247,* 561 – 565 (1972).

1.3.2 Hepatic Triacylglyceride Lipase

Triacylglycerol acylhydrolase, EC 3.1.1.3

Paavo K. J. Kinnunen and Tom Thurén

General

The presence of a heparin-releasable lipase in liver endothelial cell plasma membranes is well established [1, 2]. This lipase has an alkaline pH optimum and is found in liver plasma membrane preparations, as well as in cytosolic and microsomal fractions [3, 4]. Discrepant views exist regarding the physiological role of the hepatic lipase in the metabolism of plasma lipoproteins [5 – 8]. Recent studies indicate that a lipase with similar characteristics, and presumably identical to the lipase of liver, is present in adrenals, ovaries and testes [9]. In clinical studies the activity of the hepatic lipase is determined in plasma samples collected after the release of endothelial cell lipases into the circulation by intravenously injected heparin [10, 11]. The contribution of the above steroid-producing organs to the total 'hepatic' lipase activity of post-heparin plasma is not known.

Application of method: in clinical research and biochemistry.

Enzyme properties relevant in analysis: at least three different triacylglyceride lipase activities can be found in post-heparin plasma: hepatic lipase, lipoprotein lipase and a lipase originating from pancreas. They can be distinguished in the assay with appropriate substrates and specific inhibitors and/or activators [10 – 14]. The kinetics of the hepatic lipase, like those of other lipolytic enzymes, are rather complex [15]. The presence of a fatty acid acceptor, albumin, is mandatory in order to obtain linear kinetics.

Methods of determination: because of the rather low activities only radioisotope techniques provide sufficient accuracy. The catalytic activity is determined by measuring the liberation of radioactive fatty acids from a tracer fatty acid-labelled triacylglyceride incorporated into the substrate emulsion. No method for the estimation of enzyme mass is at present available.

International reference methods and standards: neither international level standardization nor reference standards has been established.

Enzyme effectors: no definite requirement for specific activators is known.

Assay

Method Design

Principle

$$\text{Tri[1-}^{14}\text{C]oleoylglyceride} + 3\,H_2O \xrightarrow{\text{HL*}} 3\,\text{[1-}^{14}\text{C]oleic acid} + \text{glycerol}$$

Triacylglyceride is emulsified with gum arabic to provide an oil/water interface for the lipase. Inhibitory reaction products, free fatty acids, are removed from the substrate surface into solution by previously de-fatted albumin. The reaction is terminated by an organic solvent mixture and the free fatty acid fraction is recovered by a two-phase liquid-liquid partition system [16]. Quantitation is achieved by the inclusion of radioactive fatty acids containing triacylglyceride tracer into the bulk lipid substrate. Free fatty acids are measured by liquid scintillation counting. Assay blanks must be used. In the literature the (post-heparin) plasma hepatic lipase activities are generally

* HL, hepatic endothelial cell lipase (heparin releasable), triacylglycerol acylhydrolase, EC 3.1.1.3.

expressed as µmoles free fatty acid liberated per ml of enzyme source (plasma) per hour, at 28 °C and at pH 8.6.

Optimized conditions for measurement: lipoprotein lipase (LPL, EC 3.1.1.34) present in the samples is inhibited by NaCl, 1.0 mol/l, in the assay, as well as by the absence of apolipoprotein C-II activator [11, 13]. Pancreatic lipase requires bile salts and co-lipase for maximal activity. The contribution of pancreatic lipase to the measured activities can be abolished by a proper selection of the substrate emulsifier and the lack of pancreatic colipase [12, 14]. Providing that the degree of hydrolysis is limited to less than 8% of the total triacylglyceride substrate, linear kinetics should be obtained. The final assay medium described here is as follows: trioleoylglyceride, 3.16 mmol/l, NaCl, 1.0 mol/l, Tris-HCl, 0.24 mol/l, pH 8.6, 2.5% (w/v) bovine serum albumin, and 2% (w/v) gum arabic.

Temperature conversion factors: no temperature conversion factors have been established.

Equipment

Liquid scintillation counter; *Branson* Sonifier-Cell Disruptor (*Branson Instruments Co.,* Danbury, Conn.); a thermostatted water-bath, a vortex mixer and a bench centrifuge are required. Other types of sonicators can be used, providing that optimal conditions for the substrate emulsification are established [11].

Reagents and Solutions

Purity of reagents: glycerol-tri[1-^{14}C]olein (1.11×10^{12} to $2.22 \times 10^{12}\ s^{-1} \times mol^{-1}$, CFA 258) is supplied by *Amersham International,* and unlabelled material by *Sigma.* Products of equal quality by other manufacturers can be employed. The purity of the trioleoylglyceride solutions should be checked periodically by thin-layer chromatography (TLC) on silicic acid (*E. Merck,* 5748). Develop the plate with hexane/ethyl ether/acetic acid (80:20:2 v/v/v). Stain the developed plate with iodine. Mark the spots corresponding to trioleoylglyceride and oleic acid, after evaporation of iodine, remove the silicic acid in areas corresponding to lipid spots from the plate, elute the lipid with approx. 1.0 ml methanol, and count for ^{14}C in 15 ml ACS liquid scintillation cocktail (soln. 11) for 5 min. If degradation has occurred purify the lipids on TLC as above, using preparative plates.

Bovine serum albumin is purchased from *Armour Pharmaceutical Co.,* Eastbourne, Sussex, England, heparin solution (5000 IU/ml) from *Medica,* Helsinki, Finland, and gum arabic and tris (hydroxymethyl) aminomethane buffer from *Sigma.* Sodium chloride, boric acid and sodium carbonate, all of reagent grade, and analytical grade chloroform, methanol and heptane are supplied by *E. Merck.* ACS, liquid scintillation cocktail, is obtained from *Amersham International.*

Preparation of solutions (for about 50 determinations): all solutions in re-purified water (see Vol. II, chapter 2.1.3.2).

1. Tris buffer (0.2 mol/l, pH 8.6):

   dissolve 24.2 g Tris in 200 ml water, adjust with HCl, 0.4 mol/l, to pH 8.6 and dilute to 1000 ml with water.

2. Tris (1 mol/l):

   dissolve 121.1 g Tris to 1000 ml with water.

3. 5% gum arabic:

   dissolve 10 g gum arabic in 200 ml Tris buffer (1), pass through a filter paper (*Whatman* No. 5). Store frozen at −20 °C in 10 ml batches. This solution should not be re-frozen after thawing once.

4. Bovine serum albumin, 10% (w/v):

   dissolve 14.0 g albumin in 140 ml water at 23 °C, combine with 7 g activated charcoal ([17], e.g. *Darco* M). The pH is lowered to 3.0 by adding HCl, 0.2 mol/l. Place the solution in an ice-bath and mix gently (without foaming) with a magnetic stirrer for 1 h. Remove charcoal by centrifugation at 20000 *g* for 20 min in a *Sorvall* RC 1 centrifuge with an SS-34 rotor at 4 °C. Bring the clear supernatant to pH 7.0 by adding NaOH, 0.2 mol/l. Dialyze against several volumes of water and lyophilize. Dilute 20 ml Tris solution (2) with 60 ml water. Dissolve carefully 10 g of lyophilized de-fatted bovine serum albumin in Tris solution (2) on a magnetic stirrer. Check pH and if not 8.6 adjust with HCl, 0.4 mol/l, or NaOH, 0.2 mol/l. Store solution frozen at −20 °C in 6.5 ml batches.

5. Methanol/chloroform/heptane, 1.41 : 1.25 : 1.00 (v/v/v):

   mix thoroughly 846 ml methanol, 750 ml chloroform and 600 ml heptane.

6. Borate/carbonate buffer (0.14 mol/l, pH 10.5):

   dissolve 19.35 g K_2CO_3 and 8.66 g H_3BO_3 in 900 ml water, adjust to pH 10.5 with KOH, 2.0 mol/l, and make up to 1000 ml with water.

7. NaCl/Tris solution (NaCl, 4 mol/l; Tris, 0.4 mol/l; pH 8.6):

   dissolve 58.44 g NaCl and 12.1 g Tris in 200 ml water and adjust with HCl, 0.4 mol/l, pH to 8.6. Dilute to 250 ml with water.

8. NaCl (0.9%):

   dissolve 0.9 g NaCl to 100 ml with water.

9. Trioleoylglyceride solution, unlabelled (20 mg/ml):

   dissolve 1.0 g trioleoylglyceride in 50 ml toluene to yield a concentration of 20 mg/ml. Store at $-20\,^\circ C$.

10. Tri[^{14}C]oleoylglyceride solution (2.0 mCi/l):

    dissolve $1.85 \times 10^6\ s^{-1}$ of tri[1-^{14}C]oleoylglyceride in 25 ml toluene. The final concentration is then $7.4 \times 10^4\ s^{-1} \times ml^{-1}$. Store at $-20\,^\circ C$.

11. ACS liquid scintillation cocktail:

    a commercial liquid scintillation cocktail manufactured by *Amersham International.*

12. Substrate solution (for about 50 assays):

    transfer 3.5 ml trioleoylglyceride solution (9) and 1.75 ml radioactive trioleoylglyceride solution (10) into a conical 20 ml test tube. Evaporate the solvent under a gentle stream of nitrogen, wash lipids three times with 0.5 ml heptane under N_2. Evaporate to dryness and add 10 ml gum arabic solution (3). Centre the tip of the sonicator 0.5 cm below the surface of the solution and treat the mixture in an ice-bath at setting 4 for 4 min. Add to the sonicated substrate 6.25 ml bovine serum albumin solution (4), 6.25 ml NaCl/Tris solution (7) and 2.5 ml water to a final volume of 25.0 ml. Mix thoroughly with a vortex mixer.

Stability of solutions: store buffer solutions at $+4\,^\circ C$ and organic solvents at room temperature. Buffer solutions are stable as long as no microbial contamination occurs. The emulsified substrate solution (12) should be prepared daily. Stability of trioleoylglyceride solutions (9) and (10) depends on degradation; for method of checking cf. p. 36. Working substrate solution must be prepared freshly each day.

Procedure

Collection and treatment of specimens: heparin is injected intravenously at a dose of 100 IU per kg body weight. After 15 min, a venous blood sample of approximately 2.0 ml is taken into a heparinized test tube held on ice. EDTA, citrate or fluoride should not be included. To separate plasma centrifuge at approximately 3000 *g* for 10 min. It is preferable to store the samples divided into two aliquots in order to avoid repeated freezing and thawing in case re-assay should become necessary.

Stability of the enzyme in the sample: no careful study has been carried out. However, the activity of the hepatic lipase in postheparin plasma samples remains essentially unaltered for at least six months at $-20\,^\circ C$.

Details for measurement in tissues: the enzyme can be solubilized by heparin from tissue acetone powders. Distinction from lysosomal lipase(s) is made by carrying out the assay at alkaline pH.

Assay conditions: the test tubes used in incubation and partition procedures must be free from detergents. This is achieved by rinsing the tubes with 94% or technical grade ethanol, with subsequent drying in an oven.

Final incubation conditions:

incubation time	60 min
temperature	28 °C
incubation volume	0.5 ml
volume fraction of sample	0.02
tri[1-^{14}C]oleoylglyceride	$5.17 \times 10^6 s^{-1}$/l (0.14 mCi/l)
trioleoylglyceride	3.16×10^{-3} mol/l
Tris buffer, pH 8.6	0.24 mol/l
NaCl	1.0 mol/l
albumin	25 g/l
gum arabic	20 g/l

All assays should be carried out in duplicate, taking the mean for calculations. All mixings with a vortex mixer.

Measurement

Incubation

Pipette successively into test tubes:		sample	blank	"total counts"*
substrate solution	(12)	0.49 ml	0.49 ml	–
sample		0.01 ml	–	–
NaCl solution	(8)	–	0.01 ml	0.5 ml
mix thoroughly, incubate 60 min at 28 °C in water-bath; stop reaction by adding				
methanol/chloroform heptane	(5)	3.25 ml	3.25 ml	3.25 ml
mix thoroughly; to separate the phases add				
borate/carbonate buffer	(6)	0.75 ml	0.75 ml	0.75 ml
mix thoroughly for 15 s, centrifuge 15 min at (full speed) 3000 *g;* for counting, use the upper organic phase.				

* "total counts" means reference for eliminating quench, and in addition provides reference for the exact total reference of radioactivity in each test tube.

Counting radioactivity

Pipette into counting vials:		sample	blank	"total counts"
incubation mixture, upper phase		1.0 ml	1.0 ml	1.0 ml
substrate solution	(12)	–	–	0.1 ml
ACS solution	(11)	15 ml	15 ml	15 ml
mix thoroughly, count for 5 min (settings for ^{14}C).				

Calculation: activity units are usually defined as μmoles free fatty acids liberated per hour, related to 1 ml of plasma.

Hepatic lipase activity concentration b is calculated as follows:

$$b = \frac{3 \times 2.02 \times 1.58 \times k_{vol} \times k_{time}}{0.8 \times 5 \times (C_{total\ counts} - C_{blank})} \times (C_{sample} - C_{blank})\ \mu mol \times ml^{-1} \times h^{-1}$$

$$b = 2.39 \times k_{vol} \times k_{time} \times \frac{C_{sample} - C_{blank}}{C_{total\ counts} - C_{blank}} \quad \mu mol \times ml^{-1} \times h^{-1}$$

if $k_{vol} = 100$ and $k_{time} = 1$

$$b = 239 \times \frac{C_{sample} - C_{blank}}{C_{total\ counts} - C_{blank}} = 239 \times \frac{\Delta C_{sample}}{\Delta C_{total\ counts}} \quad \mu mol \times ml^{-1} \times h^{-1}$$

C = counting rates, cpm
k_{vol} = correction factor to obtain the activity per ml of plasma
k_{time} = correction factor to obtain the activity per one hour of incubation

To obtain international units per litre ($\mu mol \times min^{-1} \times l^{-1}$) the factor 1000/60 = 16.67 has to be applied. Thus,

$$b = 3983 \times \frac{\Delta C_{sample}}{\Delta C_{total\ counts}} \quad U/l\,.$$

Results in the SI-consistent units nkat/l are obtained by a further multiplication by 16.67, since 1 U = 16.67 nkat. Thus,

$$b = 66394 \times \frac{\Delta C_{sample}}{\Delta C_{total\ counts}} \quad nkat/l\,.$$

Validation of Method

Precision, accuracy, detection limits and sensitivity: no data about accuracy are available since no standard reference material has been established. The lowest value to be measured is approximately 0.4 μmol × ml^{-1} × h^{-1} (6.7 U/l; 111 nkat/l). The imprecision near the lowest value is about ± 6%. For values of about 10.0 μmol × ml^{-1} × h^{-1} an imprecision of 2% is generally found. If hydrolysis exceeds 8% of the total amount of the substrate, linearity may not be expected. Therefore, the upper limit is approximately 38.0 μmol × ml^{-1} × h^{-1} (633 U/l; 10.55 μkat/l). If higher activities are to be measured, dilute the sample, e.g. 1:2 (v/v) with 0.9% NaCl. Sensitivity is found to be about 0.2 μmol × ml^{-1} × h^{-1}.

Sources of error: influences on the measured activity by medication are not known. Hepatic lipase appears to be inhibited by high plasma concentrations of high density lipoprotein [7].

Reference ranges: in normal human subjects the post-heparin hepatic lipase activities in plasma range between 15 – 40 μmol × ml^{-1} × h^{-1} (250 – 667 U/l; 4.17 – 11.1 μkat/l), depending on age and sex [18, 19].

References

[1] *T. Kuusi, E. A. Nikkilä, I. Virtanen, P. K. J. Kinnunen,* Localization of the Heparin-Releasable Lipase in Situ in Rat Liver, Biochem. J. *181*, 245 – 246 (1979).

[2] *P. K. J. Kinnunen,* Hepatic Endothelial Lipase: Isolation, Characteristics and Physiological Role, to appear in: *B. Borgström, H. L. Brockman* (eds.), Lipolytic Enzymes, Elsevier, North-Holland 1983.

[3] *M. Waite, P. Sisson,* Utilization of Neutral Glycerides and Phosphatidylethanolamine by the Phospholipase A_1 of the Plasma Membranes of Rat Liver, J. Biol. Chem. *248*, 7985 – 7992 (1973).

[4] *J. E. M. Groener, T. E. Knauer,* Evidence for the Existence of only one Triacylglycerol Lipase of Rat Liver Active at Alkaline pH, Biochim. Biophys. Acta *665*, 306 – 316 (1981).

[5] *E. A. Nikkilä, T. Kuusi, K. Harno, M. Tikkanen, M.-R. Taskinen,* Lipoprotein Lipase and Hepatic Endothelial Lipase are Key Enzymes in the Metabolism of Plasma High Density Lipoproteins, Particularly HDL_2, in: *A. M. Gotto, L. C. Smith, B. Allen* (eds.), Atherosclerosis V, Springer, Berlin 1980, pp. 387 – 392.

[6] *P. H. E. Groot, H. Jansen, A. van Tol,* Selective Degradation of the High Density Lipoprotein-2 Subfraction by Heparin releasable Liver Lipase, FEBS Lett. *129*, 269 – 272 (1981).

[7] *P. K. J. Kinnunen, P. Vainio, T. Thurén,* Evidence Against the Role of Hepatic Endothelial Lipase in the Metabolism of Plasma HDL_2 in Man, Atherosclerosis *40*, 377 – 379 (1981).

[8] *P. K. J. Kinnunen, J. A. Virtanen, P. Vainio,* Lipoprotein Lipase and Hepatic Endothelial Lipase. Their Roles in Plasma Lipoprotein Metabolism, in: *R. Paoletti, A. M. Gotto* (eds.) Atherosclerosis Reviews, Vol. 11, Raven Press, New York 1983, pp. 65 – 105.

[9] *H. Jansen, W. J. DeGreef,* Heparin-Releasable Lipase Activity of Rat Adrenals, Ovaries and Testes, Biochem. J. *196*, 739 – 745 (1981).

[10] *R. M. Krauss, R. I. Levy, D. S. Fredrickson,* Selective Measurement of Two Lipase Activities in Postheparin Plasma from Normal Subjects and Patients with Hyperlipoproteinemia, J. Clin. Invest. *54*, 1107 – 1124 (1974).

[11] *J. K. Huttunen, C. Ehnholm, P. K. J. Kinnunen, E. A. Nikkilä,* An Immunochemical Method for the Selective Measurement of Two Triglyceride Lipases in Human Postheparin Plasma, Clin. Chim. Acta *63*, 335 – 347 (1975).
[12] *B. Borgström, C. Erlanson-Albertsson, T. Wieloch,* Pancreatic Colipase: Chemistry and Physiology, J. Lipid Res. *20*, 805 – 816 (1979).
[13] *P. K. J. Kinnunen, R. L. Jackson, L. C. Smith, A. M. Gotto, J. T. Sparrow,* Activation of Lipoprotein Lipase by Native and Synthetic Fragments of Human Plasma ApoC-II, Proc. Natl. Acad. Sci. USA *74*, 4848 – 4851 (1977).
[14] *P. A. Verduin, J. M. H. M. Punt, H. H. Kreutzer,* Studies on the Determination of Lipase Activity, Clin. Chim. Acta *46*, 11 – 19 (1973).
[15] *R. Verger, G. deHaas,* Interfacial Enzyme Kinetics of Lipolysis, Annu. Rev. Biophys. Bioeng. *5*, 77 – 117 (1976).
[16] *P. Belfrage, M. Vaughan,* A Simple Liquid-Liquid Partition System for Isolation of Labeled Oleic Acid from Mixtures with Glycerides, J. Lipid Res. *10*, 341 – 344 (1969).
[17] *R. F. Chen,* Removal of Fatty Acids from Serum Albumin by Charcoal Treatment, J. Biol. Chem. *242*, 173 – 181 (1967).
[18] *J. K. Huttunen, C. Ehnholm, M. Kekki, E. A. Nikkilä,* Post-heparin Plasma Lipoprotein Lipase and Hepatic Lipase in Normal Subjects and in Patients with Hypertriglyceridaemia: Correlations to Sex, Age and Various Parameters of Triglyceride Metabolism, Clin. Sci. Mol. Med. *50*, 249 – 260 (1976).
[19] *P. K. J. Kinnunen, H.-A. Unnérus, T. Ranta, C. Ehnholm, E. A. Nikkilä, M. Seppälä,* Activities of Post-Heparin Plasma Lipoprotein Lipase and Hepatic Lipase During Pregnancy and Lactation, Eur. J. Clin. Invest. *10*, 469 – 474 (1980).

1.3.3 Lipoprotein Lipase (Post-heparin Lipase)

Triacylglycero-protein acylhydrolase, EC 3.1.1.34

Gerd Assmann and Hans-Ulrich Jabs

General

LPL* is a tissue-based enzyme which may be released into the circulating blood from the capillary endothelium of various organs by the action of heparin. This enzyme is absent from patients with familial hyperchylomicronaemia [1 – 4]. Lipoprotein lipase was first characterized in 1955 by *Korn* [2] following extraction from rat heart with acetone/ether. Exogenous and endogenous triglycerides of chylomicrons and very-

* Abbreviations
  FFA Free fatty acids
  HTGL Triglyceride lipase of hepatic origin, EC 3.1.1.3
  LPL Lipoprotein lipase
  PPO 2,5-Diphenyloxazole
  POPOP 1,4-Bis[2-(5-phenyloxazoyl)] benzene
  SDS Sodium dodecyl sulphate.

low-density-lipoproteins in the plasma constitute the natural substrate of LPL. The enzyme is activated by apolipoprotein C-II and inhibited by apolipoproteins C-I and C-III, protamine sulphate and high ionic strength [3 – 6]. Determination of LPL is impeded by the presence of another triglyceride lipase of hepatic origin (HTGL) in post-heparin plasma. There is no synthetic substrate for which LPL is specific. Selective determination of LPL requires isolation of the enzyme by chromatographic methods [7] or inhibition of the HTGL by means of detergents or by immunoprecipitation [8 – 10].

Application of method: in biochemistry and clinical chemistry.

Enzyme properties relevant in analysis: selective determination of LPL requires inhibition of HTGL. Inhibition of HTGL using the detergent SDS is a function of time and concentration, depending on the lipid and protein content of the sample [8].

Immunoprecipitation of the HTGL depends on the species-specificity and the selectivity of the antibodies.

Methods of determination: the HTGL-activity is inhibited by incubation of serum with SDS or, best of all, by immunoprecipitation. The lipolytic activity of LPL is measured by counting the radioactive decay after incubation with labelled substrate and lipid-extraction.

Immunoprecipitation allows selective determination of LPL, but there are limitations for application in comparative studies with regard to production of antibodies and as a consequence of the species-specificity of the antisera used in this method. The selective LPL assay following inhibition of HTGL by the detergent SDS requires only common laboratory equipment and chemicals. The method depends on the purity of the reagent SDS.

In another assay, LPL is inhibited by protamine sulphate [11, 12]. The LPL activity is calculated as the difference between the inhibited and the non-inhibited activities. This assay does not permit direct determination of LPL.

In the heparin affinity assay [7], LPL is bound to heparin-Sepharose and then eluted at an ionic strength between 1.2 and 1.5 mol NaCl/l; HTGL elutes from the affinity column at an ionic strength between 0.6 and 0.74 mol NaCl/l. However, there are losses due to non-specific adsorption on the column matrix which must be taken into account.

In addition, HTGL may be assayed by this method after inhibiting LPL with 0.75 mol NaCl/l.

International reference methods and standards: there are as yet no standards or international reference methods for determination of LPL. A comparative study of the SDS-inhibition-assay and the immunoprecipitation method has been reported [8].

Enzyme effectors: apolipoprotein C-II (or serum) activates lipoprotein lipase. The enzyme is inhibited by apolipoproteins C-I and C-III, as well as by protamine sulphate and by high ionic strength.

1.3.3.1 Determination after SDS-inhibition

The catalytic activity of LPL in postheparin plasma is determined either after SDS-inhibition or immunoprecipitation of HTGL.

Assay

Method Design

Principle

(a) Glycerol tri-[1-^{14}C]oleate + H_2O $\xrightarrow[\text{Apo C-II}]{\text{LPL}}$ [1-^{14}C]oleate + glycerol dioleate

The catalytic activity of LPL is determined from the mass of the radioactively labelled fatty acids released from the glycerol tri-[1-^{14}C]oleate substrate per unit time. Either inactivated serum or apolipoprotein C-II is added to the reaction as an activator. Extraction of lipid is performed by adsorption of oleic acid on an ion exchanger. Losses during extraction are corrected by using an external standard ([1-^{14}C]oleic acid). Non-specific adsorption of glycerol tri-[1-^{14}C]oleate on the ion exchanger is corrected by using a blank. Finally, the radioactive decay is measured after desorption of oleic acid from the ion exchanger.

Optimized conditions for measurement: the determination of lipoprotein lipase activity depends on pH, NaCl concentration and emulsification of the substrate solution. In the following methods Tris, 0.2 mol/l, pH 8.2; NaCl, 0.15 mol/l; gum arabic, 10 g/l, and glycerol trioleate, 8.475 mmol/l, are used as optimal conditions. In the SDS-inhibition assay, lipid-free albumin is added to bind the fatty acids which are set free during the reaction. Gum arabic is added as emulsifier.

Temperature conversion factors: temperature conversion factors have not been published. A temperature of 26 °C is most suitable for the SDS-inhibition assay whereas 4 °C is best for the immunoprecipitation assay.

Equipment

Sonicator (e.g. *Branson* cell disruptor), scintillation counter, water-bath.

Reagents and Solutions

Purity of reagents: all reagents must be of analytical purity. SDS is recrystallized twice from hot water and twice from hot ethanol.

The substrate has to be purified according to the following procedure: add to 4 g glycerol trioleate, glycerol tri-[1-^{14}C]oleate to yield a specific activity of 0.1 μCi/μmol. Remove free oleic acid by passing 4 g of the mixture in 50 ml benzene through a 30 × 1.6 cm Amberlite IRA-400 column which has previously been equilibrated with 500 ml benzene. Use the eluate (= stock solution) for the preparation of the substrate solution.

Preparation of solutions (for 100 determinations): all solutions in re-purified water (cf. Vol. II, chapter 2.1.3.2).

1. Tris buffer (0.5 mol/l; pH 8.2):

   dissolve 60.6 g Tris in 800 ml water. Adjust pH with HCl, 1 mol/l, to 8.2, make up with water to 1000 ml.

2. NaCl (0.15 mol/l):

   dissolve 8.77 g NaCl in 1 l water.

3. Gum arabic (5% w/v):

   dissolve 50 g gum arabic in 1 l water.

4. Lipid-free albumin (16.7% w/v):

   dissolve 167 g lipid-free albumin in 1 l water.

5. Substrate solution (glycerol tri-[1-^{14}C]oleate, 10.2 mmol/l; 1 mCi/l; Tris, 0.2 mol/l; NaCl, 0.15 mol/l; gum arabic, 10 g/l; albumin, 50 g/l, pH 8.2):

   evaporate ca. 22.6 ml glycerol trioleate stock solution (in benzene) and weigh precisely 452 mg into a sonication flask. Add 20 ml Tris buffer (1), 5 ml NaCl solution (2), 10 ml gum arabic solution (3), 15 ml lipid-free albumin solution (4) and sonicate 8 – 12 times for 10 – 15 s at 60 W and room temperature.

6. Activator apolipoprotein:

   a) Use apolipoprotein C-II isolated by means of DEAE-cellulose chromatography [13] and store at −70 °C in ammonium bicarbonate buffer, 0.05 mol/l, pH 7.4;

   b) alternatively, heat 10 ml serum from normal subjects to 56 °C for 60 min to suppress lipolytic activity, then divide into 0.1 ml aliquots and store at −70 °C.

7. SDS solution (10 mmol/l):

   dissolve 2.9 g dodecyl sulphate, sodium salt, in 1 l water.

8. Isopropanol/heptane/sulphuric acid, 2 mol/l (4 : 1 : 0.1 v/v/v):

   mix 40 ml isopropanol, 10 ml heptane and 1 ml sulphuric acid, 2 mol/l (prepared by diluting 10 ml sulphuric acid, conc., with 50 ml water).

9. NaOH (0.5 mol/l):

   dissolve 20 g NaOH in 1 l water.

9a. KOH/Triton X-100 (KOH, 1 mol/l; Triton X-100, 50% v/v):

   mix 10 ml KOH, 2 mol/l (prepared by dissolving 112.2 g KOH in 1 l methanol) with 10 ml Triton X-100 (prepared by adding 20 g Triton X-100 to 20 ml methanol).

10. [1-^{14}C]oleic acid standard solution (35.4 mmol/l; 1 mCi/l):

   dissolve 1 g oleic acid and 100 μCi [1-^{14}C]oleic acid, (X = 50 mCi/mmol = 1.85 GBq/mmol) in 40 ml Tris buffer (1), 10 ml NaCl solution (2), 20 ml gum arabic solution (3), 30 ml lipid free albumin solution (4) and sonicate 8 – 12 times for 10 – 15 s at 60 W and room temperature.

11. Scintillator solution:

   according to [14], dissolve 4 g PPO and 0.5 g POPOP in a mixture of 15 ml conc. acetic acid (96% v/v) and 1 l toluene.

Stability of solutions: store all solutions at 0 – 4 °C. The substrate solution (5) is stable for one day. The activator apolipoprotein (6) is stable for about 6 months, when stored at – 70 °C.

Procedure

Collection and treatment of specimen: blood samples are taken 15 min after injection of 60 IU heparin/kg body weight from patients who have fasted for at least 12 h. The blood is transferred to tubes containing heparin (2 U/ml blood), cooled in an ice-bath and then centrifuged for 30 min at 480 *g* to obtain the post-heparin plasma.

Stability of the enzyme in the sample: the LPL does not lose any activity when the plasma is stored at – 70 °C for up to 3 months.

Details for measurement in tissues: the determination of lipoprotein lipase activity in muscle and adipose tissue has been described by *M. R. Taskinen* [15].

Assay conditions: inhibition with SDS for 60 min at 26 °C. Pre-incubation of substrate with activator for 80 min at 37 °C and 15 min at 28 °C. Incubation of enzyme and substrate (enzyme reaction) for 60 min at 26 °C.

Run a blank which contains NaCl solution (2) instead of sample (post-heparin-SDS-plasma).

Run a standard without incubation, using standard solution (10) instead of substrate solution (5), and NaCl solution (2) instead of sample.

Measurement

SDS inhibition: add 0.05 ml SDS, solution (7) to 0.05 ml serum and incubate for 60 min at 26 °C to obtain the post-heparin-SDS-plasma. Use this solution for enzyme reaction.

Pre-incubation and enzyme reaction

Pipette into 16 × 125 mm screw-top incubation tubes:			concentration in assay mixture	
substrate solution	(5)	0.49 ml	glycerol tri[1-^{14}C]-oleate	8.6 mmol/l (843 μCi/l)
			Tris	0.17 mol/l
			NaCl	0.13 mol/l
			gum arabic	8.5 g/l
			albumin	30.6 g/l
activator apolipoprotein	(6)	0.08 ml		
incubate for 80 min at 37 °C, agitate and incubate for a further 15 min at 28 °C, add				
post-heparin-SDS-plasma (sample)		0.01 ml	volume fraction	0.017
incubate for 60 min at 26 °C.				

Lipid extraction and measurement [14]: stop the reaction by addition of 5 ml iso-propanol/heptane/sulphuric acid (8). Place the mixture in scintillation tubes and add 0.5 ml Tris-HCl buffer (1) to the reaction mixture. Mix this solution with 0.5 g Amberlite IRA 400 ion-exchange resin. Agitate the suspension and aspirate the liquid phase with a *Pasteur* pipette. Wash the exchange resin four times with heptane, after which the bound oleic acid is released from the resin by incubation for 90 min at 70 °C in 1 ml KOH/Triton X-100 solution (9a). Then add 0.1 ml conc. acetic acid and 10 ml of scintillator solution (10).

Radioactive decay is finally measured for 1 min in a scintillation counter. Solution (10) is used as standard.

Calculation: measurement of the sample assay (containing released fatty acid) yields C_{sample}. Correct for non-specific adsorption on the ion exchanger by using the blank; the difference is the net counting rate $C_{n(sample)}$. The catalytic concentration of the enzyme in the sample is (Δt in h):

$$b = \frac{C_{n(sample)}}{C_{standard}} \times c_{standard} \times \frac{1}{v} / \Delta t \qquad mmol \times l^{-1} \times h^{-1}$$

Δt in min:

$$b = \frac{C_{n(sample)}}{C_{standard}} \times c_{standard} \times \frac{10^3}{60 \times v} / \Delta t \qquad U/l$$

or

$$b = \frac{C_{n(sample)}}{C_{standard}} \times c_{standard} \times \frac{16.67 \times 10^3}{60 \times v} / \Delta t \qquad nkat/l$$

where is

C counting rate
v sample volume in assay mixture, l
$c_{standard}$ concentration of standard in assay mixture, mmol/l

1.3.3.2 Determination after Immunoprecipitation

Principle: for reaction cf. eqn. (a) in chapter 1.3.3.1, p. 44. HTGL is removed by immunoprecipitation*.

Reagents and Solutions

Purity of reagents: all reagents must be of analytical purity (cf. SDS-inhibition assay, p. 44). HTGL is prepared by heparin-Sepharose affinity-chromatography. 50 – 100 µg HTGL in 100 µl NaCl, 0.15 mol/l, is mixed with 50 µl complete *Freund*'s adjuvant and injected into a goat. The immunization is continued at intervals of 2 weeks until a good titre is obtained. The blood is collected, centrifuged and the serum is used for immunoprecipitation.

---

* *J. Augustin,* pers. comm.

Preparation of solutions (for 100 determinations):

1. Tris buffer: cf. p. 45.

2. NaCl solution: cf. p. 45.

3. Gum arabic: cf. p. 45.

4. Substrate solution (glycerol tri[1-^{14}C]oleate, 8.475 mmol/l, 0.86 mCi/l; Tris, 0.2 mol/l; NaCl, 0.15 mol/l; gum arabic, 10 g/l; pH 8.2):

   evaporate ca. 7.6 ml glycerol trioleate stock solution (in benzene) and weigh precisely 150 mg into a sonication flask. Add 8 ml Tris buffer (1), 2 ml NaCl solution (2), 4 ml gum arabic solution (3), 6 ml water and sonicate 8 – 12 times for 10 – 15 s at 60 W and room temperature.

Solutions (5), (6), and (7) cf. p. 45, (8), (9), (9a) and (11) cf. p. 46.

12. [1-^{14}C]oleic acid standard solution (35.4 mmol/l; 1 mCi/l):

    dissolve 1 g oleic acid and [1-^{14}C]oleic acid, 100 μCi (X = 50 mCi/mmol = 1.85 GBq/mmol) in 40 ml Tris buffer (1), 10 ml NaCl solution (2), 20 ml gum arabic solution (3), 30 ml water and sonicate 8 – 12 times for 10 – 15 s at 60 W and room temperature.

Stability of solutions: cf. p. 46.

Procedure

Collection and treatment of specimen: cf. p. 46.

Stability of the enzyme in the sample: cf. p. 46.

Details for measurement in tissues: cf. p. 46.

Assay conditions: inhibition by immunoprecipitation with goat anti HTGL-IgG (HTGL antiserum) for 60 min at 4 °C [16]. Incubation of serum-antiserum mixture for 30 min at 27 °C.

Run a blank for correction of non-specific adsorption on the ion exchanger after incubation of 20 μl NaCl solution (2) with 200 μl substrate solution (4).

Run a standard using standard solution (12) instead of substrate solution (5), and NaCl solution (2) instead of sample.

Measurement

Immunoprecipitation: add 10 μl HTGL antiserum to 80 μl post-heparin plasma and incubate for 60 min at 4 °C.

Enzyme reaction

Pipette into 16 × 125 mm screw-top incubation tubes:			concentration in assay mixture	
substrate solution	(4)	0.20 ml	glycerol tri[1-^{14}C]-oleate	7.7 mmol/l (781 μCi/l)
			Tris	0.18 mol/l
			NaCl	0.14 mol/l
			gum arabic	9.1 g/l
plasma-antibody mixture		0.02 ml	volume fraction	0.1
incubate for 30 min at 27 °C.				

Lipid extraction and measurement: as described on p. 47.

Calculation: the radioactive decay of the samples is converted into μmoles of fatty acids released per h per ml of post-heparin plasma, taking into account the correction for extraction losses and non-specific adsorption.

Validation of Methods

Precision, accuracy, detection limits and sensitivity

a) SDS-inhibition assay: variation within the test series is ±2–5%, day-to-day variation is ±11–12%.
b) Immunoprecipitation assay: not published.

Sources of error: in the SDS-inhibition assay, the SDS concentration must be exact and the incubation periods precisely adhered to, since LPL is inhibited at higher concentrations of SDS or with longer incubation periods. The reproducibility of the method depends on proper emulsification of the substrate solutions [8].

Specificity: the LPL activity is measured specifically with the immunoprecipitation assay. The specificity of the SDS-inhibition assay depends on the SDS concentration and incubation periods.

Reference ranges: measured with the immunoprecipitation assay [17]. Men: 9.9 ± 0.5 µmol FFA per hour and ml plasma (n = 21, 18 – 48 years). Women: 9.8 ± 0.6 µmol FFA per hour and ml plasma (n = 16, 21 – 35 years).

References

[1] *E. A. Nikkilä,* Familial Lipoprotein Lipase Deficiency and other Disorders of Chylomicron Metabolism, in: *J. B. Stanbury, J. B. Wyngaarden, D. S. Fredrickson, J. L. Goldstein, M. S. Brown* (eds.), The Metabolic Basis of Inherited Disease, Mc Graw-Hill, New York 1982, pp. 622 – 642.

[2] *E. D. Korn,* Clearing Factor, a Heparin-activated Lipoprotein Lipase. I. Isolation and Characterization of the Enzyme from Normal Rat Heart, J. Biochem. *215*, 1 – 14 (1955).

[3] *F. T. Lindgren, A. V. Nichols, N. K. Freeman,* Physical and Chemical Composition Studies on the Lipoproteins of Fasting and Heparinized Human Sera, J. Physiol. Chem. *59*, 930 – 938 (1955).

[4] *R. J. Havel, J. P. Kane, E. O. Belasse, N. Segel, L. V. Basse,* Splanchnic Metabolism of Free Fatty Acids by Production of Triglycerides of Very Low Density Lipoproteins in Normotriglyceridemic and Hypertriglyceridemic Humans, J. Clin. Invest. *49*, 2017 – 2035 (1970).

[5] *T. A. Musliner, E. C. Church, P. N. Herbert, M. J. Kingston, R. S. Shulman,* Lipoprotein Lipase Cofactor Activity of a Carboxyl-terminal Peptide of Apolipoprotein C-II, Proc. Natl. Acad. Sci. USA, *74*, 5358 – 5362 (1977).

[6] *O. Schrecker, H. Greten,* Activation and Inhibition of Lipoprotein Lipase Studies with Artificial Lipoproteins, Biochim. Biophys. Acta *572*, 244 – 256 (1979).

[7] *Ch. Ehnholm, W. Shaw, W. Harlan, V. Brown,* Separation of Two Types of Triglyceride Lipase from Human Post Heparin Plasma, Circulation (Suppl. IV), *48*, 112 (1973).

[8] *M. L. Baginsky, W. V. Brown,* A New Method for the Measurement of Lipoprotein Lipase in Postheparin Plasma using Sodium Dodecyl Sulfate for the Inactivation of Hepatic Triglyceride Lipase, J. Lipid Res. *20*, 548 – 556 (1979).

[9] *H. Greten, R. DeGrella, G. Klose, W. Rascher, J. L. de Gennes, E. Gjone,* Measurement of Two Plasma Triglyceride Lipases by an Immunochemical Method: Studies in Patients with Hypertriglyceridemia, J. Lipid Res. *17*, 203 – 210 (1976).

[10] *J. K. Huttunen, Ch. Ehnholm, P. K. J. Kinnunen, E. A. Nikkilä,* An Immunochemical Method for the Selective Measurement of Two Triglyceride Lipases in Human Postheparin Plasma, Clin. Chim. Acta *63*, 335 – 347 (1975).

[11] *J. C. LaRosa, R. I. Levy, H. G. Windmüller, D. S. Fredrickson,* Comparison of the Triglyceride Lipase of Liver, Adipose Tissue, and Postheparin Plasma, J. Lipid Res. *13*, 356 – 363 (1972).

[12] *R. M. Kraus, R. I. Levy, D. S. Fredrickson,* Selective Measurement of Two Lipase Activities in Postheparin Plasma from Normal Subjects and Patients with Hypertriglyceridemia, J. Clin. Invest. *54*, 1107 – 1124 (1974).

[13] *M. L. Baginsky, W. V. Brown,* Differential Characteristics of Purified Hepatic Triglyceride Lipase and Lipoprotein Lipase from Human Postheparin Plasma, J. Lipid Res. *18*, 423 – 437 (1977).

[14] *J. Boberg, J. Augustin, M. L. Baginsky, P. Tejada, W. V. Brown,* Quantitative Determination of Hepatic and Lipoprotein Lipase Activities from Human Postheparin Plasma, J. Lipid Res. *18*, 544 – 547 (1977).

[15] *M. R. Taskinen, E. A. Nikkilä, J. K. Huttunen, H. Hilden,* A Micromethod for Assay of Lipoprotein Lipase Activity in Needle Biopsy Samples of Human Adipose Tissue and Skeletal Muscle, Clin. Chim. Acta *104*, 107 – 117 (1980).

[16] *R. C. Pittman, J. C. Khoo, D. Steinberg,* Cholesterol Esterase in Rat Adipose Tissue and its Activation by Cyclic Adenosine-3′-5′-monophosphate-dependent Protein Kinase, J. Biol. Chem. *250*, 4505 – 4511 (1975).

[17] *J. Augustin, H. Greten,* The Role of Lipoprotein Lipase-Molecular Properties and Chemical Relevance, in: *R. Paoletti, A. M. Gotto* (eds.), Atherosclerosis Reviews, Raven Press, New York 1979, pp. 91 – 124.

1.4 Cholinesterases

Mary Whittaker

Cholinesterases, unlike other esterases, have a marked affinity for quaternary ammonium ions. These enzymes hydrolyze choline esters to form choline and the corresponding fatty acid. Two distinct types of cholinesterase are now recognized: acetylcholinesterase (AChE) and cholinesterase (ChE).

Acetylcholinesterase is usually located in membranes (e.g. of erythrocytes) of vertebrates and non-vertebrates; the enzyme controls ionic currents in excitable membranes and plays an essential role in nerve conduction processes at the neuromuscular junction [1]. It is present in the white and grey matter of brain, in amniotic fluid and placenta. It occurs in all types of excitable tissue: the electric organs from *Electrophorus electricus* and *Torpedo marmorata* are rich sources of the enzyme.

Cholinesterase occurs in many tissues of vertebrates (e.g. liver, grey matter of brain, etc.) as well as in plasma, but its physiological role remains speculative. It also occurs in plant tissues and in micro-organisms.

1.4.1 Acetylcholinesterase

(True Cholinesterase, Specific Cholinesterase)

Acetylcholine acetylhydrolase, EC 3.1.1.7

General

Application of method: in biochemistry, neurochemistry and in clinical chemistry.

Enzyme properties relevant in analysis: the active centre of acetylcholinesterase is described in terms of two "sites" – the anionic site, interacting coulombically with a positively charged quaternary nitrogen group of the substrate, and the esteratic site, where bond splitting involving the covalent bonding of the substrate to the enzyme, occurs. Substrate conversion proceeds linearly with time.

Methods of determination: a variety of techniques has been used to measure enzyme activity. These include electrometric, titrimetric, manometric, colorimetric, radiometric and potentiometric methods [2].

The introduction of radioactively labelled substrates [1-^{14}C]acetylcholine and [acetyl-1-^{14}C]-β-methylcholine has provided sensitive and accurate assay systems for acetylcholinesterase. The assay is based on the measurement of labelled acetate remaining in the aqueous phase following enzymic hydrolysis after the unhydrolyzed substrate [1-^{14}C]acetylcholine has been removed, either by extraction with sodium tetraphenylboron in higher ketones [3] or by precipitation with reineckate [4]. The method is suitable for application to sub-micro quantities of nervous tissue [5].

The demand for acetylcholinesterase assay is likely to be dramatically increased with the acceptance that the investigation of acetylcholinesterase isoenzymes present in amniotic fluid is useful in the diagnosis of neural tube defects [6, 7].

The most favoured assays used routinely for toxicological studies involve either the electrometric or a colorimetric technique. These assays are precise, cheap and quick. Furthermore the apparatus is standard equipment in most laboratories. The colorimetric method is readily adaptable for automation.

International reference method and standards: standardization at the international level is not available at present but standard preparations of AChE from the electric organ of *Electrophorus electricus* or *Torpedo marmorata* with known enzyme activity, are commercially available. Bovine and human erythrocyte AChE preparations are also commercially available.

Enzyme effectors: quinidine sulphate (2×10^{-5} mol/l) and ethopropazine (10^{-4} mol/l) selectively inhibit plasma cholinesterase whilst 3116CT (bis (3-dimethylamino-5-hydroxyphenoxy) 1,3-propane dimethiodide) selectively inhibits acetylcholinesterase. Eserine (10^{-5} mol/l) inhibits both acetylcholinesterase and plasma cholinesterase as does DFP (diisopropyl phosphofluoridate), 10^{-5} mol/l.

Assays

1.4.1.1 Electrometric Method

Method Design

Principle: acetylcholine is hydrolyzed to acetic acid and choline by acetylcholinesterase (AChE).

$$\underset{\text{Acetylcholine}}{(CH_3)_3\overset{\oplus}{N}CH_2CH_2O\text{-}COCH_3} + H_2O \xrightarrow{\text{AChE}} \underset{\text{acetic acid}}{CH_3COOH} + \underset{\text{choline}}{(CH_3)_3\overset{\oplus}{N}CH_2CH_2OH}$$

The electrometric method monitors the amount of acetic acid produced in terms of the change in pH produced by the enzyme reaction in a standard buffer solution over a definite time interval [8].

Optimized conditions for measurement: the enzyme source (erythrocyte or tissue preparation) should be sufficiently diluted so that any buffering effect is low compared with the total buffer present. A phosphate buffer of pH 8.0 is designed so that the decrease in activity of the acetylcholinesterase with pH over the range 8.1 – 6.0 is almost compensated for by a decrease in the buffer capacity. Correction factors nullify this effect as well as the decrease in pH caused by non-enzymatic hydrolysis of the substrate. To avoid precipitation of phosphate, Mg^{2+} is not added. In blood samples, heparin should be used as anticoagulant in order to avoid complex formation of citrate or oxalate with Mg^{2+} and Ca^{2+} present in blood. The substrate must always be present in excess and the temperature should be maintained at 25 ± 0.1 °C.

Temperature conversion factors: a one-degree change in temperature from 25 °C results in a 3.0% change in erythrocyte acetylcholinesterase activity [9].

Equipment

An expanded scale pH meter. Single element combination pH electrode (preferably the micro-electrode type). Water-bath set at 25 ± 0.1 °C. Stopwatch.

Reagents and Solutions

Purity of reagents: reagents should be of recognized analytical grade. Acetylcholine chloride should be stored in a desiccator.

Preparation of solutions (for about 20 determinations): all solutions in re-purified water (cf. Vol. II, chapter 2.1.3.2). All solutions should be stored in glass bottles.

1. Veronal buffer (20 mmol/l; pH 8.1 at 25 °C):

   dissolve 4.1236 g sodium barbital, 0.5446 g potassium dihydrogen phosphate (KH_2PO_4) and 44.73 g potassium chloride in 900 ml water. Add 28.0 ml HCl, 100 mmol/l, and dilute to 1000 ml with water.

2. Acetylcholine chloride (110 mmol/l):

   dissolve 0.2 g acetylcholine chloride in 10 ml water.

3. Saponin (0.01% w/v):

   dissolve 0.01 g saponin in 100 ml water.

4. Sodium chloride (0.9% w/v):

   dissolve 0.9 g sodium chloride in 100 ml water.

Stability of solutions: store all solutions at 0 – 4 °C. The buffer (1) is stable as long as no microbial contamination occurs. The substrate solution (2) is best used immediately and any unused reagent should be discarded.

Procedure

Collection and treatment of specimens: collect blood from the vein without stasis; heparin must be used as anticoagulant. Blood samples should be separated as soon as possible after collection by centrifugation at 1000 g for 15 min. Separate the plasma and buffy layer. Wash the cells twice with an equal volume of sodium chloride solution (4). Centrifuge at 1000 *g* for 15 min, and discard the washings. Dilute the packed, washed red cells with an equal volume of saline solution (4) and mix thoroughly. Measure the haematocrit of the suspension.

Stability of the enzyme in the sample: after centrifugation the samples can be stored for 2 – 3 days at 4 °C without appreciable loss of enzymic activity. Lysates should be analyzed within 4 h of preparation (lysates or whole cells stored deep frozen are suitable material for long term quality control [10]). AChE in erythrocytes of other species are less stable [11].

Details of measurement in tissues: the tissue should be thoroughly homogenized in a *Virtis* homogenizer* prior to enzymic assay. The optimum substrate concentration may vary considerably, not only for different tissues but also for different species and should be determined in preliminary experiments. Assays using brain homogenates have been reported [12].

Assay conditions: final volume 2.22 ml; water-bath at 25 °C ± 0.1 °C; incubation period 60 min. All determinations must be duplicated. A blank reaction is mandatory and is prepared by substituting sodium chloride solution (4) instead of the sample to measure the spontaneous hydrolysis of substrate.

* *Virtis* Co., Yonkers, N.Y.

Measurement

Pipette successively into the reaction vessel (5 ml tubes are very suitable):		concentration in assay mixture	
saponin solution (3) sample (mixed cell suspension)	1 ml 0.02 ml	saponin volume fraction	45 μg/ml 0.009
mix thoroughly to give complete hydrolysis			
buffer solution (1)	1 ml	veronal phosphate KCl	9 mmol/l 1.82 mmol/l 270 mmol/l
equilibrate for 10 min at 25 °C and determine pH_1			
acetylcholine chloride solution (2)	0.20 ml	acetylcholine	9.91 mmol/l
shake thoroughly, note time; after 60 min read pH_2. The readings are repeated for the blank.			

The initial pH_1 is generally between 7.97 and 8.03. If the reading is below 7.97 it is essential to discard the test and start again using freshly prepared buffer solution (1).

The electrode is rinsed with de-ionized water and gently blotted dry with soft adsorbent tissue.

Calculation: $\delta pH = pH_1 - pH_2$, $\delta pH_{blank} = pH_{1blank} - pH_{2blank}$.

Subtract δpH for blank from δpH for sample, yielding ΔpH per unit time ($\Delta pH/\Delta t$). The catalytic activity concentration is usually expressed as $pH \times h^{-1} \times l^{-1}$ erythrocytes. $\Delta pH/\Delta t$ cannot be directly converted to substrate conversion in μmol/min (U). Considering the volume fraction of the sample in the assay mixture the catalytic activity concentration for 1 litre erythrocytes is

$$b = \frac{f \times 100}{0.009 \times H} \times 10^3 \times \Delta pH/\Delta t \qquad \text{arbitr. units/l}$$

where H is the haematocrit of erythrocyte suspension (%)
f is the correction factor for variations in ΔpH/h with pH corresponding to pH_2. The factor increases linearly from 0.94 at pH 7.9 to 1.00 at pH 7.3. It remains constant at 1.00 until the pH falls to 6.6 after which it becomes 0.97 until pH 6.2.

Validation of Method

Precision, accuracy, detection limits and sensitivity: for values of about 0.90 ΔpH/h in human erythrocytes a precision of 0.02 ΔpH/h was found. The coefficient of variation for human erythrocytes is ±2.8% [13] and for brain homogenates ±7.7% [12]. The relative standard deviation referred to the mean value of a series of interlaboratory measurements was 4%. The lowest practical value to be measured is approximately 0.03 ΔpH/h using a pH meter accurate to ±0.001 pH units. The imprecision of this low catalytic activity is $< \pm 20\%$.

Sources of error: the therapeutic use of anticholinesterases may result in an abnormally low enzymatic activity in an individual. Interference in the assay technique due to uneven sampling of the suspended red cells will result in poor reproducibility in duplicate samples. Deterioration of the electrode is a frequent source of error.

Specificity: although cholinesterase hydrolyzes acetylcholine, it does so at a slower rate than acetylcholinesterase. Substrate and buffer capacity are optimal for AChE hydrolysis but the inclusion of a selective inhibitor of ChE may be necessary.

Reference ranges: in human erythrocytes of healthy probands ΔpH/h at 25 °C is 0.81 [13]. Is seems probable that men have a higher erythrocyte acetylcholinesterase activity than women (0.84 ± 0.11 for men and 0.75 ± 0.07 for women [14]).

1.4.1.2 Colorimetric Method

Method Design

Principle: acetylthiocholine is hydrolyzed by AChE to acetic acid and thiocholine.

$$\text{(a)} \quad (CH_3)_3\overset{\oplus}{N}CH_2CH_2SCOCH_3 + H_2O \xrightarrow{\text{AChE}} CH_3COOH + (CH_3)_3\overset{\oplus}{N}CH_2CH_2SH$$

Acetylthiocholine — acetic acid — thiocholine

The catalytic activity is measured by following the increase of the yellow anion, 5-thio-2-nitrobenzoate, produced from thiocholine when it reacts with DTNB* [15].

* DTNB = 5.5′-dithio-bis-2-nitrobenzoic acid.

(b)

$$(CH_3)_3\overset{\oplus}{N}CH_2CH_2SH + O_2N\text{-}C_6H_3(COO^-)\text{-}S\text{-}S\text{-}C_6H_3(COO^-)\text{-}NO_2$$

Thiocholine DTNB

$$\longrightarrow (CH_3)_3\overset{\oplus}{N}CH_2CH_2SS\text{-}C_6H_3(COO^-)\text{-}NO_2 + O_2N\text{-}C_6H_3(^-OOC)\text{-}S^-$$

2-nitrobenzoate-5-mercaptothiocholine 5-thio-2-nitrobenzoate

Optimized conditions for measurement: AChE activity is at a maximum between pH 7.5 and 9 [1]. At high pH there is considerable non-enzymatic hydrolysis of the substrate, acetylthiocholine, which would yield a high blank in the assay procedure. It is customary to use phosphate buffer, 100 mmol/l, at pH 8 for the assay but a reagent blank is essential. In some studies it may be desirable to lower the pH to reduce the blank to negligible proportions. In such cases it is necessary to confirm a linear relationship of the assay procedure at the selected temperature. Cholinesterase also hydrolyzes acetylthiocholine, although at a slower rate, and so it is customary to include a specific inhibitor for this enzyme in the assay. Quinidine sulphate (0.02 mmol/l) or ethopropazine (0.1 mmol/l) are commonly used for this purpose. The assay is valid at 20 °C, 25 °C, 30 °C or 37 °C.

Temperature conversion factors: the relative activity of human erythrocyte acetylcholinesterase activity at 25 °C is 0.65 and at 30 °C is 0.79 times the activity at 37 °C [10].

Equipment

Spectrophotometer for exact measurement at 410 nm preferably with a thermostatted cuvette holder, water-bath, stopwatch or recorder.

Reagents and Solutions

Purity of reagents: analytical grade reagents must be used. All reagents should be stored in glass bottles. Solid acetylthiocholine iodide must be stored in a desiccator.

Preparation of solutions (for about 50 determinations): all solutions in re-purified water (cf. Vol. II, chapter 2.1.3.2).

1. Phosphate buffer (100 mmol/l; pH 8.0):

   dissolve 15.6 g $Na_2HPO_4 \cdot 2\ H_2O$ in 750 ml water. Adjust the pH to 8.0 by adding sodium hydroxide solution (100 mmol/l). Check pH at 25 °C. Make up to 1 l with water.

2. Phosphate buffer (100 mmol/l; pH 7.0):

   dissolve 15.6 g $Na_2HPO_4 \cdot 2\ H_2O$ in 750 ml water. Adjust to pH 7.0 by addition of sodium hydroxide solution (100 mmol/l). Check pH at 25 °C. Make up to 1 l with water.

3. Buffered *Ellman's* reagent, DTNB (DTNB, 10 mmol/l; $NaHCO_3$, 17.85 mmol/l):

   dissolve 39.6 mg DTNB in 10 ml phosphate buffer (2) and add 15 mg sodium bicarbonate.

4. Acetylthiocholine iodide (75 mmol/l):

   dissolve 108.35 mg acetylthiocholine iodide in 5 ml water.

Stability of solutions: store all solutions at 4 °C. The phosphate buffers (1) and (2) are stable as long as no microbial contamination occurs. Solution (3) is made up in phosphate buffer (2) in which it is more stable than in phosphate buffer (1) of pH 8.0. Solution (3) is stable for 4 weeks if stored in dark bottles. Solution (4) should not be kept for more than 7 days.

Procedure

Collection and treatment of specimens: as described on p. 55. It is necessary to estimate either the haemoglobin mass concentration (g/l) or the red cell count of the original blood sample prior to assay. Dilute the saline suspension of red cells with water 1: 50 before assay.

Stability of the enzyme in the sample: as described on p. 55.

Details for measurement in tissues: as described on p. 55.

Assay conditions: wavelength 410 nm; light path 10 mm; final volume 3.14 ml; water-bath 25 ± 0.1 °C; thermostatted cuvette holder at 25 °C; measure against air. A blank reaction must be set up to estimate the non-enzymatic hydrolysis of the substrate. It is prepared by substituting phosphate buffer solution (1) in place of the sample.

Measurement

Pipette successively into the cuvette:			concentration in assay mixture	
buffer	(1)	3.00 ml	phosphate	100 mmol/l
acetylthiocholine solution	(4)	0.02 ml	acetyl-thiocholine	0.47 mmol/l
DTNB solution	(3)	0.10 ml	DTNB	0.315 mmol/l
			phosphate	100 mmol/l
			$NaHCO_3$	0.56 mmol/l
incubate at 25 °C for 10 min				
sample		0.02 ml	volume fraction	0.0064
mix by inversion, read absorbance and start stopwatch. Repeat the readings at intervals of 0.5 min or monitor the reaction on a recorder for 5 min to verify that a linear reaction occurs.				

The volume quoted can be adjusted to the instrumentation available, or to the size or nature of the sample. If whole blood is used it is necessary to include quinidine in the assay in order to inhibit the plasma cholinesterase present.

Calculation: acetylcholinesterase activity is expressed as micromoles of substrate hydrolyzed at 25 °C per g haemoglobin per litre whole blood; it is also expressed as micromoles of substrate hydrolyzed at 25 °C per litre of packed red cells. The absorption coefficient of the yellow anion is $\varepsilon = 1.36\ l \times mmol^{-1} \times mm^{-1}$ [11]. Subtract the change in absorbance per min for the blank from the change in absorbance per min for the sample yielding $\Delta A/\Delta t$.

Calculate the catalytic concentration of acetylcholinesterase in the erythrocyte suspension according to eqns. (k) or (k_1), cf. "Formulae", Appendix 3. With t in minutes

$$b = \frac{F}{1.36} \times \Delta A/\Delta t \quad U/l$$

$$b = \frac{16.667 \times F}{1.36} \times \Delta A/\Delta t \quad nkat/l$$

where F is the dilution factor of erythrocytes in saline and water.

If ρ_{Hb} is the mass concentration of haemoglobin (g/l) then the acetylcholinesterase content (μmoles substrate hydrolyzed per min per g haemoglobin) is according to eqns. (l) or (l_1) – cf. "Formulae", Appendix 3:

$$z_c/m_s = \frac{F}{1.36 \times \rho_{Hb}} \times \frac{\Delta A}{\Delta t} \quad U/g$$

$$z_c/m_s = \frac{16.667 \times F}{1.36 \times \rho_{Hb}} \times \frac{\Delta A}{\Delta t} \quad nkat/g$$

Validation of Method

Precision, accuracy, detection limits and sensitivity: correlation coefficients of 4% and 7.9% have been reported for the acetylcholinesterase activity of human erythrocytes [10, 15] and brain homogenates [12] respectively. The correlation coefficient between the calculated and experimental activities of two human erythrocyte populations mixed in five different proportions was 0.99. Limit of detection is approximately 20 U/l. Sensitivity is equivalent to $\Delta A/\Delta t$ = 0.002 per 5 min.

Reference ranges: erythrocyte acetylcholinesterase activities in healthy adults, newborn infants and patients with disorders characterized by a reduction of AChE activity are given in Table 1.

Table 1. Human erythrocyte acetylcholinesterase activity at 25 °C

Subjects	Number	Relative activity	Activity, U/l RBC	Ref.
Healthy adults	23	100	6956 – 11673	[10]
Healthy adults	10	100	7556 – 12222	[19]
Healthy adults	NG	100	6469 – 9854	[20]
Healthy adults	60	100	6689	[17]
Paroxysmal nocturnal haemoglobinuria	5	41, 39, 29, 19, 14	2742, 2609, 1940, 1271, 936	[17]
Autoimmune haemolytic anaemia	2	28, 30	1873, 2007	[17]
Healthy full term neonates	253	62	4147	[17]
Full term neonates with haemolytic disease				
Rh disease	65	64	4280	[17]
ABO disease	57	37	2475	[17]
Infants of low birthweight for gestational age	120	58	3880	[17]

Conversion factors for different red cell measurements [20]:

U/gHb to U/ml packed red cells, multiply by 0.01 × mean corpuscular haemoglobin concentration (NV = 0.34).

U/10^{10} RBC to U/ml RBC, multiply by 100/mean corpuscular volume (NV = 1.15).

Sources of error

Effect of drugs: as described on pp. 53,57.

Interference in the assay technique: uneven sampling as described on p. 57. The assay of erythrocyte cholinesterase of dogs, rats and rabbits cannot be recommended by this method since the low activity of the erythrocyte enzyme in these species necessitates the use of large volumes of haemolysates which interfere with the spectrophotometric readings.

Specificity: it may be necessary to use a selective inhibitor of acetylcholinesterase such as DFP (cf. p. 53) or eserine in order to use the method for the specific assay of AChE. For example, most tissue contain non-specific esterases and their contribution to the hydrolysis of the substrate an be determined by incorporating physostigmine in an extra blank to be subtracted from the observed rate of change of absorbance in the assay [18].

References

[1] *D. Nachmansohn, E. Neumann,* in: Chemical and Molecular Basis of Nerve Activity, Academic Press, New York 1975.

[2] *K. B. Augustinsson,* Determination of Activity of Cholinesterases, in: *D. Glick* (ed.), Methods of Biochemical Analysis, Vol. 19, John Wiley and Sons Inc., New York 1971, pp. 217 – 273.

[3] *F. Fonnum,* Radiochemical Microassays for the Determination of Choline Acetyltransferase and Acetylcholinesterase Activities, Biochem. J. *115*, 465 – 472 (1969).

[4] *M. W. McCaman, L. R. Tomey, R. E. McCaman,* Radiometric Assay of Acetylcholinesterase Activity in Submicrogram Amounts of Tissue, Life Science *7*, 233 – 244 (1968).

[5] *J. L. Maderdrut,* A Radiometric Anion-Exchange Method for Acetylcholinesterase, Neurochem. Res. *2*, 717 – 721 (1977).

[6] *A. D. Smith, N. J. Wald, H. S. Cuckle, G. M. Stirrat, M. Bobrow, H. Lagercrantz,* Amniotic-fluid Acetylcholinesterase as a Possible Diagnostic Test for Neural-tube Defects in Early Pregnancy, Lancet *I*, 685 – 688 (1979).

[7] *R. D. Barlow, H. S. Cuckle, N. J. Wald,* A Simple Method for Amniotic Fluid Gel-acetylcholinesterase Determination, Suitable for Routine Use in the Antenatal Diagnosis of Open Neural-tube Defects, Clin. Chim. Acta *110*, 137 – 142 (1982).

[8] *H. O. Michel,* An Electrometric Method for the Determination of Red Blood Cell and Plasma Cholinesterase Activity, J. Lab. Clin. Med. *34*, 1564 – 1568 (1949).

[9] *R. I. Ellin, P. P. Vicario,* ΔpH Method for Measuring Blood Cholinesterase, Arch. Environ. Health *30*, 263 – 265 (1975).

[10] *P. J. Lewis, R. K. Lowing, D. Gompertz,* Automated Discrete Kinetic Method for Erythrocyte Acetylcholinesterase and Plasma Cholinesterase, Clin. Chem. *27*, 926 – 929 (1981).

[11] *H. J. Mersmann, M. C. Sanguinetti,* Automated Determination of Plasma and Erythrocyte Cholinesterase in Various Species, Am. J. Vet. Res. *35*, 579 – 583 (1974).

[12] *K. I. Hawkins, C. E. Knittle,* Comparison of Acetylcholinesterase Determinations by the *Michel* and *Ellman* Methods, Anal. Chem. *44*, 416 – 417 (1972).

[13] *R. I. Ellin, B. H. Burkhardt, R. D. Hart,* A Time-modified Method for Measuring Red Blood Cell Cholinesterase Activity, Arch. Environ. Health *27*, 48 – 49 (1973).

[14] *H. Hecht, E. Stillger,* Normalwerte und individuelle Schwankungsbreite der Acetylcholinesterase des Blutes, Z. Klin. Chem. Klin. Biochem. *5*, 156 – 160 (1967).

[15] *G. L. Ellman, K. D. Courtney, V. Andres Jr., R. M. Featherstone,* A New and Rapid Colorimetric Determination of Acetylcholinesterase Activity, Biochem. Pharmacol. *7*, 88 – 95 (1961).
[16] *G. L. Ellman,* Tissue Sulfhydryl Groups, Arch. Biochem. Biophys. *82*, 70 – 77 (1959).
[17] *F. Herz, E. Kaplan,* A Review: Human Erythrocyte Acetylcholinesterase, Pediatr. Res. *7*, 204 – 214 (1973).
[18] *G. B. Koelle,* The Histochemical Identification of Cholinesterase in Cholinergic, Adrenergic and Sensory Neurons, J. Pharmacol. Exp. Ther. *114*, pp. 167 – 184 (1955).
[19] *E. Reiner, A. Buntic, M. Trdak, V. Simenn,* Effect of Temperature on the Activity of Human Blood Cholinesterase, Arch. Toxicol. *32*, 347 – 350 (1974).
[20] *E. Beutler,* in: Red Cell Metabolism, Grune and Stratton, New York 1975.

1.4.2 Cholinesterase

(Pseudocholinesterase, Plasma or Serum Cholinesterase, Non-specific Cholinesterase, Butyrylcholinesterase or Benzoylcholinesterase)

Acylcholine acylhydrolase, EC 3.1.1.8

General

Applications of method: in biochemistry and in clinical chemistry.

Enzyme properties relevant in analysis: the catalytic activity of serum cholinesterase, ChE, is due to several isoenzymes with different activities [1]. Substrate conversion proceeds linearly with time.

Methods of determination: enzyme activity can be measured by a wide variety of techniques, e.g. electrometric, titrimetric, potentiometric, manometric, colorimetric, conductimetric, radiometric or fluorimetric. An excellent review of many of these assays has been produced by *Augustinsson* [2]. An update of some of these assays together with the range of human serum enzymatic activity has been tabulated [3]. Determination of enzyme immunological mass [4, 5] as well as active-site titration [1, 6, 7] are steadily making an impact on studies of the enzyme and its genetic variants. The most favoured assays, which are used routinely, involve a colorimetric technique. Such assays are precise, cheap and quick. They are also adaptable for either manual or automated continuous monitoring systems.

International reference method and standards: standardization at the international level is not available at present but standard sera of known enzymic activity are commercially available.

Enzyme effectors: eserine, 10^{-5} mol/l, and E600 (diethyl-4-nitrophenyl phosphate), 10^{-5} mol/l, inhibit both acetylcholinesterase and cholinesterase. 62C47, 1:5-bis-(4-trimethyl ammonium phenyl)pentan-3-one dibromide, and 284C51, 1:5-bis-(4-allyl dimethyl ammonium phenyl)pentan-3-one dibromide, are specific inhibitors of acetylcholinesterase (10^{-5} mol/l) and iso-OMPA, tetraisopropylpyrophosphoramide, specifically inhibits cholinesterase (10^{-5} mol/l). The specificity should be checked for each species investigated.

Assay

Method Design

Principle: acylthiocholines are hydrolyzed by cholinesterase (ChE) to the corresponding fatty acid and thiocholine. The rate of formation of the thiocholine can be monitored by the continuous reaction of the thiol group with 5,5′-dithio-bis-(2-nitrobenzoic acid) – DTNB – to form a yellow anion, 5-thio-2-nitrobenzoate, and other products. The rate of formation of the yellow anion can be measured spectrophotometrically at 410 nm [8].

(a) $(CH_3)_3\overset{\oplus}{N}CH_2CH_2SCOR + H_2O \xrightarrow{ChE} RCOOH + (CH_3)_3\overset{\oplus}{N}CH_2CH_2SH$

Acylthiocholine — fatty acid — thiocholine

(b) $(CH_3)_3\overset{\oplus}{N}CH_2CH_2SH$ + DTNB (O_2N–$C_6H_3(COO^-)$–S–S–$C_6H_3(COO^-)$–NO_2) ⟶

Thiocholine — DTNB

5-thio-2-nitrobenzoate (O_2N–$C_6H_3(COO^-)$–S^-) + 2-nitrobenzoate-5-mercaptothiocholine ($(CH_3)_3\overset{\oplus}{N}CH_2CH_2S$–S–$C_6H_3(COO^-)$–$NO_2$)

Acetyl-, propionyl- and butyryl-thiocholines have been used as substrates in modifications of the method which can readily be adapted to automated equipment. The affinity of butyrylthiocholine for human serum ChE is twice that of acetylthiocholine and additionally it is more stable with respect to pH and temperature than acetylthiocholine [9]. Butyrylthiocholine is therefore the substrate of choice for either human or horse serum ChE using the *Ellman* method of assay. Other species have different specificities for various acylthiocholines as substrate [1].

Optimized conditions for measurement: maximum activity of the enzyme is obtained at pH 8.5 – 9.0 using concentrations of butyrylthiocholine of 10 mmol/l at 37 °C [10].

However there is considerable non-enzymatic hydrolysis under these conditions which would produce high reagent blank readings in the assay procedure. Therefore, despite a pH optimum of 8.5, the use of *Sørensen's* phosphate buffer, 67 mmol/l, pH 7.4 is preferred. This reduces the non-enzymatic hydrolysis to negligible proportions and still allows the linear relationship of the assay procedures to obtain at 37 °C. A true index of serum ChE activity can be obtained only when the assay is conducted at 37 °C [10]. The assay given below is valid at 20 °C, 25 °C or 30 °C.

Temperature conversion factors: *Das & Liddell* [11] have shown that temperature correction can be made by adding or subtracting 9% for each 1 °C variation of the reaction mixture from 25 °C.

Equipment

Spectrophotometer capable of exact measurement at 410 nm preferably with a thermostatted cuvette holder, water-bath, recorder or stopwatch.

Reagents and Solutions

Purity of reagents: reagents should be of recognized analytical grade. Solid butyrylthiocholine iodide should be stored under desiccation.

Preparation of solutions (for about 30 determinations): all solutions in re-purified water (cf. Vol. II, chapter 2.1.3.2). All solutions should be stored in glass bottles.

1. *Sørenson's* phosphate buffer (67 mmol/l; pH 7.4 at 37 °C):

   dissolve 9.595 g $Na_2HPO_4 \cdot 2\,H_2O$ and 1.74 g KH_2PO_4 in 750 ml water. Check the pH = 7.4 at 37 °C. Cool to room temperature and make up to 1 l.

2. Buffered *Ellman's* reagent, DTNB (0.27 mmol/l):

   dissolve 10.7 mg DTNB in 100 ml buffer solution (1). Store in dark bottle at least overnight at 4 °C.

3. Butyrylthiocholine iodide (90 mmol/l):

   dissolve 149 mg butyrylthiocholine iodide in 5 ml water. Store in a dark bottle.

Stability of solutions: store all solutions at 0 – 4 °C. The phosphate buffer (1) is stable as long as microbial contamination is absent. Solutions (2) and (3) are stable for 4 weeks if stored in dark bottles. Storage of other acylthiocholines should be limited to 1 week.

Procedure

Collection and treatment of specimen: collect blood from the vein without stasis. Serum or heparinized plasma may be used for assay and blood samples should be separated as soon as possible after collection. Haemolysis should be avoided.

Stability of the enzyme in the specimen and sample: ChE is extremely stable when undiluted. Unrefrigerated blood can be sent by airmail or first class domestic post without significant change in activity. Samples may be kept for many years without deterioration at $-20\,^{\circ}C$ if they are not thawed and re-frozen [12].

Details for measurement in tissues: the total ChE activity of a cell is only quantitatively extracted after complete homogenization of the tissue. The optimum substrate concentration may vary considerably for different tissues and should be determined in preliminary experiments.

Assay conditions: wavelength 410 nm; light path 10 mm; final volume 3.02 ml; water-bath $37\,^{\circ}C$ (thermostatted cuvette holder); measure against air. Blank reactions are unnecessary since the non-enzymatic hydrolysis is virtually negligible under these conditions.

Measurement

Pipette successively into the cuvette:			concentration in assay mixture	
DTNB solution	(2)	2.9 ml	phosphate	67 mmol/l
			DTNB	0.259 mmol/l
butyrylthiocholine solution	(3)	0.1 ml	butyrylthio-choline	2.98 mmol/l
sample (serum or other sample)		20 µl	volume fraction	0.0066
mix by inversion, read absorbance and start stopwatch. Repeat readings at intervals of 0.5 min or monitor the reaction on a recorder.				

The volumes quoted can be scaled down if required to accommodate the available instrumentation or size or nature of sample.

Calculation: serum cholinesterase activity is expressed as micromoles of substrate hydrolyzed at $37\,^{\circ}C$ per l plasma or serum per minute.

The absorption coefficient of the product of the chemical reaction, 5-thio-2-nitrobenzoate is $\varepsilon = 1.36\ l \times mmol^{-1} \times min^{-1}$ [13]. For calculation of the catalytic concentration of cholinesterase in serum or plasma use eqns. (k) or (k_1) from "Formulae", Appendix 3. With t in minutes

$$b = 11.1 \times 10^3 \times \Delta A/\Delta t \quad \text{U/l}$$

$$b = 185 \times 10^3 \times \Delta A/\Delta t \quad \text{nkat/l}$$

Validation of Method

Precision, accuracy, detection limits and sensitivity: the coefficient of variation is of the order of −4%. Two sera with high (5.00 kU/l) and low (0.9 kU/l) enzymatic activity were mixed in six different proportions and the correlation coefficient between the calculated and experimental activities was 0.986. Limit of detection is approximately 10 U/l. Sensitivity is equivalent to $\Delta A/\Delta t$ 0.002 per 5 min.

Sources of error: in general haemolyzed samples do not yield a greater net activity of plasma cholinesterase but less consistent results were obtained for multiple assays of such samples. A plasma blank is essential with heavily jaundiced samples.

Specificity: butyrylthiocholine is the substrate of choice for human or horse plasma cholinesterase but the specificity of rat plasma is highest for propionylthiocholine. Other sources of the enzyme exhibit different specificity for the substrate. There is little variation in substrate specificity for the enzyme when it occurs in different tissues within a species.

Reference ranges: the range for human plasma cholinesterase activity is 1.8 – 5.0 kU/l plasma for women and 2.0 – 5.2 kU/l plasma for men at 25 °C [9]. Similar ranges of catalytic concentration have been reported by other workers [3, 11].

Suxamethonium sensitivity: suxamethonium is widely used as a muscle relaxant in anaesthesia. The drug, a dicholine ester of succinic acid, acts by effecting chemical neuromuscular block causing excessive depolarization which spreads to the surrounding muscle fibres making them unexcitable. It competes with acetylcholine for the cholinergic receptors at the neuromuscular junction producing an acetylcholine-like but prolonged depolarization of the end plate. Suxamethonium, unlike acetylcholine, is not metabolized by acetylcholinesterase and its action persists until it is hydrolyzed by plasma cholinesterase. This hydrolysis is usually rapid in most individuals but a few patients metabolize the drug only slowly and a prolonged apnoea occurs. *Kalow* and his colleagues showed that suxamethonium sensitivity is mainly due to a genetically determined variant of plasma cholinesterase with low enzymatic activity. Several genetic variants of the enzyme have now been recognized by the use of suitable differential inhibitors of the enzyme [12].

1.4.3 Phenotyping of Genetic Variants

General

The method, which is suitable both for routine assay or as a standard procedure, estimates cholinesterase activity in plasma or serum at 25 °C. Benzoylcholine chloride, a specific substrate for ChE, has been widely used in studying the genetic variants [14]. The variants are recognized by use of dibucaine and fluoride incorporated in the assay system [15, 16]. Silent gene phenotypes are assigned on the basis of enzymic activity measurements and family studies [17].

Method Design

Principle

$$\text{Benzoylcholine} + H_2O \xrightarrow{\text{ChE}} \text{benzoic acid} + \text{choline}\,.$$

The use of the two differential inhibitors dibucaine and sodium fluoride enables the probable genotype of an individual to be ascribed. This is confirmed by family studies [12].

Optimized conditions for measurement: benzoylcholine and benzoate ion absorb in the UV. Although the maximum difference in the absorption spectra of the substrate and its hydrolysis product occurs at 235 nm, measurements are made at 240 nm because of the optical properties of diluted plasma. The change in absorbance, at 25 °C, in the UV at 240 nm during the hydrolysis of benzoylcholine is proportional to the decrease in concentration of benzoylcholine. Fluoride inhibition has been shown to be very temperature dependent and the rigorous control of temperature at 25 °C is mandatory.

Temperature conversion factors: *Kalow & Lindsay* [14] investigated the effect of temperature on the rate of hydrolysis of benzoylcholine and have shown that results may be converted from 25 °C to 37 °C or 30 °C by multiplication by appropriate factors. These are given in Table 2.

Equipment

A manual or recording spectrophotometer, with a thermostatically controlled cuvette holder, capable of absorption measurements at 240 nm. Silica cuvettes 10 mm light path. Water-bath at 25 °C ± 0.05 °C.

Table 2. Temperature conversion factors for cholinesterase [18]

Assay temperature °C	To convert to 25 °C multiply by:	To convert to 37 °C multiply by:
16	1.75	3.2
17	1.64	3.0
18	1.54	2.82
19	1.45	2.65
20	1.36	2.48
21	1.28	2.33
22	1.2	2.19
23	1.12	2.05
24	1.055	1.93
25	1.00	1.83
26	0.95	1.74
27	0.9	1.65
28	0.855	1.57
29	0.815	1.5
30	0.775	1.42
31	0.74	1.35
32	0.70	1.29
33	0.67	1.23
34	0.635	1.17
35	0.605	1.11
36	0.575	1.055
37	0.545	1.00

Reagents and Solutions

Purity of reagents: use only reagents of analytical grade. Dry reagents should be kept under desiccation.

Preparation of solutions (for about 150 determinations): all solutions in re-purified water (cf. Vol. II, chapter 2.1.3.2).

Stock solutions

1. *Sørensen's* phosphate buffer (67 mmol/l; pH 7.4 at 25 °C):

   dissolve 37.92 g Na_2HPO_4 (anhydrous) and 9.09 g KH_2PO_4 (anhydrous) in 2 l water. Check the pH and make up to 5 l.

2. Benzoylcholine chloride (0.2 mmol/l):

   dissolve 24.4 mg benzoylcholine chloride in 500 ml buffer (1).

3. Dibucaine hydrochloride (0.04 mmol/l):

   dissolve 15.3 mg dibucaine hydrochloride in 1 l buffer (1).

4. Sodium fluoride (0.2 mmol/l):

   dissolve 42 mg in 100 ml buffer (1) yielding a solution of 10 mmol/l. Dilute 5 ml of this solution to 250 ml with buffer (1) yielding a solution of 0.2 mmol/l.

Working solutions (for about 24 determinations)

For the assay of cholinesterase

5. Benzoylcholine (0.1 mmol/l):

   mix equal volumes (say 25 ml) of phosphate buffer (1) and benzoylcholine solution (2).

For the dibucaine inhibited reaction

6. Benzoylcholine (0.1 mmol/l) and dibucaine (0.02 mmol/l):

   mix equal volumes of benzoylcholine solution (2) and dibucaine solution (3).

For the fluoride inhibited reaction

7. Benzoylcholine (0.1 mmol/l) and sodium fluoride (0.1 mmol/l):

   mix equal volumes of benzoylcholine solution (2) and sodium fluoride solution (4).

Stability of solutions: benzoylcholine solution (2) should be prepared daily. Dibucaine hydrochloride and sodium fluoride solutions (3) and (4) are stable for one week at 4°C. Phosphate buffer (1) is stable unless contaminated with micro-organisms. All working solutions must be prepared daily.

Procedure

Collection and treatment of specimens: as described on p. 66.

Stability of enzyme in sample: as described on p. 66.

Assay conditions: wavelength 240 nm; light path 10 mm; final volume 4.0 ml; water-bath at 25 ± 0.05°C; measure against air in blank reaction. Blank solutions are usually unnecessary since non-enzymatic hydrolysis is virtually negligible under these conditions. If appreciable non-enzymatic hydrolysis occurs, a blank solution is prepared by substituting buffer solution (1) for benzoylcholine solution (5). All solutions must be equilibrated to 25°C before measurement occurs. Dilute all plasma or serum 1 + 99 by making 0.1 ml of plasma up to 10 ml with buffer (1).

Measurement

Pipette into the cuvette:			concentration in assay mixture
benzoylcholine solution	(5)	2 ml	phosphate 67 mmol/l benzoylcholine 0.05 mmol/l volume fraction 0.005
sample (diluted)		2 ml	
mix, read absorbance and start stopwatch, repeat the reading at 0.5 min intervals or monitor the reaction on a recorder for at least 4 min until a series giving equal increments is obtained.			

If the dibucaine number is required, repeat the procedure with the benzoylcholine and dibucaine solution (6) instead of benzoylcholine/phosphate solution (5) in order to determine the rate of reaction in the presence of dibucaine (0.01 mmol/l). The fluoride number is measured by repeating the assay procedure using the benzoylcholine/fluoride solution (7) instead of solution (5) in order to determine the rate of reaction in the presence of fluoride (0.05 mmol/l). The presence of either inhibitor in the assay solution causes higher absorbance readings than the uninhibited reaction.

Calculation: plot the absorbance values as a function of time and draw the best straight line through the points. Calculate the change in absorbance per minute ($\Delta A/\Delta t$). The absorption coefficient of the substrate benzoylcholine at 25 °C is $\varepsilon = 0.66 \text{ l} \times \text{mmol}^{-1} \times \text{mm}^{-1}$ [14].

For calculation of the catalytic concentration of plasma cholinesterase use eqns. (k) or (k_1) from "Formulae", Appendix 3. With t in minutes

$$b = 30.3 \times 10^3 \times \Delta A/\Delta t \quad \text{U/l}$$

$$b = 505 \times 10^3 \times \Delta A/\Delta t \quad \text{nkat/l}$$

If $(\Delta A/\Delta t)_S$, $(\Delta A/\Delta t)_D$ and $(\Delta A/\Delta t)_F$ are the changes in absorbance per min occurring in the uninhibited, dibucaine and fluoride reactions then

$$\text{Dibucaine number DN} = \frac{(\Delta A/\Delta t)_S - (\Delta A/\Delta t)_D}{(\Delta A/\Delta t)_S} \times 100$$

and

$$\text{Fluoride number FN} = \frac{(\Delta A/\Delta t)_S - (\Delta A/\Delta t)_F}{(\Delta A/\Delta t)_S} \times 100$$

Validation of Method

Precision, accuracy, detection limits and sensitivity: the correlation coefficient of the enzymatic activity is ±2.4%. For measurements at 240 nm with a sufficiently sensitive spectrophotometer the lowest possible value to be measured is approximately 63 U/l with an imprecision < ±25%. Sensitivity is found to be $\Delta A/\Delta t$ = 0.002/5 min.

A mean value for DN of 80.58 with a standard deviation of 0.76 was found for 31 assays in one individual; the mean value for FN, measured at the same time, was 61.0 with a standard deviation of 0.97. The relative standard deviation of the DN and FN are 0.94% and 1.59% respectively for the genotype $E_1^uE_1^u$ (for nomenclature cf. Tab. 4). The relative standard deviation of the DN and FN are 7% and 4% respectively for the genotype $E_1^uE_1^a$. The corresponding values for the genotype $E_1^aE_1^a$ are 5% for both the DN and FN.

Sources of error: an incorrect genotype may be assigned to an individual if the blood specimen is taken following a recent blood transfusion. In such an event the enzymatic activity will also be incorrect. Some drug therapy or exposure to organophosphorous compounds will also effect the cholinesterase activity (Table 3). Fluoride numbers are very temperature sensitive and decrease with rise in temperature.

Table 3. Conditions producing changes in human plasma cholinesterase (ChE) activity

Type	Decreased ChE activity	Increased ChE activity
Inherited	Rare cholinesterase variants.	Electrophoretic and *Cynthiana* variants.
Physiological	Pregnancy and the puerperium. Newborns and infants. Liver disease (acute hepatitis and hepatic metastasis). Myocardial infarction. Hyperpyrexia. Malnutrition. Collagen diseases (progressive muscular dystrophy, congenital myotonia and dermatomyositis). Myxoedema. Shock. Tuberculosis. Acute infections. Carcinomas. Uraemia. Chronic debilitating diseases. Chronic anaemias.	Young children (6/12 → 5 years). Obesity. Hyperlipidaemia. Nodular goitre. Psoriasis. Essential hypertension. Thyrotoxicosis. Nephrosis. Asthma. Anxiety states. Alcoholism. Schizophrenia. *Alzheimer's* disease.
Iatrogenic	X-ray therapy. Anti-cancer drugs. Contraceptive pills. Monoamine oxidase inhibitors. Ecothiopate iodide. Propanidid. Neostigmine. Chlorpromazine chloride. Pancuronium. Organophosphorus insecticides. Cyclophosphamide. Burned patients. Kwashiorkhor. Propranolol. Tetanus.	

Specificity: benzoylcholine is a specific substrate for cholinesterase.

Reference ranges: Table 4 summarizes the mean relative activities of plasma cholinesterase and inhibition characteristics in the most common genotypes. Family studies are desirable and are of great importance in assigning the correct genotype. In spite of

Table 4. Distribution, suxamethonium sensitivity and biochemical characteristics of the plasma cholinesterase variants in a Caucasian population

Genotype	Enzymatic activity		Dibucaine number		Fluoride number		Frequency	Suxamethonium sensitivity
	Range kU/l	Relative activity	Average	Range	Average	Range		
$E_1^uE_1^u$	0.650 – 1.450	100	80	78 – 83	62	56 – 69	96%	? 1 in 2500 moderately sensitive
$E_1^uE_1^s$	0.455 – 1.015	70	80	78 – 83	62	56 – 69	1 in 190	? 1 in 1000 moderately sensitive
$E_1^uE_1^f$	0.559 – 1.247	86	74	70 – 83	52	46 – 54	1 in 200	? 1 in 100 moderately sensitive
$E_1^uE_1^a$	0.500 – 1.116	77	62	48 – 69	50	44 – 54	1 in 25	* ? 1 in 500 moderately sensitive
$E_1^aE_1^f$	0.384 – 0.856	59	53	45 – 59	33	28 – 39	1 in 20000	All moderately sensitive
$E_1^fE_1^f$	0.358 – 0.798	55	66	64 – 69	35	30 – 42	1 in 154000	All moderately sensitive
$E_1^fE_1^s$	0.195 – 0.435	30	67	64 – 69	35	30 – 42	1 in 150000	All sensitive
$E_1^aE_1^a$	0.345 – 0.769	43	21	8 – 28	19	10 – 28	1 in 2000	All sensitive
$E_1^aE_1^s$	0.163 – 0.363	22	21	8.28	19	10 – 28	1 in 29000	All very sensitive
$E_1^sE_1^s$	enzymic activity nil or too low to measure →						1 in 100000	All very sensitive

* Many patients (? 1 in 10) with this genotype become moderately sensitive when pregnant.

Nomenclature: E designates the gene for plasma cholinesterase. Subscript 1 represents the first found locus (subscript 2 is used for the electrophoretic variants). E_1^u denotes the gene responsible for the formation of the usual cholinesterase, E_1^a, E_1^f, and E_1^s represent the genes responsible for the biosynthesis of the atypical, fluoride resistant, and 'silent' or an enzymatic cholinesterase respectively. The four genes are allelic and thus there are 10 genotypes. $E_1^uE_1^u$ denotes the homozygote for the usual enzyme. $E_1^uE_1^a$ represent the heterozygote for the usual and atypical enzyme etc.

Over rare variants, all confined to single or few family studies have been reported [12] but in many cases there is uncertainty whether the responsible gene is active at locus 1 or another locus.

family studies the assignment of the correct genotype may be difficult, especially in individuals having silent gene heterozygotes. The figures quoted in Table 4 are valid only when benzoylcholine is used as substrate. Fluoride numbers decrease with rise in temperature and it is essential to adhere to the experimental conditions described [10, 14–16].

References

[1] *S. S. Brown, W. Kalow, W. Pilz, M. Whittaker, C. L. Woronick,* The Plasma Cholinesterases: A New Perspective, in: *A. L. Latner, M. K. Schwartz,* Advances in Clinical Chemistry, Vol. *22*, Academic Press, New York 1981, pp. 1–123.

[2] *K. B. Augustinsson,* Determination of Activity of Cholinesterases, in: *D. Glick* (ed.), Methods of Biochemical Analysis, Vol. *19*, John Wiley and Sons Inc., New York 1971, pp. 217–273.

[3] *M. Whittaker,* Plasmacholinesterase Abnormalities: Laboratory Investigations, in: *F. R. Ellis* (ed.),Inherited Disease and Anaesthesia, Excerpta Medica, Amsterdam 1981, pp. 105–126.

[4] *H. M. Rubinstein, A. A. Dietz, T. Lubrano, P. J. Garry,* E_1^J, a Quantitative Variant at Cholinesterase Locus 1: Immunological Evidence, J. Med. Genet. *13*, 43–45 (1976).

[5] *J. C. Bog-Hansen, I. Lorenc-Kubis, O. J. Bjerrum,* in: *B. J. Radola* (ed.), Quantitative Immunoelectrophoresis. A Survey and some Applications in Electrophoresis '79, Walter de Gruyter and Co., Berlin 1980, pp. 173–192.

[6] *A. R. Main,* Kinetic Evidence of Multiple Reversible Cholinesterases Based on Inhibition by Organophosphates, J. Biol. Chem. *244*, 829–840 (1969).

[7] *O. Lockridge, B. N. La Du,* Comparison of Atypical and Usual Human Serum Cholinesterase. Purification, Number of Active Sites, Substrate Affinity and Turnover Number, J. Biol. Chem. *253*, 361–366 (1978).

[8] *G. L. Ellman, K. D. Courtney, V. Andres Jr., R. M. Featherstone,* A New and Rapid Colorimetric Determination of Acetylcholinesterase Activity, Biochem. Pharmacol. *7*, 88–95 (1961).

[9] *G. Szasz,* Cholinesterase-Bestimmung im Serum mit Acetyl- und Butyrylthiocholin als Substrat, Clin. Chim. Acta *19*, 191–204 (1968).

[10] *E. Silk, J. King, M. Whittaker,* Assay of Cholinesterase in Clinical Chemistry, Ann. Clin. Biochem. *16*, 57–75 (1979).

[11] *D. K. Das, J. Liddell,* Value of Butyrylthiocholine Assay for Identification of Cholinesterase Variants, J. Med. Genet. *7*, 351–355 (1970).

[12] *M. Whittaker,* Plasma Cholinesterase Variants and the Anaesthetist, Anaesthesia *35*, 174–197 (1980).

[13] *G. L. Ellman,* Tissue Sulfhydryl Groups, Arch. Biochem. Biophys. *82*, 70–77 (1959).

[14] *W. Kalow, H. A. Lindsay,* A Comparison of Optical and Manometric Methods for the Assay of Human Serum Cholinesterase, Can. J. Biochem. *33*, 568–574 (1955).

[15] *W. Kalow, K. Genest,* A Method for the Detection of Atypical Forms of Human Cholinesterase; Determination of Dibucaine Numbers, Can. J. Biochem. *35*, 339–346 (1957).

[16] *H. Harris, M. Whittaker,* Differential Inhibition of Human Serum Cholinesterase with Fluoride: Recognition of Two New Phenotypes, Nature *191*, 496–498 (1961).

[17] *J. Liddell, H. Lehmann, E. Silk,* A "Silent" Pseudocholinesterase Gene, Nature *193*, 561–562 (1962).

[18] *J. King,* The Hydrolases in Practical Clinical Enzymology, Van Nostrand, London 1965, p. 180.

1.5 Alkaline Phosphatases

Orthophosphoric monoester phosphohydrolase (alkaline optimum), EC 3.1.3.1

1.5.1 Routine Method

Jean-Pierre Bretaudiere and Thomas Spillman

General

Alkaline phosphatase occurs in cell membranes throughout most of the organs and tissues of the body. The relative distribution of the activity of the enzyme has been reported by *Fernley* [1] as: intestinal mucosa = placenta > kidney = bone > liver = lung = spleen. Three distinct forms of the enzyme have been recognized and the available evidence strongly supports the view that these forms are the products of three distinct genes [2, 3]. The intestinal enzyme, the placental enzyme, and the bone/liver/kidney enzymes have been distinguished from one another on the basis of enzymatic properties, electrophoretic mobility and immunological properties. Additionally, these enzymes exist in differentially glycosylated forms characteristic of the tissue of origin [4]. Alkaline phosphatase determination is of importance in the diagnosis of liver diseases associated primarily with pathology of the biliary tract. Even in cases of partial biliary obstruction in which serum bilirubin is not appreciably elevated, alkaline phosphatase from the biliary canalicular membranes will often appear in the serum. Characterization of the forms of the enzyme in serum is useful in distinguishing elevations in liver disease from those originating from placental release, intestinal release in steatorrhea, or conditions characterized by increased osteoblastic activity, such as osteomalacia, osteogenic sarcoma, hyperparathyroidism with bone involvement, or *Paget*'s disease (osteitis deformans).

Application of method: in clinical chemistry.

Enzyme properties relevant in analysis: alkaline phosphatase is a dimeric molecule of about 160000 molecular weight [5]. The two subunits interact with negative co-operativity and consequently the enzyme shows "half of the sites" reactivity at physiological pH [6, 7], but displays *Michaelis* kinetics at alkaline pH (above 9). Each subunit contains a tightly-bound atom of zinc which is essential for the structural integrity of the enzyme and a second, less tightly-bound zinc atom which is involved in the catalytic process [8].

Methods of determination: phosphomonoesterase activity can be determined with a number of substrates but most simply with 4-NPP* which provides an intense, spontaneous absorbance signal upon hydrolysis at alkaline pH. In the presence of high concentrations of certain amino alcohols, the enzyme will function as a phosphotransferase, generating a phosphorylated alcohol. The presence of such acceptors, for example 2A2M1P, accelerates the dephosphorylation of the substrate [9].

International reference method and standards: a reference method is being developed by the International Federation of Clinical Chemistry which employs the assay principle described here [10] and is described in detail in chapter 1.5.2 (p. 83). The conditions have been the subject of controversy in that the recommended phosphate acceptor, as commercially available, is very often contaminated by an impurity which acts as a chelator, resulting in a time-dependent and irreversible inactivation of the enzyme by removal of both structural and catalytic zinc [11]. For standard reference material cf. Vol. II, chapter 2.3; e.g. SRM-909 from U.S. National Bureau of Standards.

Enzyme effectors: alkaline phosphatase activity is stimulated by magnesium ions [12, 13]. This ion binds to an effector site on each subunit which is different from the site for zinc. The magnesium ion exerts an allosteric effect on the enzyme to stimulate the dephosphorylation process. Zinc ions can also bind to the magnesium site with greater affinity than the magnesium itself, resulting in a loss of activity. Common inhibitors of alkaline phosphatase are chelators of divalent cations, such as EDTA [14]. Alkaline phosphatase activity is markedly affected by its electrostatic environment (ionic strength, solvent effect) [15]. It has been shown that amino acids such as L-phenylalanine and L-homoarginine selectively inhibit some molecular forms of alkaline phosphatase [16]. They have been used clinically for the determination of tissue of origin of alkaline phosphatase in serum.

Assay

Method Design

Principle

$$\text{4-NPP} + \text{X-OH} \xrightarrow{\text{ALP}} \text{4-NP} + \text{X-OPO}_3\text{H}_2$$

* Non standard abbreviations: 4-NPP, 4-nitrophenylphosphate; 4-NP, 4-nitrophenol; 2A2M1P, 2-amino-2-methyl-1-propanol; ALP, alkaline phosphatase.

The amount of 4-NP generated per unit time, monitored by its absorbance, is a measure of the catalytic activity of alkaline phosphatase. The phosphate acceptor (X-OH) in most common use is 2A2M1P [9].

Optimized conditions for measurement: although the principle of measurement of enzyme activity with 4-NPP as a substrate is generally accepted, optimization of the other reaction conditions has proved controversial. 2A2M1P is widely used as a phosphate acceptor but commercial preparations have been shown to be contaminated by a chelating agent which complicates the control of zinc and magnesium ion concentrations [11]. Metal buffers have recently been employed which minimize the influence of chelating contaminants, but a more satisfactory approach to the problem would seem to be the use of an effectively pure acceptor reagent. Such a method is described here.

Dependence of alkaline phosphatase activity on pH, and on the concentrations of substrate and phosphate acceptor, is complex [17, 18]. The pH optimum of the reaction increases with substrate concentration and thus must be optimized for the selected conditions. The phosphate acceptor concentration must also be optimized for the reaction conditions. While such phosphate acceptors increase activity by stimulating the dephosphorylation of the enzyme and the transfer of phosphate, they also promote an enzyme conformation which is unfavourable to the binding of substrate. Although they stimulate the activity at lower concentrations, they act as competitive inhibitors at higher concentration [19]. For a given initial choice of 4-NPP or 2A2M1P there is a single corresponding optimum pH and protagonist concentration.

The ultimate maximal activity that theoretically could be obtained would be at such an alkaline pH that the resulting reaction conditions would be impractical to control. Furthermore, inconveniently high absorbance of the substrate and the viscosity of highly concentrated solutions of the phosphate acceptors dictate limitations of these parameters. Consequently, an arbitrary judgement is required as to the "best" reaction conditions for measurement of the enzyme. We have chosen the method recommended by the "Societé Française de Biologie Clinique" because it has been used in clinical practice and reference ranges have been determined with it [20].

Temperature conversion factors: even though no significant differences have been found between the slopes of the *Arrhenius* plots between 25 °C and 37 °C for various purified alkaline phosphatases [21], utilization of the proposed technique is not recommended at a temperature other than 30 °C. Temperature conversion factors are yet to be determined experimentally using these reaction conditions.

Equipment

Any spectrophotometer capable of providing an accurate absorbance measurement at 405 nm is suitable. The spectrophotometer must have sufficient thermostatic capability to maintain the temperature of the reaction mixture at 30 °C ± 0.05 °C.

Reagents and Solutions

Purity of reagents: 2A2M1P must be tested for the presence of potential alkaline phosphatase inhibitors, particularly for zinc chelators. These chelators have been found to develop during the synthesis of 2A2M1P [11]. Since the inhibition by these chelators is time-dependent, their presence can be detected by measuring the loss in activity of a specimen containing alkaline phosphatase after a 10 minute incubation with 2A2M1P in the assay conditions. A test mixture is prepared as described in the protocol below, which shows the standard assay mixture, and is incubated for 10 minutes at 30 °C before the addition of 4-NPP. A control is prepared in which the 4-NPP is substituted for the specimen in the 10 minute incubation and the specimen added last, to initiate the reaction. A reduction in activity after pre-incubation with the 2A2M1P/magnesium reagent (solution 1) will reflect an inactivation of the alkaline phosphatase by the contaminating chelator. If the activity difference is over 2%, the 2A2M1P should be discarded. *Aldrich Chemical,* Milwaukee, WI 53201 (U.S.A) produces a special 2A2M1P (reference 23,260-2) which is claimed to have a level of contaminating metal chelator of less than 0.01 mol %. This product is suitable for alkaline phosphatase determination. The 4-NPP should meet the criteria of purity proposed by *Bowers et al.* [22]. Magnesium sulphate heptahydrate must be of analytical grade.

Preparation of solutions: all solutions are prepared in re-purified water (cf. Vol. II, chapter 2.1.3.2).

1. 2-Amino-2-methyl-1-propanol/magnesium solution (2A2M1P 964.3 mmol/l; Mg^{2+}, 1.071 mmol/l; pH 10.5):

   warm the 2A2M1P at about 35 °C until it is completely liquefied. Weigh exactly 86 g 2A2M1P in a 1 litre beaker (caution: 2A2M1P is viscous and awkward to manipulate). Add approximately 500 ml water and mix. Add approximately 120 ml HCl, 1 mol/l, and mix. In a separate beaker weigh exactly 264 mg $MgSO_4 \cdot 7\ H_2O$ and dissolve with approximately 200 ml water. Pour the magnesium solution into the 2A2M1P solution. Rinse several times with water. While constantly monitoring temperature at 30 °C ± 1 °C, adjust pH at 10.5 ± 0.02 with HCl, 1 mol/l. Transfer the content of the beaker to a 1 litre volumetric flask, rinse the beaker with water and dilute to exactly 1 litre with water.

2. 4-Nitrophenyl phosphate solution (4-NPP 480 mmol/l):

   dissolve exactly 8.9 g 4-NPP-$Na_2 \cdot 6\ H_2O$ in about 40 ml water. Dilute to 50 ml with water.

Stability of solutions: the 2A2M1P/Mg^{2+} solution (1) is stable for at least one month at room temperature if it is protected from contact with atmospheric carbon dioxide.

The 4-NPP solution (2) is stable for only one hour at room temperature. During the assay, protect the solution from light by wrapping the container with aluminum foil and keep the solution cool by placing the container in crushed ice. The 4-NPP solution can be stored frozen at $-20\,^{\circ}\mathrm{C}$ for at least a month. It is usually convenient to freeze suitable aliquots.

Procedure

Collection and treatment of specimen: collect blood with a minimum of venous stasis. The preferred specimen is freshly collected serum. Haemolysis or lipaemia do not affect the determination of the activity but may cause difficulties in the spectrophotometric measurements if pronounced. Plasma obtained using heparin as an anticoagulant is equally acceptable. Anticoagulants such as EDTA, oxalate, or citrate which chelate divalent cations should not be used since they would result in inhibition of the enzyme.

Stability of the enzyme in the specimen: the determination of the activity should be performed within 6 h of collection if the specimen is kept at room temperature. The stability at room temperature depends upon the origin of the alkaline phosphatase. Serum alkaline phosphatase activity derived from bone tissue disappears rapidly while that from placenta may remain unchanged for several days. Serum alkaline phosphatase activity of hepatic or intestinal origin is stable for at least one day.

The activity of lyophilized quality control specimens has been shown to increase gradually to a plateau after reconstitution. The time necessary to reach this plateau depends on both the temperature at which the specimen is incubated and the origin of the alkaline phosphatase [23]. Therefore lyophilized quality control specimens should be used at a fixed time after reconstitution. The use of liquid quality control specimens is preferred since they do not present this anomaly.

Details for measurements in tissue: because alkaline phosphatase is a membrane enzyme, it is associated with the particulate matter of the organ homogenate [1]. Therefore, most preparation techniques employ an extraction of alkaline phosphatase by disruption of the membrane remnants using n-butanol [24, 25]. Once this operation is complete and the homogenate has been centrifuged, alkaline phosphatase is located in the aqueous phase. This aqueous phase cannot be used directly for activity measurments since it is saturated with n-butanol which inhibits the enzyme. A dialysis step is required to eliminate n-butanol before activity measurement.

Assay conditions: wavelength 405 nm; light path 10 mm; final volume 3.00 ml; temperature $30\,^{\circ}\mathrm{C}$. It is good practice to perform a reagent blank with each new reagent.

Measurement

Pipette successively into the cuvette:			concentration in assay mixture	
2A2M1P/Mg^{2+} solution	(1)	2.8 ml	2A2M1P	900 mmol/l
			Mg^{2+}	1 mmol/l
sample (serum or diluted tissue homogenate)		0.1 ml	volume fraction	0.033
mix thoroughly without removing any of the solution from the cuvette; incubate at 30 °C for 10 min or until temperature equilibration has occurred,				
4-NPP solution	(2)	0.1 ml	4-NPP	16 mmol/l
mix thoroughly; monitor change in absorbance with a recorder or manually record absorbance readings at fixed time intervals using an appropriate timer. The reaction should be monitored for about 3 min.				

The value of $\Delta A/\Delta t$ should not exceed 0.250/min; otherwise dilute specimen with sodium chloride, 9.0 g/l, and repeat the assay.

The pre-incubation conditions and particularly the value of pre-incubating the sample with magnesium has been extensively studied [21, 26]. Unfortunately, most of the results reported are of little value since, when they were obtained, the critical requirement of purity of 2A2M1P was not recognized. In addition, the use of a higher concentration of magnesium (i.e. 1 mmol/l) [21] leads to an almost instantaneous association of this cation to the enzyme. It is therefore acceptable to trigger the reaction with either 4-NPP or the sample. Therefore, an equally appropriate protocol would be one where 2.8 ml of the 2A2M1P/Mg^{2+} reagent (1) is incubated with 0.1 ml of 4-NPP reagent (2) and the reaction is triggered by 0.1 ml of sample.

Calculation: use eqn. (k) or (k_1), cf. "Formulae", Appendix 3. Under the conditions described here, $\varepsilon = 1.88\ l \times mmol^{-1} \times mm^{-1}$. Correct $\Delta A/\Delta t$ readings for blank. The following relation is valid for the catalytic concentration in the sample (Δt in minutes):

$$b = 1596 \times \Delta A/\Delta t \quad \text{U/l}$$

$$b = 26600 \times \Delta A/\Delta t \quad \text{nkat/l.}$$

Validation of Method

Precision, accuracy, detection limits and sensitivity: a typical manual analysis showed an imprecision of about 10 U/l for a patient pool specimen with a catalytic activity of 100 U/l. (The standard deviation is 5 U/l). The magnitude of a reagent blank has been found to be stable with various batches of reagents with an activity level lower than 2 U/l.

Data regarding the accuracy of the proposed method can be generated by using a reference material produced by the U.S. National Bureau of Standards (SRM-909). The collaborative study, involving 6 laboratories, which has led to the determination of the target value for SRM-909, shows a within-laboratory repeatability of 2.5% and a between-laboratory reproducibility of 2.9% at a level of 75.4 U/l*. The method used differed from the proposed method only in the 2A2M1P concentration (1 mol/l). The lowest value which can be reliably measured is 4 U/l with an imprecision of about 25%.

Sources of error: there are no known influences of therapeutic agents on activity measurement [27]. Since alkaline phosphatase is readily adsorbed onto the surface of many plastics, the use of plastic tubing or components in conjunction with the measuring chamber may introduce an undesired activity background.

Specificity: the method is specific for alkaline phosphatase.

Reference ranges: the reference range for human serum varies with age and sex. The following data (30 °C) are from reference [20].

up to 2 months	99 – 228 U/l
from 3 months to 6 months	78 – 276 U/l
from 7 months to 3 years	96 – 228 U/l
from 4 years to 15 years	90 – 300 U/l
males over 15 years	90 – 300 U/l
females from 15 to 40 years	30 – 90 U/l
females over 40 years	30 – 102 U/l.

References

[1] *H. N. Fernley*, Mammalian Alkaline Phosphatases, in: *P. D. Boyer* (ed.), The Enzymes, Vol. *IV*, Academic Press, New York 1971, pp. 417 – 447.

[2] *M. J. McKenna, T. A. Hamilton, H. H. Sussman*, Comparison of Human Alkaline Phosphatase Isoenzymes, Biochem. J. *181*, 67 – 73 (1979).

* *G. N. Bowers, R. Alvarez, J. P. Cali et al.*, Enzymes/SRM-909. Standard reference material: the measurement of the catalytic (activity) concentration of seven enzymes in N.B.S. human serum SRM-909. NBS Special Publication 260-83. National Bureau of Standards, Dept. of Commerce, Washington D.C. 20234, USA.

[3] *R. A. Stinson, L. E. Seargeant,* Comparative Studies of Pure Alkaline Phosphatases from five Human Tissues, Clin. Chim. Acta *110*, 261 – 272 (1981).
[4] *P. M. Crofton,* Biochemistry of Alkaline Phosphatase Isoenzymes, CRC Crit. Rev. Clin. Lab. Sci. *16*, 161 – 194 (1982).
[5] *J. M. Trepanier, L. E. Seargeant, R. A. Stinson,* Affinity Purification and Some Molecular Properties of Human Liver Alkaline Phosphatase, Biochem. J. *155*, 653 – 660 (1976).
[6] *D. Chappelet-Tordo, M. Fosset, M. Iwatsubo, C. Gache, M. Lazdunski,* Intestinal Alkaline Phosphatase. Catalytic Properties and Half of the Sites Reactivity, Biochemistry *13*, 1788 – 1795 (1974).
[7] *M. Lazdunski,* "Half-of-the sites" Reactivity and the Role of Subunit Interactions in Enzyme Catalysis, in: *E. T. Kaiser, F. J. Kezdy* (eds.), Progress in Bioorganic Chemistry, Vol. *3*, John Wiley & Sons, New York 1974, pp. 81 – 140.
[8] *K.-D. Gerbitz,* Human Alkaline Phosphatases II. Metalloenzyme Properties of the Enzyme from Human Liver, Hoppe-Seyler's Z. Physiol. Chem. *358*, 1491 – 1497 (1977).
[9] *R. B. McComb, G. N. Bowers Jr., S. Posen,* Alkaline Phosphatase, Plenum Press, New York 1979.
[10] *N. W. Tietz, C. Burtis, K. Ervin et al.,* Progress in the Development of a Recommended Method for Alkaline Phosphatase Activity Measurements, Clin. Chem. *26*, 1023 (1980). Abstract.
[11] *R. Rej, J.-P. Bretaudiere, R. W. Jenny, K. Y. Jackson,* Measurement of Alkaline Phosphatase Activity: Characterization and Identification of an Inactivator in 2-Amino-2-methyl-1-propanol, Clin. Chem. *27*, 1401 – 1409 (1981).
[12] *G. Cathala, D. Chappelet-Tordo, M. Lazdunski, C. Brunel,* Bovine Kidney Alkaline Phosphatase. Catalytic Properties, Subunit Interactions in the Catalytic Process, and Mechanism of Mg^{2+} Stimulation, J. Biol. Chem. *250*, 6046 – 6053 (1975).
[13] *G. Linden, D. Chappelet-Tordo, M. Lazdunski,* Milk Alkaline Phosphatase. Stimulation by Mg^{2+} and Properties of the Mg^{2+} Site, Biochim. Biophys. Acta *483*, 100 – 106 (1977).
[14] *B. P. Ackermann, J. Ahlers,* Kinetics of Alkaline Phosphatase from Pig Kidney, Influence of Complexing Agents on Stability and Activity, Biochem. J. *153*, 151 – 157 (1976).
[15] *M. Lazdunski, J. Brouillard, L. Ouellet,* Etude des effets electrostatiques sur le mecanisme de la phosphatase alcaline intestinale de veau. Can. J. Chem. *43*, 2222 – 2235 (1965).
[16] *W. H. Fishman,* Perspectives on Alkaline Phosphatase Isoenzymes, Am. J. Med. *56*, 617 – 650 (1974).
[17] *I. Hinberg, K. J. Laidler,* Influence of pH on the Kinetics of Reactions Catalyzed by Alkaline Phosphatase, Can. J. Biochem. *51*, 1096 – 1103 (1973).
[18] *I. Hinberg, K. J. Laidler,* The Kinetics of Reactions Catalyzed by Alkaline Phosphatase: the Effects of Added Nucleophiles, Can. J. Biochem. *50*, 1360 – 1368 (1972).
[19] *J.-P. Bretaudiere,* Influence of Nucleophiles on the Kinetic Properties at Steady-state of Human Alkaline Phosphatases, D. Sc. dissertation, University of Paris-Sud 1978.
[20] Société Française de Biologie Clinique, Enzymology Commission, Recommendations for Measuring the Catalytic Concentration of Alkaline Phosphatase in Human Serum at 30 °C, Ann. Biol. Clin. *40*, 150 – 154 (1982).
[21] *J.-P. Bretaudiere, A. Vassault, L. Amsellem et al.,* Criteria for Establishing a Standardized Method for Determining Alkaline Phosphatase Activity in Human Serum, Clin. Chem. *23*, 2263 – 2274 (1977).
[22] *G. N. Bowers Jr., R. B. McComb, A. Upretti,* 4-Nitrophenyl Characterization of High-purity Materials for Measuring Alkaline Phosphatase Activity in Human Serum, Clin. Chem. *27*, 135 – 143 (1981).
[23] *A. F. Smith, B. A. Fogg,* Possible Mechanism for the Increase in Alkaline Phosphatase Activity of Lyophilized Control Material, Clin. Chem. *18*, 1518 – 1523 (1972).
[24] *R. K. Morton,* The Purification of Alkaline Phosphatases of Animal Tissues, Biochem. J. *57*, 595 – 603 (1954).
[25] *H. H. Sussman, P. A. Small Jr., E. Cotlove,* Human Alkaline Phosphatase, Immunological Identification of Organ-specific Isoenzymes, J. Biol. Chem. *243*, 160 – 166 (1968).
[26] *R. Rej,* Effect of Incubation with Mg^{2+} on the Measurement of Alkaline Phosphatase Activity, Clin. Chem. *23*, 1903 – 1911 (1977).
[27] *D. S. Young, L. C. Pestaner, V. Gibberman,* Effects of Drugs on Clinical Laboratory Tests, Clin. Chem. *21*, 1D – 432D (1975).

1.5.2 IFCC Reference Method (provisional)

Jean-Pierre Bretaudiere and Thomas Spillman

General

The provisional IFCC reference method is based on conditions developed by the working group on alkaline phosphatase of the Subcommitee on Standards of the American Association of Clinical Chemistry (AACC) [1]. This method has not yet been used in general routine practice and therefore data on analytical performance and reference ranges are not available. The technique proposed here is adapted from Draft 3, stage 1 of a document circulated by the Expert Panel on Enzymes of the IFCC and dated November 1981.

Assay

Method Design

Principle: the method principle is identical to that of the routine method described on pp. 76, 77. The assay conditions differ in the concentrations of the reagents and in the addition of a metal buffer in the IFCC method to maintain constant concentrations of free zinc and magnesium ions. These conditions are shown in Table 1.

Table 1. Optimized reaction conditions of IFCC reference method (provisional).

Reaction temperature	30 °C
pH	10.4
Volume fraction of specimen	0.0196
4-Nitrophenyl phosphate	16 mmol/l
2-Amino-2-methyl-1-propanol	350 mmol/l
Mg^{2+}	2 mmol/l
Zn^{2+}	1 mmol/l
N-(2-hydroxyethyl)ethylenediamine triacetate	2 mmol/l

Establishing optimized conditions

Buffer

2-Amino-2-methyl-1-propanol has been found to be a suitable buffer because of both its buffer capacity near the optimum pH of the enzymatic reaction and its ability to function as a phosphate acceptor. 2-Amino-2-methyl-1-propanol has been widely accepted in routine practice provided its purity can be assured (cf. chapter 1.5.1, pp. 77, 78 for further discussion).

Zinc, Magnesium

The addition of an otherwise excess of zinc and magnesium ions in the presence of an appropriate metal buffer allows an optimum concentration of free metal ions to be maintained [2]. It has been shown that alkaline phosphatase in serum is depleted of these ions [3]. The depletion is more dramatic with lyophilized quality control specimens [4]. This system has the additional advantage of alleviating the chelating effect of a possible contaminant in 2-amino-2-methyl-1-propanol which results in an undesired inactivation of the alkaline phosphatase [5]. Since the relative proportions of zinc, magnesium, and metal buffer were determined empirically with a reasonably pure 2-amino-2-methyl-1-propanol screened as described on page 78, it cannot be assumed that the addition of a metal buffer will protect the enzyme against indefinitely increasing amounts of a contaminant chelator in 2-amino-2-methyl-1-propanol. Therefore, it is an absolute necessity to screen the 2-amino-2-methyl-1-propanol for such an impurity.

pH and substrate

As discussed on p. 77, pH and the concentrations of 4-nitrophenyl phosphate and 2-amino-2-methyl-1-propanol are intimately related. Therefore, by keeping one factor constant, corresponding optimal conditions can be always derived for the other two factors. The AACC working group chose to determine the optimal reaction conditions by processing the experimental results obtained by a phenomenological statistical procedure, response surface methodology [6]. This method of analysis finds the optimum within the range of conditions studied. Nevertheless, this optimum is still based on the initial choice of the condition ranges, i.e. 10.3 to 10.5 for pH, 14 to 18 mmol/l for 4-nitrophenyl phosphate and 300 to 400 mmol/l for 2-amino-2-methyl-1-propanol. A different result would be obtained if different initial choices were made. The conditions for the metals and the metal buffer have been determined similarly by response surface methodology.

Equipment, Reagents and Solutions

The equipment and reagents are described on p. 77 and78, respectively. See p. 78 for purity of 2-amino-2-methyl-1-propanol, 4-nitrophenyl phosphate and magnesium sulphate heptahydrate. Zinc sulphate heptahydrate and N-(2-hydroxyethyl)ethylenediamine triacetic acid trisodium salt dihydrate (HEDTA) must be of analytical grade.

1. Buffer solution (2A2M1P, 372 mmol/l; Zn^{2+}, 1.062 mmol/l; Mg^{2+}, 2.125 mmol/l; HEDTA, 2.125 mmol/l):

   warm the 2-amino-2-methyl-1-propanol at about 35 °C until it is completely liquefied. Weigh exactly 33.16 g 2-amino-2-methyl-1-propanol in a one litre beaker and add approximately 500 ml water. Add 45 ml HCl, 1 mol/l and mix. Adjust the pH to 10.4. In separate 100 ml beaker weigh exactly 808 mg HEDTA-$Na_3 \cdot 2H_2O$

and dissolve it with approximately 50 ml water. Then add in this order, allowing each to dissolve, 306 mg $ZnSO_4 \cdot 7\ H_2O$ and 524 mg $MgSO_4 \cdot 7\ H_2O$. Transfer quantitatively the contents of the 100 ml beaker to the one litre beaker containing 2-amino-2-methyl-1-propanol. Adjust pH to 10.4 ± 0.05 at 30°C and dilute to one litre with water.

2. Substrate solution (4-NPP, 408 mmol/l):

   dissolve exactly 7.57 g 4-nitrophenyl phosphate, disodium salt, hexahydrate, with about 40 ml water in a 100 ml beaker. Then dilute to exactly 50 ml with water.

Procedure

Conditions for collecting and processing of specimens are identical to those given in p. 79.

Stability of the enzyme in the specimen: cf. p. 79.

Details for measurement in tissues: see p. 79.

Assay conditions: see p. 79.

Measurement

Pipette successively into the cuvette:			concentration in assay mixture	
buffer solution	(1)	2.40 ml	2A2M1P Mg^{2+} Zn^{2+} HEDTA	350 mmol/l 2 mmol/l 1 mmol/l 2 mmol/l
sample		0.05 ml	volume fraction	0.0196
mix thoroughly without removing any of the solution from the cuvette; incubate at 30°C for 10 min or until temperature equilibration has occurred,				
substrate solution	(2)	0.10 ml	4-NPP	16 mmol/l
mix thoroughly; monitor change in absorbance with a recorder or manually record absorbance readings at fixed time intervals using an appropriate timer. The reaction should be monitored for about 3 min.				

The value of $\Delta A/\Delta t$ should not exceed 0.250/min; otherwise dilute specimen with sodium chloride, 9.0 g/l, and repeat the assay.

It is acceptable to trigger the reaction with either 4-nitrophenyl phosphate or the specimen. However, the IFCC recommendation calls for triggering with the sample. Therefore, an equally appropriate protocol would be one where 2.4 ml solution (1) is incubated with 0.1 ml solution (2) and the reaction triggered by 0.05 ml of sample.

Calculation: use calculation formula (k) from "Formulae", Appendix 3, $\varepsilon = 1.88$ $l \times mmol^{-1} \times mm^{-1}$. The following relation is valid for the catalytic concentration in the serum (Δt in s):

$$b = 45267 \times \Delta A/\Delta t \quad \text{nkat/l.}$$

Validation of method: cf. Routine Method, p. 81.

References

[1] *N. W. Tietz, C. Burtis, K. Ervin et al.,* Progress in the Development of a Recommended Method for Alkaline Phosphatase Activity Measurements, Clin. Chem. *26*, 1023 (1980), Abstract.
[2] *H. U. Wolf,* Divalent Metal Ion Buffers with Low pH-Sensitivity, Experientia *29*, 241 – 249 (1973).
[3] *J.-P. Bretaudiere, A. Vassault, L. Amsellem et al.,* Criteria for Establishing a Standardized Method for Determining Alkaline Phosphatase Activity in Human Serum, Clin. Chem. *23*, 2263 – 2274 (1977).
[4] *R. Rej, J.-P. Bretaudiere,* Effects of Metal Ions on the Measurement of Alkaline Phosphatase Activity, Clin. Chem. *26*, 423 – 428 (1980).
[5] *R. Rej, J.-P. Bretaudiere, R. W. Jenny, K. Y. Jackson,* Measurement of Alkaline Phosphatase Activity: Characterization and Identification of an Inactivator in 2-Amino-2-methyl-1-propanol, Clin. Chem. *27*, 1401 – 1409 (1981).
[6] *W. J. Hill, W. G. Hunter,* Response Surface Methodology: a Review, Technometrics *8*, 571 – 590 (1966).

1.5.3 Isoenzymes of Alkaline Phosphatase

Determination After Heat Inactivation of Labile Forms

Discrimination among hepatobiliary, bone, and placental/Regan isoenzymes

Jean-Pierre Bretaudiere and Thomas Spillman

General

Multiple forms of alkaline phosphatase exist between species, between different tissues of the same species, and within single tissues of the same species.

In some cases, multiple forms of the enzyme found within single tissues have been shown to differ with respect to carbohydrate moiety or primary structure [1]. For some other reported multiple forms, artefactual generation during extraction or purification has not been ruled out. In any case, the existence of multiple forms within a single tissue source has not found clinical application.

However, well-established and clinically-important differences have been recognized between alkaline phosphatases from different organs and tissues, namely: those from bone, kidney, liver, biliary tract, intestine, placenta and neoplastic tissue. The most common means of distinguishing among the forms when found in the serum are evaluation of resistance to heat, urea, or L-phenylalanine; resolution by electrophoresis; or precipitation by tissue-specific antibodies, though many other differences in the kinetic properties of the isoenzymes have been reported [2].

The alkaline phosphatase in human serum can be a mixture of the intestinal, bone, and hepatobiliary forms. In adolescents, the bone form constitutes the majority of the alkaline phosphatase activity, and in late pregnancy the placental form predominates. Pathological conditions which change the distribution of the enzymes are: hepatobiliary obstruction, in which liver (parenchymal) or biliary or both forms may be elevated; osteoblastic proliferation in osteomalacia or osteosarcoma, in which the bone form is elevated; and organ tumours in which the *Regan* or pseudo-placental form is elevated. Thus, clinical interest lies in distinguishing among elevations of serum activity due to the liver/biliary, bone, and *Regan*/placental forms.

Applications of method: in clinical chemistry.

Enzyme properties relevant in analysis: although not dramatic enough for routine diagnostic discrimination, some differences in the catalytic activities of alkaline phosphatase isoenzymes have been reported. The human placental enzyme has a pH optimum almost a pH unit higher than the other forms [3]. The more structurally stable (i.e., with heat or urea) forms are less sensitive to magnesium concentration and chelators, possibly due to tighter retention of endogenous magnesium within their more stable secondary structures. Some differences of substrate specificity [4] and amount of change in activity with change in temperature [2] have also been noted. Although these differences are recognized, correction is generally not made for them. Otherwise, all isoenzymes conform to the general description of alkaline phosphatase, chapter 1.5.1, p. 75.

Methods of determination: many observed differences in enzyme properties have been employed to resolve multiple forms of alkaline phosphatase and have been reviewed extensively [2]. In clinical practice, two basic principles have been widely employed: electrophoretic separation of the isoenzymes and selective inactivation of specific forms followed by measurement of the remaining activity. Immunochemical techniques have been of some value but are often equivocal.

Electrophoretic resolution of isoenzymes (cf. Vol. III, chapter 1.3.2, p. 71) has been performed on many supporting media but the most satisfactory separations have been obtained with polyacrylamide gel or cellulose acetate membranes. Bone, liver, bile, intestine, and placental enzymes have been resolved from one another by use of a combination of acrylamide gel systems, although only the biliary enzyme migrates far from the others [5]. On cellulose acetate, under conditions which resolve the usual classes of serum proteins, all the alkaline phosphatases run roughly in the alpha 2-beta region. The bone enzyme gives a broad distribution and bridges this entire region, but is readily heat-inactivated to allow identification of the other isoenzymes [6, 7]. An activity stain is used to detect alkaline phosphatase zones after electrophoresis and the forms are estimated visually or densitometrically.

Limitations of the electrophoretic technique for quantitation of the multiple forms are the possibility of preferential loss of activity of the less stable forms during electrophoresis, reported alterations in the electrophoretic patterns with storage of the specimen, and variability of migration rate of a given form following different tissue extraction techniques. Other sources of quantitative error are the non-linearity of the activity stain due to precipitation of the dephosphorylated substrate in some cases, or slow diffusion of the substrate into the gel. Nevertheless, large isoenzyme imbalances can be recognized and many workers find the techniques satisfactory for clinical purposes.

Inhibition and inactivation studies have repeatedly shown large differences between the sensitivities of the isoenzymes. Several amino acids inhibit different forms differentially, but two are particularly effective: L-(but not D) phenylalanine and L-homoarginine. Enzyme forms which are more stable to heat and urea (placental, intestinal) are also the more resistant to L-homoarginine, 10 mmol/l, but the more sensitive to L-phenylalanine, 5 mmol/l. The anti-helminthic compound levamisole and derivatives of it are also inhibitors which discriminate between the placental/intestinal and non-placental/non-intestinal alkaline phosphatases [8]. Enzyme inactivation due to disruption of secondary structure by heat or molar concentrations of urea discriminates effectively among the tissue types of enzyme [9]. *Regan*/placental forms are highly stable, whereas enzyme from bone, liver, or kidney is readily inactivated. Intestinal and biliary forms show intermediate stability. For the advantage of simplicity of performance, the heat inactivation technique (cf. also Vol. III, chapter 1.3.4, p. 101) will be described here.

International reference method and standards: there is no international reference method or standard for alkaline phosphatase isoenzyme determination at this time.

Enzyme effectors: aside from those mentioned in "Methods of Determination", these are the same as described in chapter 1.5.1, p. 76.

Assay

Method Design

Principle: aliquots of serum are incubated in parallel at room temperature, 56 °C, or 65 °C. After 10 min, the heated aliquots are chilled rapidly to 0 °C and the activity is determined in each aliquot as for alkaline phosphatase.

Optimized conditions for measurement: as for alkaline phosphatase, chapter 1.5.1, p. 77.

Temperature conversion factors: see remarks under alkaline phosphatase, chapter 1.5.1, p. 77.

Equipment

Constant temperature water-baths capable of maintaining 56 °C ± 0.05 °C and 65 °C ± 0.10 °C and also the equipment described for measurement of alkaline phosphatase activity, chapter 1.5.1, p. 77.

Reagents and Solutions

As specified and described for alkaline phosphatase, chapter 1.5.1, p. 78.

Procedure

Collection and treatment of specimen: as described for alkaline phosphatase, chapter 1.5.1, p. 79.

Stability of enzyme in the specimen: since the isoenzymes are differently stable, specimens for determination of relative isoenzyme composition are subject to the conditions described for total alkaline phosphatase activity, p. 79. Additionally, heat-inactivation rates are influenced by pH during heating. Loss of dissolved carbon dioxide over a period of hours may increase the pH of the specimen by a unit and cause faster inactivation [10]. Therefore, the specimen should be analyzed promptly after venepuncture.

Details for measurement in tissues: extraction can be performed as described for alkaline phosphatase, chapter 1.5.1, p. 79, but the sensitivity to heat of activity extracted from tissues has been reported to be heavily influenced by the degree of purification: thus, appropriate confirmation of isoenzyme identity by other methods should be considered. For the same reason, a tissue extract does not function adequately as a calibrator for evaluation of the tissue of origin of serum alkaline phosphatase by heat inactivation [11].

Assay conditions: following the heat-inactivation step, assay is performed as described for alkaline phosphatase, chapter 1.5.1, p. 79, 80.

Measurement

1. *Division of aliquots:* transfer 0.2 ml aliquots of serum to each of 3 thin-walled glass tubes at room temperature.
2. *Incubation of aliquots:* start a timer and transfer one aliquot into each water-bath, leaving the third at room temperature. Swirl the tubes for 5 s to facilitate heat transfer to the specimen. The rack used to hold the tubes for the remainder of the incubation period should be equilibrated with the water-bath prior to use.
3. *Termination of inactivation period:* after precisely 10 min, remove the aliquots and swirl in an ice-water mixture for 15 s to chill quickly. Leave the specimens on ice for 3 min before returning them to room temperature.
4. *Assay of alkaline phosphatase activity:* assay as described for alkaline phosphatase, chapter 1.5.1, p. 80.

Calculation: percent remaining alkaline phosphatase activity for each aliquot is calculated as follows:

$$b_{\text{remaining}} = \frac{b_{\text{heated}}}{b_{\text{unheated}}} \times 100 \quad \%$$

Validation of Method

Precision, accuracy, detection limits and sensitivity: standard deviations of 9 – 11% of the original activity have been reported [12, 13] for 10 min inactivation of serum alkaline phosphatase isoenzymes at 56 °C. Evaluation of the accuracy and sensitivity of isoenzyme determination suffers from a lack of availability of standards. The only means of confirming tissue-specificity is extraction of the enzyme from tissue. However, addition of tissue extracts to serum does not provide a standard which is equivalent to the isoenzyme naturally occurring in serum [11]. Consequently, the most clinically useful reference material is serum from patients known to suffer from diseases which have been recognized to introduce extreme isoenzyme biases. Such reference material, though useful, does not contain a precisely known isoenzyme composition and does not allow rigorous evaluation of the accuracy or detection limits of the method. The precision, accuracy, detection limits and sensitivity of the alkaline phosphatase assay itself are discussed under alkaline phosphatase, p. 81.

Sources of error: the control of temperature is a major potential source of error. An increase of 0.5 °C during a short incubation has been reported to decrease the remaining activity by 30% [14]. Obviously, the use of thick-walled or plastic tubes, or of larger specimen volume, in the heat inactivation step will delay equilibration and affect the results. The increase of heat stability with increased protein concentration of the serum is reported to be about 3% within the normal range of 6.5 – 8.0 g/100 ml protein. The loss of carbon dioxide from serum or storage at 4 °C for long periods (e.g., overnight) may increase the pH as much as 1 unit and reduce the fraction of activity remaining by 7% [10]. Large meals before sampling are reported to increase the intestinal isoenzyme which will be measured by heat inactivation as an apparent elevation in the liver form [15]. Sources of error in the alkaline phosphatase reaction are discussed under alkaline phosphatase, p. 81.

Specificity: considerable overlap exists between percent inactivation of alkaline phosphatase from normal subjects and patients with bone disease. However, the isoenzyme determination is most valuable in patients already defined as abnormal by elevated levels of total alkaline phosphatase activity. The overlap of residual percentage activities between patients with skeletal disease and hepatobiliary disease is small. *Cadeau & Malkin* [12], for example, chose thresholds which mis-classified only 2 of 65 cases of liver isoenzyme elevation and none of 33 cases of bone disease. Of these 98 cases, 20 were categorized as "indefinite" on the basis of the percentage inactivation. It must be kept in mind that, while the use of patient sera addresses the issue of clinical predictivity, it does not ensure the presence of single tissue isoenzymes. In the case of hepatobiliary complications, both biliary and liver-parenchymal enzymes may be released. The biliary enzyme is more stable to heat inactivation than the liver enzyme [14], but will still be scored as "liver" by the thresholds given below. *Regan*/placental isoenzyme is readily recognized as activity remaining after heating at 65 °C.

Reference ranges [12, 16]

	percent alkaline phosphatase activity remaining after incubation at	
	56 °C	65 °C
Normal adult serum	<26%	<1%
Bone predominant	<26%	<1%
Mixed or indeterminate	26% – 34%	<1%
Hepatobiliary predominant	34% – 60%	<1%
Regan/placental predominant	>90%	>90%

References

[1] *W. H. Fishman,* Perspectives on Alkaline Phosphatase Isoenzymes, Am. J. Med. *56,* 617 – 650 (1974).

[2] *R. B. McComb, G. N. Bowers Jr., S. Posen,* Alkaline Phosphatase, Plenum Press, New York 1979.

[3] *J.-P. Bretaudiere, A. Vassault, L. Amsellem et al.,* Criteria for Establishing a Standardized Method for Determining Alkaline Phosphatase Activity in Human Serum, Clin. Chem. *23,* 2263 – 2274 (1977).

[4] *P. M. Crofton,* Biochemistry of Alkaline Phosphatase Isoenzymes, CRC Crit. Rev. Clin. Lab. Sci. *16,* 161 – 194 (1982).

[5] *R. B. Johnson Jr., K. Ellingboe, P. Gibbs,* A Study of Various Electrophoretic and Inhibition Techniques for Separating Alkaline Phosphatase Isoenzymes, Clin. Chem. *18*, 110–115 (1972).
[6] *H. A. Fritsche Jr., H. R. Adams-Park,* Cellulose Acetate Electrophoresis of Alkaline Phosphatase in Human Serum and Tissue, Clin. Chem. *18*, 417–421 (1972).
[7] *W. H. Siede, U. B. Seiffert,* Quantitative Alkaline Phosphatase Isoenzyme Determination by Electrophoresis on Cellulose Acetate Membranes, Clin. Chem. *23*, 28–34 (1977).
[8] *H. van Belle,* Alkaline Phosphatase. I. Kinetics and Inhibition by Levamisole of Purified Isoenzymes from Humans, Clin. Chem. *22*, 972–976 (1976).
[9] *D. W. Moss,* Alkaline Phosphatase Isoenzymes, Clin. Chem. *28*, 2007–2016 (1982).
[10] *D. W. Moss, M. J. Shakespeare, D. M. Thomas,* Observations on the Heat-stability of Alkaline Phosphatase Isoenzymes in Serum, Clin. Chim. Acta *40*, 35–41 (1972).
[11] *L. G. Whitby, D. W. Moss,* Analysis of Heat Inactivation Curves of Alkaline Phosphatase Isoenzymes in Serum, Clin. Chim. Acta *59*, 361–367 (1975).
[12] *B. J. Cadeau, A. Malkin,* A Relative Heat Stability Test for the Identification of Serum Alkaline Phosphatase Isoenzymes, Clin. Chim. Acta *45*, 235–242 (1973).
[13] *D. W. Moss, G. L. Whitby,* A Simplified Heat-inactivation Method for Investigating Alkaline Phosphatase Isoenzymes in Serum, Clin. Chim. Acta *61*, 63–71 (1975).
[14] *S. P. Posen, F. C. Neale, J. S. Clubb,* Heat Inactivation in the Study of Human Alkaline Phosphatase, Ann. Int. Med. *62*, 1234–1243 (1965).
[15] *R. O. Briere,* Alkaline Phosphatase Isoenzymes, CRC Crit. Rev. Clin. Lab. Sci. *10*, 1–30 (1979).
[16] *D. W. Moss, E. J. King,* Properties of Alkaline-phosphatase Fractions Separated by Starch-gel Electrophoresis, Biochem. J. *84*, 192–195 (1962).

1.6 Acid Phosphatases

Orthophosphoric monoester phosphohydrolase, EC 3.1.3.2

Donald W. Moss

1.6.1 General

Non-specific acid phosphatase activity is widely distributed throughout the living world. Acid phosphatase occurs in the lysosomes of human and animal tissues. Extra-lysosomal acid phosphatases are also present in many cells. The acid phosphatase secreted by the human prostate gland has attracted most attention, because of its clinical importance, and extensive characterization and structural studies have now been carried out on it. Acid phosphatases from other human tissues have also been the subject of clinical and genetic studies, but on a less extensive scale, while numerous studies of acid phosphatases from some plants such as potato have also been published. Acid phosphatase has also been used as an analytical reagent, e.g. in structural studies on nucleotides, because the purified enzyme is free from diesterase activity.

Application of method: the majority of determinations of acid phosphatase activity are carried out on blood serum in clinical laboratories, as an aid to the detection and monitoring of metastatic carcinoma of the prostate. Levels of activity in serum are typically normal in benign prostatic diseases, or when a cancer of the prostate remains confined to the gland itself. Increases in activity are observed when metastasis occurs and elevations are seen in approximately 75% of patients with metastatic spread. Acid phosphatase activity in serum falls on successful treatment.

The introduction in recent years of immunologically-based methods of determination of prostatic acid phosphatase, particularly radioimmunoassay, has re-awakened and stimulated interest in the determination of this enzyme in prostatic cancer. The greater sensitivity of these methods allows an increased acid phosphatase in serum to be demonstrated in a significant proportion (about 20%) of patients in whom metastasis has not yet occurred.

Slight or moderate elevations in total acid phosphatase activity occur in *Paget*'s disease of bone (osteitis deformans), in hyperparathyroidism with skeletal involvement, and in the presence of malignant invasion of the bones by cancers such as female breast cancer. Unlike the acid phosphatase of prostate, the enzyme in these cases is not inhibited by tartrate and is thought to come from osteoclasts. The osteoclasts are also the probable source of the increased tartrate-resistant acid phosphatase activity of growing children. Elevated levels of non-prostatic acid phosphatase have also been observed in patients with the lysosomal storage diseases, *Gaucher*'s disease and *Niemann-Pick* disease, in leukaemic reticuloendotheliosis, and in some other haematological disorders.

Assays of acid phosphatase activity are also of general biochemical interest, e.g. in the study of lysosomes and in developmental studies in plants.

Multiple forms of acid phosphatase: several forms of acid phosphatase exist in human tissues. Some of these forms result from the existence of different genetic loci for acid phosphatase. As with other isoenzymes determined by multiple gene loci, acid phosphatases derived from different loci differ to some extent in their catalytic properties, such as relative rates of hydrolysis of various orthophosphate ester substrates and response to inhibitors.

The exact number of genes responsible for determining acid phosphatases in man is not yet certain. The acid phosphatase of erythrocytes is certainly the product of a unique gene locus which is subject to considerable allelic variation, giving rise to electrophoretically distinguishable isoenzymes [1]. Erythrocyte acid phosphatase is a monomer with a molecular weight of 16500.

Lysosomal acid phosphatases are also the products of at least one separate gene locus, as is shown by reports of an inherited generalized deficiency of lysosomal acid phosphatase [2]. Lysosomal isoenzymes have molecular weights of the order of 100000 – 120000 and a dimeric structure. Two further loci appear to account for three forms of the enzyme in human placenta, two forms being homodimers of the products of the respective loci and the third consisting of a hybrid dimer of the two distinct subunits. One or both of these loci are probably expressed in tissues other than placenta.

It seems possible that prostatic acid phosphatase is also the product of a unique genetic locus. It is a dimer with molecular weight about 100000 at pH 5 – 6 which readily undergoes aggregation or dissociation.

The acid phosphatases are able to undergo post-translational modifications which give rise to heterogeneous patterns on electrophoresis. For example, oxidation of the sulphydryl groups *in vitro* gives rise to modified forms of red cell acid phosphatase [3], while prostatic acid phosphatase displays multiple zones on electrophoresis due to the presence of variable amounts of sialic acid in the molecule [4].

Enzyme properties relevant in analysis: acid phosphatase hydrolyzes a wide range of orthophosphate monoesters, and also catalyzes phosphate transfer. Pyrophosphates or polyphosphates are acted on to only a slight extent by prostatic phosphatase, but an acid phosphatase from leukaemic human spleen has been shown to have considerable pyrophosphatase activity [5]. *Michaelis* constants for the hydrolysis of phenolic orthophosphates by prostatic acid phosphatase are low. Values of 1.7×10^{-4} mol/l for 4-nitrophenyl phosphate [6] and 0.9×10^{-4} mol/l for 1-naphthyl phosphate [7] at pH 5 have been reported. Hydrolysis proceeds linearly with respect to time.

Acid phosphatase activity is not dependent on or activated by metal ions. The presence of hydroxyl compounds, e.g. 1,5-pentanediol, which can act as phosphate acceptors, stimulates activity [8, 9]. The differential effects of various inhibitors on the multiple forms of acid phosphatase from human tissues have played a significant part in attempts to increase the specificity of assay methods. Formaldehyde or cupric ions inhibit red cell acid phosphatase more than the prostatic isoenzyme. However, the most useful selective inhibitor is L-(+)-tartrate, which is a fully competitive inhibitor of prostatic and lysosomal acid phosphatases, but does not inhibit erythrocytic phosphatase or certain other forms of the enzyme [10]. Values of K_I for the inhibition of prostatic acid phosphatase by dextrorotatory tartrate at pH 5 are of the order of $3 - 4 \times 10^{-5}$ mol/l [6, 7].

Histidine and arginine residues have been shown to be essential for the activity of prostatic acid phosphatase and the enzyme is inactivated by reagents specific for these residues [11, 12].

Acid phosphatase is unstable, particularly at alkaline pH. It is also readily inactivated at liquid interfaces, e.g. in foams. Detergents protect the enzyme against surface inactivation.

Methods of determination: determination of catalytic activity can be carried out conveniently with a wide range of synthetic substrates. The choice of substrate has been made on the grounds of analytical convenience, e.g. liberation of a product which can readily be determined, and on grounds of specificity for the isoenzyme to be determined, particularly when this is prostatic acid phosphatase. The β-glycerophosphate and phenyl phosphate used in earlier clinical studies have now almost entirely fallen into disuse. Assays at present in use can be divided into two groups: those in which 4-nitrophenyl phosphate is the substrate, and those in which 1-naphthyl phosphate [13] or thymolphthalein monophosphate [14] is chosen because these substrates are much more rapidly hydrolyzed by prostatic acid phosphatase than by the various non-prostatic acid phosphatases which may occur in serum.

Assays based on 4-nitrophenyl phosphate can be applied generally to all acid phosphatases and, for this reason, a 4-nitrophenyl phosphate procedure is described here. When additional specificity for prostatic acid phosphatase is required, it is obtained in 4-nitrophenyl phosphate assays by the introduction of L-(+)-tartrate as a selective inhibitor. Fluorimetry of 1-naphthol [7] or 4-methyl umbelliferone [15] liberated by the enzyme from the corresponding orthophosphate ester has been used to increase the sensitivity of measurement of the low levels of acid phosphatase that are of clinical interest.

In contrast with corresponding assays for alkaline phosphatase, the hydrolysis of potentially chromogenic substrates such as 4-nitrophenyl phosphate cannot be monitored continuously since the differences in absorption spectra between the phosphate ester and its parent phenol are only developed in alkaline solution. Therefore, acid phosphatase assays usually have to be carried out on a fixed incubation basis.

Hillmann [16] has shown that naphthol released by the enzyme from 1-naphthyl phosphate will couple with Fast Red TR at a sufficiently rapid rate at acid pH to allow the formation of the resulting red dye to be monitored continuously during the enzymatic reaction. This procedure has been used as the basis of several manual and automated methods for continuous monitoring of acid phosphatase activity. However, it is not without its difficulties. For example, the rate of coupling is dependent on pH, so that it may be necessary to carry out the assay at a slightly more alkaline pH than the pH optimum of the enzyme, in order to ensure that the hydrolysis of the phosphate ester is rate-limiting. The coloured product also undergoes time-dependent changes in spectral characteristics, apparently due to interaction with components of the reaction mixture. Nevertheless, the *Hillmann* reaction represents the best approach to continuous monitoring of acid phosphatase activity at present available; therefore, a version of the method suitable for automated analysis is included in this chapter.

Several immunochemical methods for the determination of prostatic acid phosphatase have become available as the result of the preparation of antisera with a high degree of specificity for the prostatic isoenzyme. Such antisera typically do not cross-react with acid phosphatases from erythrocytes, spleen, liver, kidney, intestine, bladder or bone.

Radioimmunoassay is entirely independent of acid phosphatase activity. Part of the increased ability of radioimmunoassay to detect increases in prostatic acid phosphatase in serum compared with methods based on enzyme activity may result from the ability of radioimmunoassay to measure catalytically-inactive but immunologically-recognizable acid phosphatase molecules in the circulation. Several radioimmunoassays have been described (e.g. [17, 18, 19]) and reagents are available commercially.

Other immunoassays depend on the enzymatic activity retained by the acid phosphatase-antibody complex, either for the detection and measurement of the peaks in a *Laurell* "rocket" electro-immunoassay [20], or for quantitation of the amount of isoenzyme bound to antibody in solid-phase immuno-adsorbent assays [21], or double-antibody immunoprecipitation assays [22]. Colorimetric and fluorimetric determinations of the product of the acid phosphatase reaction have been used in these assays. Reagents for an immunoprecipitation method are available commercially. Counter-immunoelectrophoretic assays of prostatic acid phosphatase have also been described [23, 24].

An antiserum raised against tartrate-resistant acid phosphatase purified from the spleen of a patient with leukaemic reticuloendotheliosis has been used in a double-antibody immunoprecipitation assay to investigate this type of acid phosphatase activity in serum [25].

Apart from the use of specific antisera or inhibitors to determine the prostatic isoenzyme, other methods of isoenzyme analysis have not been applied extensively to acid phosphatases. However, electrophoresis is used in the study of allelozymes of erythrocyte acid phosphatase [1, 3] and electrophoresis of serum acid phosphatase in polyacrylamide gel has been advocated as an aid to diagnosis [26].

International reference methods and standards: no standard acid phosphatase preparation is yet available*, nor has international agreement been reached on a reference assay method. A recommended assay method (based on the *Hillmann* method) has been published by the Association of Clinical Biochemists in Great Britain [27], and national recommendations are being drawn up by working parties in other countries.

1.6.2 Fixed-time Method

Assay

Method Design

Principle

(a) 4-Nitrophenyl phosphate + H_2O $\xrightarrow[\text{pH 4.9}]{\text{acid phosphatase}}$ 4-nitrophenol + P_i

(b) 4-Nitrophenol $\xrightarrow[\text{dissociation and rearrangement}]{\text{pH 11}}$ 4-nitrophenolate ion

* Human Serum SRM 909 (NBS, USA) contains some human acid phosphatase. However, the activity (determined with thymolphthalein monophosphate as substrate) is so low that the material can not be recommended as a reference material for acid phosphatase.

Hydrolysis of 4-nitrophenyl phosphate by acid phosphatase at pH 4.9 liberates 4-nitrophenol. The reaction is stopped by raising the pH to ca. 11 by addition of NaOH. At this pH the strongly-coloured quinonoid 4-nitrophenolate ion is produced and its absorbance is measured at 405 nm. Tartrate-sensitive (e.g. prostatic) acid phosphatase activity is measured by the difference in activity observed when the assay is carried out in the absence and presence of the competitive inhibitor, dextrorotatory tartrate.

A blank is essential to correct for the small amount of non-enzymatic hydrolysis of the substrate during incubation and, more particularly, for the absorbance of serum samples at the wavelength of measurement.

Optimized conditions for measurement: prostatic acid phosphatase activity in serum has a pH optimum close to 5 in citrate buffer, with little variation in activity over a range of 4.8 – 5.3. At the concentration of 4-nitrophenyl phosphate of 6.3 mmol/l in the assay mixture, the reaction velocity is 97% of the theoretical V. Inhibition of tartrate-sensitive acid phosphatase (e.g. from prostate) is 95% complete at the final concentrations of substrate and inhibitor.

Temperature conversion factors: almost all assays for clinical purposes have been carried out at 37 °C, in the interests of increased sensitivity. Data on relative activities at other temperatures are correspondingly few. However, comparison of reference ranges at 30 °C and 37 °C indicates that activities are lower by a factor of 0.85 at the lower temperature.

Equipment

Spectrophotometer or spectral-line photometer providing accurate absorbance measurements at 405 nm; thermostatted water-bath capable of maintaining 37 ± 0.1 °C; stopwatch.

Reagents and Solutions

Purity of reagents: preparations of disodium 4-nitrophenyl phosphate of acceptable purity are now available commercially. Other reagents should be of analytical grade.

Preparation of solutions: make all solutions in re-purified water (cf. Vol. II, chapter 2.1.3.2).

1. Stock citrate buffer (45.9 mmol/l; pH 4.9):

   dissolve 13.5 g trisodium citrate dihydrate in about 700 ml water, adjust pH to 4.9 with HCl, 1 mol/l, and dilute to 1 litre with water.

2. Tartrate solution (1 mol/l):

   dissolve 15 g L-(+)tartaric acid (i.e. the dextrorotatory form) in about 70 ml water, adjust pH to 4.9 with NaOH (10 mol/l), then make up volume to 100 ml with water. Tartrate may precipitate initially during the addition of NaOH but redissolves on further addition.

3. Buffer/substrate solution for up to 25 total activity estimations (4-nitrophenyl phosphate, 7.6 mmol/l; citrate, 45 mmol/l; pH 4.9):

   dissolve 144 mg disodium 4-nitrophenyl phosphate hexahydrate in 50 ml stock buffer (1) and adjust volume to 51 ml with water.

4. Buffer/substrate/tartrate solution for up to 25 estimations of tartrate-resistant activity (4-nitrophenyl phosphate, 7.6 mmol/l; citrate, 45 mmol/l; pH 4.9; tartrate, 19.6 mmol/l):

   dissolve 72 mg disodium 4-nitrophenyl phosphate hexahydrate in 25 ml stock buffer (1) and add 0.5 ml tartrate solution (2).

   Note: when both total and tartrate-sensitive activities are to be determined on each sample, as is usually the case in clinical laboratories, it is convenient to dissolve 216 mg disodium 4-nitrophenyl phosphate hexahydrate in 75 ml solution (1), divide into 50 ml and 25 ml portions, add 1 ml water to the larger volume to give solution (3) and 0.5 ml tartrate solution (2) to the smaller volume to make solution (4).

5. Sodium hydroxide (0.1 mol/l):

   dissolve 4 g NaOH in water and dilute to 1 litre.

Stability of solutions: store all solutions at 0 – 4 °C. Solution (5) is stable indefinitely. Solutions (1) and (2) are stable as long as there is no microbial contamination. Solutions (3) and (4) can be stored frozen; however, it is preferable to add the substrate only on the day of analysis.

Procedure

Collection and treatment of specimen: obtain blood for acid phosphatase determination by venepuncture. Serum is separated as soon as possible and acidified by addition of 20 µl acetate buffer, 5 mol/l, pH 5.0, per ml serum. Plasma from blood prevented from clotting with EDTA can be used. Discard haemolyzed samples. Store frozen at – 20 °C until analyzed. Earlier reports that serum acid phosphatase is elevated following rectal examination of the prostate have not been confirmed by more recent studies.

Stability of the enzyme in the sample: acid phosphatase is rapidly inactivated at alkaline pH, e.g. in serum samples which become alkaline by loss of carbon dioxide. Acidification greatly increases the stability of the enzyme. Although immunological reactivity is lost at a rather slower rate than enzymatic activity, samples for radio-immunoassay should also be preserved by freezing.

Details for measurement in samples other than serum: acid phosphatase is readily extracted from prostate by homogenization in water, or from erythrocytes by lysing them with water (1 vol cells + 9 vol water). Release of the enzyme from lysosomes is facilitated by adding a detergent, e.g. Triton X-100. Acid phosphatase can be extracted by grinding plant materials with an acid buffer (e.g. acetate, 0.1 mol/l, pH 5.8) in a mortar.

Assay conditions: wavelength 405 nm; light path 10 mm; final volume (incubation) 1.2 ml, (measurement) 5.2 ml; 37 °C (water-bath). Incubation time usually 30 min. Measure against water or blank.

Measurement

Pipette into test tubes:	total activity	blank activity	tartrate resistant activity	concentration in incubation medium
buffer/substrate (3)	1.00 ml	1.00 ml	–	4-nitrophenyl phosphate 6.3 mmol/l citrate 37.5 mmol/l
buffer/substrate/tartrate (4)	–	–	1.00 ml	tartrate 16.3 mmol/l
allow to attain 37 °C in water-bath				
sample (e.g. serum)	0.20 ml	–	0.20 ml	volume fraction 0.167
mix, start stopwatch; incubate for exactly 30 min in a 37 °C water-bath				
NaOH (5)	4.00 ml	4.00 ml	4.00 ml	
mix				
sample (e.g. serum)	–	0.20 ml	–	
read absorbances at 405 nm against water or against blank.				

Calculation: use eqn. (k) or (k_1), cf. "Formulae", Appendix 3. Correct readings for total acid phosphatase activity and for tartrate-resistant activity by blank readings (if readings are not made against the blank); yielding ΔA_{total} and $\Delta A_{inh.}$, respectively. With a value of $\varepsilon_{405} = 1.85\ l \times mmol^{-1} \times mm^{-1}$ for 4-nitrophenol, the respective catalytic activity concentrations (b) are (Δt in minutes),

total acid phosphatase:	$b_1 = 1405 \times \Delta A_{total}/\Delta t$	U/l
tartrate-resistant phosphatase:	$b_2 = 1405 \times \Delta A_{inh.}/\Delta t$	U/l
tartrate-inhibited phosphatase:	$b_3 = b_1 - b_2$	U/l

For conversion to nkat/l multiply all values by 16.67.

Validation of Method

Precision, accuracy, detection limits and sensitivity: the relative standard deviation of determinations of total activity is of the order of 6% for elevated activities and 12% for values within the normal range, on a between-run basis. For tartrate-inhibited activity the value is 6% at 14 U/l. No independent evidence of accuracy can be given, since no standard acid phosphatase preparation is available. However, the results for tartrate-inhibited acid phosphatase activity in serum are strongly correlated with results of an immunoprecipitation assay using an antiserum specific for the prostatic isoenzyme. An activity of 0.4 U/l can be detected. For activities greater than 40 U/l, dilute the sample with NaCl, 9 g/l.

Sources of error: drugs do not generally interfere with the determination, although increased activities in patients treated with androgens or clofibrate have been reported. As already mentioned, routine digital examination of the prostate is now considered to be without effect on serum enzyme activity, but transient increases have been observed after trans-urethral resection of the prostate [28], and after cystoscopy or prostatic biopsy in patients with benign prostatic hyperplasia [29].

Specificity: the assay without added tartrate measures the activity of all acid phosphatases. Acid phosphatases which contribute to this activity in normal serum probably originate from platelets, erythrocytes and other blood cells, and from general tissue wear and tear. Tartrate-inhibited acid phosphatase in normal serum probably derives from platelets and other cells; however, raised tartrate-inhibited activity is almost entirely confined to disease of the prostate so that with the addition of tartrate the test becomes essentially specific for the prostatic isoenzyme.

Reference ranges: total acid phosphatase activity in serum from adult males ranges from 2.5 to 11.5 U/l at 37 °C; up to 4 U/l is inhibited by tartrate. Total activity in sera of females is up to 9 U/l, of which less than 1 U/l is inhibited by tartrate. Total serum acid phosphatase in children may be up to twice the normal adult activity. This is mainly due to an increased contribution of tartrate-resistant acid phosphatase, presumably derived from osteoclasts of bone.

1.6.3 Continuous-monitoring Method

Assay

Method Design

Principle

(a) 1-Naphthyl phosphate + H_2O $\xrightarrow{\text{acid phosphatase}}$ 1-naphthol + P_i

(b) 1-Naphthol + Fast Red TR $\longrightarrow$ chromophore

The method [30] is modified from *Hillmann* [16]. Enzymatic hydrolysis of 1-naphthyl phosphate at pH 5.6 liberates 1-naphthol. At this pH, 1-naphthol combines rapidly with the stabilized diazonium salt, Fast Red TR, present in the incubation mixture to form a red dye. The appearance of the dye is monitored continuously at 410 nm.

There is some increase in absorbance due to combination of the diazonium salt with components of the system other than enzymatically-liberated 1-naphthol. However, the blank rate thus generated is equivalent to less than 1 U/l under the present conditions and can be neglected for most purposes. The progress of reaction first shows an enhanced rate after addition of substrate due to chromophore formation with free 1-naphthol. Absorbance increases linearly with time after about 1 min under the conditions described.

The method can readily be adapted to automatic analyzers working on serum-start or substrate-start principles.

Optimized conditions for measurement: the chosen conditions represent a compromise between the requirements of the enzymatic and chromogenic reactions. The buffer pH of 5.6 is somewhat above the central value of maximum activity, but this allows more rapid coupling between 1-naphthol and Fast Red TR, reducing the interval before zero-order kinetics are established, without a significant reduction in enzymatic activity [30, 31, 32]. The substrate concentration of 3 mmol/l provides for a reaction velocity which is approximately 90% of the theoretical V; higher concentrations prolong the delay before a linear progress curve is obtained because of the presence of free naphthol, and also increase the concentration of tartrate required to inhibit prostatic acid phosphatase. Increased concentrations of Fast Red TR produce an unacceptably high blank rate.

When tartrate is added at the concentration indicated, inhibition of prostatic acid phosphatase is 95% complete.

A detergent, Triton X-100, is added to enhance the solubility of the dye. The chromophore has an absorption maximum at 420 nm in the presence of Triton X-100. The spectrum of the chromophore undergoes time-dependent changes. For this reason, measurement at the isosbestic point at 390 nm has been recommended, since absorbance remains constant with time at this wavelength [32]. However, measurements are made at 410 nm in the present method since filters for this wavelength are generally available for automatic analyzers.

The concentrations in the complete assay mixture are: citrate buffer, pH 5.6, 44 mol/l; 1-naphthyl phosphate, 3 mmol/l; Fast Red TR, 0.8 mmol/l; dextrorotatory tartrate (when present), 39 mmol/l.

Temperature conversion factors: see above, p.97.

Equipment

Spectrophotometer or spectral-line photometer capable of accurate absorbance measurement at 410 nm (or Hg 405 nm) with thermostatted cuvette holder; recorder or stopwatch.

Reagents and Solutions

Purity of reagents: 1-naphthyl disodium orthophosphate should be freed from traces of 1-naphthol by washing with diethyl ether on a *Buchner* funnel or extraction in a *Soxhlet* apparatus. Free naphthol in the substrate solution causes the phase of accelerated reaction by coupling with Fast Red TR; as the free naphthol content rises, this effect becomes more pronounced and the delay before zero-order kinetics are established is lengthened. Within-laboratory preparation of the Fast Red TR salt has been recommended by some authors; however, commercial preparations are of acceptable purity. All other reagents should be of analytical grade.

Preparation of solutions: make all solutions in re-purified water (cf. Vol. II, chapter 2.1.3.2).

1. Citrate buffer (50 mmol/l; pH 5.6):

   dissolve 14.7 g trisodium citrate dihydrate in about 700 ml water, adjust pH to 5.6 with HCl, 1 mol/l, and dilute to 1 litre with water.

2. Tartrate solution (1 mol/l):

   dissolve 15 g L-(+)tartaric acid (i.e. the dextrorotatory form) in about 70 ml water, adjust pH to 5.6 with NaOH (10 mol/l), then make up volume to 100 ml with water. Tartrate may precipitate initially during the addition of NaOH but redissolves on further addition.

3. Buffer/substrate solution for about 20 estimations of total activity (1-naphthyl phosphate, 37.5 mmol/l; citrate, 50 mmol/l, pH 5.6):

   dissolve 28 mg disodium 1-naphthyl phosphate in 3 ml citrate buffer (1).

4. Buffer/Fast Red TR solution:

   not longer than 10 min before each batch of 20 analyses, dissolve 10 mg Fast Red TR (diazotized 2-amino-5-chlorotoluene) and 25 µl (20 mg) Triton X-100 in 25 ml citrate buffer (1).

Stability of solutions: store all solutions at 0 – 4 °C. Solutions (1) and (2) are stable as long as there is no microbial contamination. Solutions (3) and (4) should be prepared immediately before use.

Procedure

Collection and treatment of specimen: obtain venous blood, separate and acidify serum as above (p. 98). Heparinized or EDTA-treated plasma can be used, but is not recommended because it may cause turbidity. Although red-cell acid phosphatase acts only slowly on 1-naphthyl phosphate haemolyzed samples should be rejected.

Stability of the enzyme in the sample: see p. 99.

Details for measurement in other samples: because 1-naphthyl phosphate is hydrolyzed at a low rate by tartrate-resistant acid phosphatase, the method is of less general application than the 4-nitrophenyl phosphate procedure. The present method is intended principally for the determination of prostatic acid phosphatase in serum.

Assay conditions: wavelength 410 (Hg 405) nm; light path 10 mm; final volume 1.25 ml (total activity), 1.30 ml (tartrate-resistant activity); 37 °C (thermostatted cuvette holder). Measure against air.

Measurement

Pipette successively into cuvettes:		total activity	tartrate-resistant activity	concentration in incubation medium	
buffer/Fast Red TR	(1)	1.00 ml	1.00 ml	citrate Fast Red TR	44 mmol/l 0.8 mmol/l
tartrate	(2)	–	0.05 ml	tartrate	39 mmol/l
sample (serum)		0.15 ml	0.15 ml	volume fractions	0.12 and 0.115
mix; incubate each cuvette for at least 20 min at 37 °C					
buffer/substrate	(3)	0.10 ml	0.10 ml	1-naphthyl phosphate	3 mmol/l
mix, record absorbance; determine slope of linear portion of progress curve, beginning at least 1 min after start of reaction.					

Calculation: use eqn. (k) or (k_1), cf. "Formulae", Appendix 3. The value of ε at 410 nm for the reaction product of 1-naphthol and Fast Red TR is $1.56\ \text{l} \times \text{mmol}^{-1} \times \text{mm}^{-1}$. Catalytic concentrations (b) in the sample are therefore (Δt in minutes),

total acid phosphatase:	$b_1 = 534 \times \Delta A_{\text{total}}/\Delta t$	U/l
tartrate-resistant phosphatase:	$b_2 = 556 \times \Delta A_{\text{inh.}}/\Delta t$	U/l
tartrate-inhibited phosphatase:	$b_3 = b_1 - b_2$	U/l

The value of ε at Hg 405 nm is $1.29\ \text{l} \times \text{mmol}^{-1} \times \text{mm}^{-1}$, and the corresponding factors are 646 (total activity) and 672 (tartrate-resistant activity), respectively.

For conversion to nkat/l multiply all values by 16.67.

Validation of Method

Precision, accuracy, detection limits and sensitivity: within-run values of the relative standard deviation for the determination of total activity are 3.6% at the 6 U/l level and 2.5% at 13 U/l; between-run values are 7% at 5 U/l and 3% at 19 U/l. The imprecision of measurement of tartrate-inhibited activity is 2% at the 2 U/l level (within-run), and 5% and 10% at 17 and 2 U/l, respectively (between-run). The detection limit is 0.5 U/l.

Sources of error: see also p. 100. Errors in this method may arise from excessive contamination of the substrate with free naphthol, which may result in incorrect identification of the portion of the progress curve corresponding to enzymatic hydrolysis.

Specificity: the specificity of the method for tartrate-inhibited acid phosphatases such as the prostatic isoenzyme is much higher than that of the 4-nitrophenyl phosphate method. However, non-prostatic, tartrate-resistant acid phosphatases do act on the substrate to some extent; therefore, tartrate is added to increase the specificity of the assay for prostatic acid phosphatase.

Reference ranges: the normal range of total acid phosphatase activity in men is up to 7.5 U/l, of which 2.5 U/l is inhibited by tartrate.

References

[1] *D. A. Hopkinson, N. Spencer, H. Harris,* Red Cell Acid Phosphatase Variants: a New Human Polymorphism, Nature *199*, 969 – 971 (1963).

[2] *H. L. Nadler, T. J. Egan,* Deficiency of Lysosomal Acid Phosphatase. A New Familial Metabolic Disorder, New Engl. J. Med. *282*, 302 – 307 (1970).

[3] *R. A. Fisher, H. Harris,* Studies on the Purification and Properties of the Genetic Variants of Red Cell Acid Phosphohydrolase in Man, Ann. N.Y. Acad. Sci. *166*, 380 – 391 (1969).

[4] *J. K. Smith, L. G. Whitby,* The Heterogeneity of Prostatic Acid Phosphatase, Biochim. Biophys. Acta *151*, 607 – 618 (1968).

[5] *K.-W. Lam, L. T. Yam,* Biochemical Characterization of the Tartrate-Resistant Acid Phosphatase of Human Spleen with Leukemic Reticuloendotheliosis as a Pyrophosphatase, Clin. Chem. *23*, 89 – 94 (1977).

[6] *K. Jacobsson,* On the Inhibition of Prostatic Phosphatase by Tartrate, Scand. J. Clin. Lab. Invest. *11*, 358 – 360 (1959).

[7] *D. M. Campbell, D. W. Moss,* Spectrofluorimetric Determination of Acid Phosphatase Activity, Clin. Chim. Acta *6*, 307 – 315 (1961).

[8] *H. Gallati, M. Roth,* Aktivierung der sauren Prostataphosphatase durch 1-Pentanol, J. Clin. Chem. Clin. Biochem. *14*, 581 – 587 (1976).

[9] *C. Poindexter, K. Ervin, E. G. Rice,* A Kinetic Acid Phosphatase (Hillmann) Procedure of Improved Sensitivity, Clin. Chem. *26*, 1009 (Abstr.) (1980).

[10] *M. A. M. Abul-Fadl, E. J. King,* Properties of the Acid Phosphatases of Erythrocytes and of the Human Prostate Gland, Biochem. J. *45*, 51 – 60 (1949).

[11] *J. J. McTigue, R. L. Van Etten,* An Essential Active-site Histidine Residue in Human Prostatic Acid Phosphatase. Ethoxyformylation by Diethyl Pyrocarbonate and Phosphorylation by a Substrate, Biochim. Biophys. Acta *523*, 407 – 421 (1978).

[12] *J. J. McTigue, R. L. Van Etten,* An Essential Arginine Residue in Human Prostatic Acid Phosphatase, Biochim. Biophys. Acta *523*, 422 – 429 (1978).

[13] *A. L. Babson, P. A. Read,* A New Assay for Prostatic Acid Phosphatase in Serum, Am. J. Clin. Path. *32*, 89 – 91 (1959).

[14] *A. V. Roy, M. E. Brower, J. E. Hayden,* Sodium Thymolphthalein Monophosphate: a New Acid Phosphatase Substrate with Greater Specificity for the Prostatic Enzyme in Serum, Clin. Chem. *17*, 1093 – 1102 (1971).

[15] J. P. Chambers, L. Aquino, R. H. Glew, R. E. Lee, L. R. McCafferty, Determination of Serum Acid Phosphatase in *Gaucher*'s Disease using 4-Methylumbelliferyl Phosphate, Clin. Chim. Acta *80*, 67 – 77 (1977).

[16] *G. Hillmann,* Fortlaufende photometrische Messung der sauren Prostataphosphatase-Aktivität, Z. Klin. Chem. Klin. Biochem. *9*, 273 – 274 (1971).

[17] A. G. Foti, H. Herschman, J. F. Cooper, A Solid-phase Radioimmunoassay for Human Prostatic Acid Phosphatase, Cancer Res. *35*, 2446 – 2452 (1975).

[18] *P. Vihko, E. Sajanti, O. Janne, L. Pettonen, R. Vihko,* Serum Prostate-specific Acid Phosphatase: Development and Validation of a Specific Radioimmunoassay, Clin. Chem. *24*, 1915 – 1919 (1978).

[19] *D. E. Mahan, B. P. Doctor,* A Radioimmune Assay for Human Prostatic Acid Phosphatase – Levels in Prostatic Disease, Clin. Biochem. *12*, 10 – 17 (1979).

[20] *V. Milisauskas, N. R. Rose,* Immunochemical Quantitation of Prostatic Phosphatase, Clin. Chem. *18*, 1529–1531 (1972).
[21] *C. L. Lew, C. S. Killian, G. P. Murphy, T. M. Chu,* A Solid-phase Immunoadsorbent Assay for Serum Prostatic Acid Phosphatase, Clin. Chim. Acta *101*, 209–216 (1980).
[22] *B. K. Choe, E. J. Pontes, M. K. Dong, N. R. Rose,* Double-antibody Immunoenzyme Assay for Human Prostatic Acid Phosphatase, Clin. Chem. *26*, 1854–1859 (1980).
[23] *A. G. Foti, J. F. Cooper, H. Herschman,* Counterimmunoelectrophoresis in Determination of Prostatic Acid Phosphatase in Human Serum, Clin. Chem. *24*, 140–142 (1978).
[24] *T. M. Chu, M. C. Wang, W. W. Scott, R. P. Gibbons, D. E. Johnson, J. D. Schmidt, S. A. Loening, G. R. Prout, G. P. Murphy,* Immunochemical Detection of Serum Prostatic Acid Phosphatase. Methodology and Clinical Evaluation, Invest. Urol. *15*, 319–323 (1978).
[25] *K.-W. Lam, M. Siemens, T. Sun, C.-Y. Li, L. T. Yam,* Enzyme Immunoassay for Tartrate-resistant Acid Phosphatase, Clin. Chem. *28*, 467–470 (1982).
[26] *T. Sun, K.-W. Lam, L. Narukar,* Clinical Applicability of Acid Phosphatase Isoenzyme Assay, Clin. Chem. *27*, 1742–1744 (1981).
[27] Association of Clinical Biochemists, News Sheet No. 202, Supplement (1980).
[28] *C. D. Johnson, D. Costa, J. E. Castro,* Acid Phosphatase after Examination of the Prostate, Brit. J. Urol. *51*, 218–223 (1979).
[29] *P. Vihko, O. Lukkarinen, M. Kontturi, R. Vihko,* The Effect of Manipulation of the Prostate Gland on Serum Prostate-specific Acid Phosphatase Measured by Radioimmunoassay, Invest. Urol. *18*, 334–336 (1981).
[30] *R. J. Warren, D. W. Moss,* An Automated Continuous-monitoring Procedure for the Determination of Acid Phosphatase Activity in Serum, Clin. Chim. Acta *77*, 179–188 (1977).
[31] *R. Bais, J. B. Edwards,* An Optimized Continuous-monitoring Procedure for the Semiautomated Determination of Serum Acid Phosphatase Activity, Clin. Chem. *22*, 2025–2028 (1976).
[32] *G. Gundlach, B. Mülkausen,* Untersuchungen zur Kupplung des 1-Naphthols mit Fast Red TR, Z. Klin. Chem. Klin. Biochem. *18*, 603–610 (1980).

1.7 5′-Nucleotidase

5′-Ribonucleotide phosphohydrolase, EC 3.1.3.5

1.7.1 UV-method

Fritz Heinz and Rainer Haeckel

General

5′-Nucleotidase catalyzes the hydrolysis of most ribonucleoside 5′-monophosphates and deoxynucleoside 5′-monophosphates to the corresponding nucleosides and orthophosphate. 5′-Nucleotidase activity has been detected in a great number of human and animal tissues and is localized predominantly in the plasma membrane [1, 2]. In mature erythrocytes the activity is extremely low or absent [3].

Application of method: in biochemistry as a membrane marker-enzyme and in clinical chemistry, especially in diseases affecting the hepatobiliary tract [1].

Enzyme properties relevant in analysis: the existence of isoenzymes has been described. The enzyme activity present in human sera is released mainly from the membrane of liver cells by bile salts. Substrate conversion proceeds linearly with time.

Methods of determination: a number of different methods have been described for the determination of catalytic activity. Radioactive methods make use of ^{14}C, ^{3}H, and ^{32}P-labelled nucleotides, particularly AMP [4 – 6]. One of the oldest methods is the determination of liberated orthophosphate according to *Fiske & Subbarow* [7, 8]. With AMP as substrate, the reaction product adenosine can be deaminated to inosine by adenosine deaminase. Inosine thus produced can be measured at 265 nm [6, 9], or the ammonia liberated can be determined by the *Berthelot* reaction [10 – 12] or with glutamate dehydrogenase [13 – 16]. With IMP as substrate, the reaction product inosine is converted to uric acid by nucleoside phosphorylase and xanthine oxidase. The appearance of uric acid is measured at 295 nm [17].

International reference method and standards: neither standardization at the international level nor the existence of reference standard materials is known up to now.

Enzyme effectors: 5′-nucleotidase is inhibited by ADP, NAD, NADH, phosphate, α,β-methylene adenosine 5′-diphosphate, concanavalin A and Ni ions. Mg^{2+} or Mg^{2+} are needed for maximal activity [17]. A special problem in 5′-nucleotidase assays derives from interference by non-specific phosphatases, especially alkaline phosphatase. Several methods have been described to avoid this difficulty. 5′-Nucleotidase may be inhibited by nickel ions [8], concanavalin A [11, 12] or α,β-methylene-adenosine 5′-diphosphate [4]; 5′-nucleotidase activity is represented by the difference between the activities in the presence and absence of the inhibitor.

Alternatively, the activity of alkaline phosphatase may be inhibited by amino acids such as L-histidine, L-cysteine or glycine [18]. Another possibility for the elimination of non-specific nucleoside 5′-monophosphate hydrolysis is the simultaneous determination of alkaline phosphatase at pH 7.5 with A-2-MP and A-3-MP as substrates, or at pH 9.8 with 4-nitrophenyl phosphate, from which the interference of alkaline phosphatase with the 5′-nucleotidase assay can be calculated using an empirical correction factor [19].

In our hands the technique of "enzyme diversion" according to *Belfield & Goldberg* [20] gives the best results. High concentrations of glycerol 2-phosphate are offered to the alkaline phosphatase resulting in a competitive displacement of the nucleoside 5′-monophosphate from the active centre; phenyl phosphate may substitute for glycerol 2-phosphate, but there is no advantage.

Assay

Method Design

Principle

(a) $IMP + H_2O \xrightarrow{5'\text{-nucleotidase}} \text{inosine} + P_i$

(b) $\text{Inosine} + P_i \xrightarrow{NP^*} \text{hypoxanthine} + \text{ribose-1-P}$

(c) $\text{Hypoxanthine} + 2\,H_2O + 2\,O_2 \xrightarrow{XOD^{**}} \text{urate} + 2\,H_2O_2$

(d) $2\,H_2O_2 + 2\,\text{ethanol} \xrightarrow{\text{catalase}^+} 2\,\text{acetaldehyde} + 4\,H_2O$

(e) $2\,\text{Acetaldehyde} + 2\,NADP^+ + 2\,H_2O \xrightarrow{AlDH^{++}} 2\,\text{acetate} + 2\,NADPH + 2\,H^+$

Sum:

(f) $IMP + 2\,NADP^+ + 2\,\text{ethanol} + 2\,O_2 + H_2O \longrightarrow \text{urate} + \text{ribose-1-P} + 2\,\text{acetate} + 2\,NADPH + 2\,H^+$

The amount of inosine 5-phosphate (IMP) hydrolyzed per min, detected by the increase of absorbance due to the production of NADPH, is a measure of the catalytic activity of 5′-nucleotidase. One mole of inosine 5′-phosphate hydrolyzed corresponds to the formation of 2 moles of NADPH. The equilibrium of the complete reaction sequence is in favour of acetate production, because the 5′-nucleotidase-, xanthine oxidase- and aldehyde dehydrogenase reactions are practically irreversible under the assay conditions. Blanks are necessary for routine assays. A mechanized version of the following method has been described [21].

Optimized conditions for measurement: for optimization it was necessary to establish the influence of phosphate: purine nucleoside phosphorylase needs phosphate as a substrate, but phosphate inhibits 5′-nucleotidase. The concentration of 0.002 mol/l present in the assay shows no inhibitory effect on the 5′-nucleotidase reaction and is sufficient for optimal reaction rate.

The high ethanol concentration of 1.14 mol/l is necessary to prevent the normal catalase reaction with two moles of hydrogen peroxide.

* Nucleoside phosphorylase, Purine-nucleoside: orthophosphate ribosyltransferase, EC 2.4.2.1.

** Xanthine oxidase, Xanthine: oxygen oxidoreductase, EC 1.2.3.2.

\+ Catalase, Hydrogen peroxide: hydrogen peroxide oxidoreductase, EC 1.11.1.6.

\++ Aldehyde dehydrogenase, Aldehyde: $NAD(P)^+$ oxidoreductase, EC 1.2.1.5.

NADP cannot be replaced by NAD, because NAD inhibits 5′-nucleotidase. The K ions are activators for aldehyde dehydrogenase from yeast. The activity of 5′-nucleotidase is dependent on divalent cations; manganese and magnesium are reported to be the best activators. With magnesium ions a plateau of maximal activity is reached at concentrations higher than 10 mmol/l.

Manganese is even more effective and would increase the sensitivity of the method, but difficulties occur with the solubility of manganese phosphate. In order to add manganese in concentrations up to the optimum of 1 mmol/l, it is necessary to lower the phosphate concentration to 0.1 mmol/l. In this case, all indicator enzymes have to be freed from phosphate by dialysis and the amount of nucleoside phosphorylase has to be increased six-fold to reach the activity which can be maximally determined with the assay conditions described below. The sensitivity with magnesium is high enough for the determination of 5′-nucleotidase activity in sera, even those of healthy persons.

To prevent the formation of magnesium ammonium phosphate precipitates it is necessary to remove ammonium sulphate from the xanthine oxidase suspension by dialysis.

Temperature conversion factors: factors for the conversion of the results determined at 30 °C to other temperatures have not been reported.

Equipment

Spectrophotometer or spectral-line photometer capable of exact measurements at 340 nm or 334 nm, and with a thermostatted cuvette holder; recorder or stopwatch.

Reagents and Solutions

Purity of reagents: in the commercially available enzymes aldehyde dehydrogenase from yeast, catalase and nucleoside phosphorylase, the NADPH-oxidation activities together should not exceed 0.01% of that of aldehyde dehydrogenase; NADP dependent alcohol dehydrogenase should be absent. If nucleoside phosphorylase from manufacturers other than those cited are used, the specificity against inosine should be determined, because specificity varies depending on the material used for isolation. Commercial preparations of enzymes, NADP and IMP are available. Other chemicals should be of analytical grade.

Preparation of solutions: all solutions in re-purified water (cf. Vol. II, chapter 2.1.3.2).

1. Triethanolamine buffer (0.2 mol/l; pH 7.5):

   dissolve 18.57 g triethanolamine hydrochloride, 7.456 g potassium chloride and 0.372 g EDTA-$Na_2H_2 \cdot 2H_2O$ in 900 ml water and adjust the pH to 7.5 with NaOH (2 mol/l), then dilute to 1 l.

2. Phosphate solution (1 mol/l):

   dissolve 22.82 g $KH_2PO_4 \cdot 3\ H_2O$ in water, dilute to 100 ml.

3. Magnesium chloride solution (1 mol/l):

   dissolve 20.33 g $MgCl_2 \cdot 6\ H_2O$ in water, dilute to 100 ml.

4. Xanthine oxidase:

   for elimination of ammonium sulphate, dialyze a sample of xanthine oxidase (*Boehringer Mannheim* 110442), 10 mg/ml, ca. 0.4 U/mg, overnight at 4°C against 2 l potassium chloride solution (0.1 mol/l; 7.456 g KCl/l).

5. Inosine 5′-monophosphate (0.04 mol/l):

   dissolve 20 mg IMP-$Na_2 \cdot 6\ H_2O$ (*Boehringer Mannheim* 106704) in 1 ml water.

6. Working solution (for about 20 determinations):

   to a mixture of 9 ml buffer (1) and 1 ml ethanol add 0.01 ml catalase (*Boehringer Mannheim* 106810: 20 mg/ml, ca. 13000 U), 5 mg NADP, Na salt (*Boehringer Mannheim* 128058), 135 mg glycerol 2-phosphate, Na salt (*Sigma* 6-6251), 0.02 ml phosphate solution (2), 0.15 ml magnesium chloride solution (3), aldehyde dehydrogenase (lyophilized powder, *Boehringer Mannheim* 171832) equivalent to 35 U, 1 ml xanthine oxidase (4) and 0.14 ml nucleoside phosphorylase (*Boehringer Mannheim* 107964; 5 mg/ml).

Stability of solutions: the dialyzed xanthine oxidase is stable for one week at 4°C. The working solution (6) is stable at least for 6 hours at room temperature. The other solutions are stable for several months at 4°C if growth of micro-organisms is prevented.

Procedure

Collection and treatment of specimen: collect blood from the vein. Addition of oxalate, citrate, fluoride and heparin are without effect. Centrifuge for 10 min at about 3000 *g* in order to obtain plasma or serum.

Stability of enzymes in the sample: sera may be stored for 3 days at 4°C without change in activity.

Details for measurement in tissues: the 5′-nucleotidase is localized in the plasma membrane and has to be solubilized before determination. Experience with lymphocytes shows that sonication (40 s, 25 watts at 4°C) and the addition of triton X-100 (0.5%

w/w) results in maximal solubilization. After this treatment, the enzyme can no longer be sedimented at 100000 *g* for 1 h [22].

Assay conditions: wavelength 339 nm or Hg 334 nm; light path 10 mm; final volume 0.61 ml; 30 °C (thermostatted cuvette holder). Before starting the assay adjust the temperature of the working solution (6) and the sample to 30 °C.

Measurement

Pipette successively into a semi-microcuvette:			concentration in the assay mixture	
working solution	(6)	0.45 ml	triethanolamine	60 mmol/l
			KCl	60 mmol/l
			EDTA	0.6 mmol/l
			ethanol	1.14 mol/l
			NADP	0.42 mol/l
			glycerol-2-P	41.7 mmol/l
			phosphate	1.33 mmol/l
			Mg^{2+}	10 mmol/l
			AlDH	2300 U/l
			NP	933 U/l
			XOD	267 U/l
			catalase	866 kU/l
sample		0.15 ml	volume fraction	0.248
mix thoroughly with a plastic spatula; wait for ca. 5 min, until a linear change in absorbance can be registered (blank),				
IMP	(5)	0.005 ml	IMP	0.33 mmol/l
mix, wait 2 min till the change in absorbance becomes linear, then read the absorbance; repeat the reading after exactly 5 and 10 min or monitor the reaction on a recorder.				

The value $\Delta A/\Delta t$ must be <0.04/min, otherwise dilute samples correspondingly with triethanolamine buffer (0.1 mol/l) pH 7.5.

Calculation: values of $\Delta A/\Delta t$ are taken from the linear part of the conversion curve. Calculate $\Delta A/\Delta t$ for 1 min. Subtract $\Delta A/\Delta t$ of the blank from $\Delta A/\Delta t$ of the sample assay. Calculate the catalytic concentration *b* of the sample according to eqn. (k) or (k_1), cf. "Formulae", Appendix 3. Note that 2 mol NADPH are produced for each mole of inosine 5-phosphate hydrolyzed by 5′-nucleotidase. For ε values cf. Appendix 4. The following relations are valid:

	Hg 334 nm	339 nm	
$b =$	$326 \times \Delta A/\Delta t$	$320 \times \Delta A/\Delta t$	U/l
$b =$	$5433 \times \Delta A/\Delta t$	$5333 \times \Delta A/\Delta t$	nkat/l

Validation of Method

Precision, accuracy, detection limits and sensitivity: for values of 12.8 U/l in serum an imprecision of ±0.85 U/l and a coefficient of variation of 6.6% has been found for manual performance; with 3.05 U/l the values were ±0.3 U/l and 9.7%. For the mechanized procedure with a mean value of 12.75 U/l, an imprecision of ±0.38 U/l and a coefficient of variation of 3.0% were obtained; for 3.25 U/l the imprecision was 0.11 U/l and the coefficient of variation 3.3% (each case $n = 10$). Data on accuracy are not available since standard reference materials is not established yet. The minimal enzyme activity detectable is 0.18 U/l. Sensitivity is found to be $\Delta A/\Delta t = 0.001/$ 10 min (*Eppendorf* photometer, Hg 334 nm).

Sources of error: influences on the measured activity by therapeutics are unknown, as described by *Haeckel* [23] for the determination of uric acid, which uses the identical indicator system.

Specifity: 5′-nucleosidase activity is measured specifically by the method described here (see enzyme effectors, p. 107).

Reference ranges: in sera of normal adults ($n = 95$) the activity ranges from 2.1–9.5 U/l (24).

References

[1] *D. M. Goldberg,* 5′-Nucleotidase, Recent Advances in Cell Biology, Methodology and Clinical Significance, Digestion *8*, 87–99 (1973).

[2] *G. I. Drummond, Y. Masanobu,* Nucleotide Phosphomonoesterases, *P. D. Boyer* (ed.), The Enzymes, 3rd ed., Vol. *IV*, Academic Press, New York 1971, pp. 337–352.

[3] *J. Delaunay, S. Fischer, M. Tortolero, J. P. Piau, G. Schapira,* Absence of any Detectable Activity of the Membrane Marker Enzyme 5′-Nucleotidase in Human Red Blood Cells, Biomedicine (Express) *29*, 173–175 (1978).

[4] *M. K. Gentry, R. A. Olson,* A Simple Specific Radioisotopic Assay for 5′-Nucleotidase, Anal. Biochem. *64*, 624–627 (1975).

[5] *B. Glastaris, S. E. Pfeiffer,* Mammalian Membrane Marker Enzymes: Sensitive Assay for 5′-Nucleotidase and Assay for Mammalian 2′,3′-Cyclic-Nucleotidase-3′-Phosphohydrolase, in: *S. P. Colowick, N. O. Kaplan* (eds.), Methods of Enzymology, Vol. *XXXII*, Academic Press, New York 1974, pp. 124–131.

[6] *P. L. Ipata,* A Coupled Optical Enzyme Assay for 5′-Nucleotidase, Anal. Biochem. *20*, 30–36 (1967).

[7] *V. G. Bethune, M. Fleisher, M. K. Schwartz,* Automated Method for Determination of Serum 5′-Nucleotidase Activity, Clin. Chem. *18*, 1525 – 1526 (1972).
[8] *S. V. Rieder, M. Otero,* A Simplified Procedure for the Assay of 5′-Nucleotidase, Clin. Chem. *15*, 727 – 729 (1969).
[9] *J. Beckmann, O. Beckmann,* Verbesserte Bestimmung der 5′-Nucleotidaseaktivität im Serum, Z. Klin. Chem. Klin. Biochem. *9*, 277 (1971).
[10] *A. Belfield, G. Ellis, D. M. Goldberg,* A Specific Colorimetric 5′-Nucleotidase Assay Utilizing the Berthelot-Reaction, Clin. Chem. *16*, 396 – 401 (1970).
[11] *E. R. Zygowicz, F. W. Sunderman, E. Horak,* Concanavalin-Assay for Serum 5′-Nucleotidase, Clin. Chem. *23*, 1171 (1977).
[12] *E. R. Zygowicz, F. W. Sunderman, E. Horak, J. F. Dooley,* Inhibition by Concanavalin A as the Basis for a Specific Assay of Serum 5′-Nucleotidase Activity, Clin. Chem. *23*, 2311 – 2323 (1977).
[13] *C. L. Arkesteijn,* A Kinetic Method for Serum 5′-Nucleotidase Using Stabilized Glutamate Dehydrogenase, J. Clin. Chem. Clin. Biochem. *14*, 155 – 158 (1976).
[14] *J. Bootsma, B. G. Wolthers, A. Groen,* Determination of Serum 5′-Nucleotidase by Means of a NADH Linked Reaction, Clin. Chem. Acta *41*, 219 – 222 (1972).
[15] *G. Ellis, D. M. Goldberg,* An Improved Kinetic 5′-Nucleotidase Assay, Anal. Letters *5*, 65 – 73 (1972).
[16] *A. A. Adel, A. Ismail, D. G. Williams,* Scope and Limitations of a Kinetic Assay for Serum 5′-Nucleotidase Activity, Clin. Chim. Acta *55*, 211 – 216 (1974).
[17] *E. Fioretti, G. Caulini, G. Magni, R. A. Felicioli,* Spectrophotometric Assays for 5′-Nucleotidase Using IMP, GMP and CMP as Substrates, Ital. J. Biochem. *21*, 102 – 112 (1972).
[18] *O. Bodansky, M. K. Schwartz,* 5′-Nucleotidase, Adv. Clin. Chem. *11*, 277 – 328 (1968).
[19] *J. Beckmann, K. Leybold, L. Weisbecker,* Zur Bestimmung der 5′-Nucleotidase im Serum, Z. Klin. Chem. Klin. Biochem. *7*, 18 – 24 (1969).
[20] *A. Belfield, D. M. Goldberg,* Comparison of Sodium β-Glycerolphosphate and Disodium Phenylphosphate as Inhibitors of Alkaline Phosphatase in Determination of 5′-Nucleotidase Activity of Human Serum, Clin. Biochem. *3*, 105 – 110 (1970).
[21] *F. Heinz, R. Pilz, S. Reckel, J. R. Kalden, R. Haeckel,* A New Spectrophotometric Method for the Determination of 5′-Nucleotidase, J. Clin. Chem. Clin. Biochem. *18*, 781 – 788 (1980).
[22] *R. Pilz,* Bestimmung der Aktivität von Purinstoffwechselenzymen in menschlichen peripheren Blutlymphocyten: Entwicklung von Methoden und ihre exemplarische Anwendung, Dissertation, Med. Hochschule Hannover (1980).
[23] *R. Haeckel,* The Use of Aldehyde Dehydrogenase to Determine H_2O_2-Producing Reactions, J. Clin. Chem. Clin. Biochem. *14*, 101 – 107 (1976).
[24] *M. Jablonski,* Die diagnostische Bedeutung der katalytischen Aktivität der 5′-Nucleotidase im Serum bei Lebererkrankungen, Dissertation, Med. Hochschule Hannover (1983).

1.7.2 Colorimetric Method

Norbert van Husen and Ulrich Gerlach

General

5′-Nucleotidase was discovered by *Reis* [1] in 1933. The enzyme specifically catalyzes the dephosphorylation of nucleoside phosphates with phosphate groups attached at the C5 position of the ribose ring. It is found in various plant, animal and human

tissues [2]. *Heppel & Hilmoe* isolated 5′-nucleotidase from bull semen in 1951 [3]. Since then several methods of preparation have been described [2, 4]. In human tissues, 5′-nucleotidase is found (in descending order of concentration) in organs such as prostate, pituitary gland, thyroid, testicle, aorta, kidney, lung, liver etc. [2]. Subcellular distribution studies show the enzyme to be localized to some extent in the microsomal fraction of the liver and, in even higher activity, in the plasma membrane fraction that is rich in bile canaliculi [4]. From this source, 5′-nucleotidase can be solubilized by deoxycholate [5]. In addition, a sex-dependency of intracellular enzyme distribution has been shown in rat liver [6], female rats presenting more 5′-nucleotidase in the sinusoidal border while male rats exhibit greater activity near the bile ducts. Liver lysosomes contain 5′-nucleotidase in two different forms, one probably originating from the plasma membrane by pinocytosis [7].

The function of 5′-nucleotidase is not yet fully understood; speculations include a role in nucleic acid biosynthesis, in RNA degrading systems or in ATP metabolism [4]. In hepatobiliary disease, elevation of 5′-nucleotidase activity in serum is not due to *de novo* synthesis but to depletion of the plasma membrane 5′-nucleotidase in liver [8]. Considering that 5′-nucleotidase can be solubilized by detergents and bile salts, the high concentration of the latter in cholestatic liver may explain why 5′-nucleotidase activity increases in serum with an impaired bile flow [4].

Application of method: in clinical chemistry and biochemistry. Because of its substrate specificity, 5′-nucleotidase is of importance in nucleic acid analysis. Various publications describe the differential diagnostic value of the assay of 5′-nucleotidase activity in addition to the assay of non-specific alkaline phosphatase in patients' sera [4, 9–13].

Like γ-glutamyltransferase, determination of 5′-nucleotidase in serum may be helpful in cases of elevated activity of total alkaline phosphatase to differentiate between hepatobiliary disease and disease of other sources of alkaline phosphatase, such as bone. However, unlike γ-glutamyltransferase, 5′-nucleotidase is not so easily affected by moderately hepatotoxic substances such as ethanol, thus reflecting more truly severe hepatic disorders. Determination of 5′-nucleotidase is considered to be of particular clinical value in the detection of liver metastases [13].

Enzyme properties relevant in analysis: 5′-nucleotidase is a glycoprotein. Estimation of the molecular weight of 5′-nucleotidase revealed M_r 130000 for pig lymphocytes and M_r 140000 or 150000, respectively, for mouse liver [14]. In human placenta a subunit of 5′-nucleotidase with a molecular weight of 95000 was detected [15]. The pH optima and the *Michaelis-Menten* constants of 5′-nucleotidase depend on the source of the enzyme. *Hardonk* [16] demonstrated in rat and mouse tissues two different pH optima for 5′-nucleotidase of 5.0 and 7.0–7.5, respectively. At least 5 zones of 5′-nucleotidase activity were identified by electrophoresis [16]. 7 distinct forms were demonstrated in human placental preparations by isoelectric focusing [17]. With adenosine 5′-monophosphate as substrate, hydrolysis follows *Michaelis* kinetics up to 3 mmol/l [2]. In human serum, the pH optimum of 5′-nucleotidase is 7.5 [2, 4]. For adenosine 5′-monophosphate the K_m is 1.1×10^{-5} mol/l in the presence of Mn^{2+} [18].

Methods of determination: the catalytic activity can be determined, among other methods, by colorimetric measurement of the liberated phosphate [19], by methods based on radiolabelled substrates [20], or by so-called coupled methods using the product of a following reaction for the assessment of 5′-nucleotidase activity [21, 22]. Correction for alkaline phosphatase activity has been achieved by inhibition of 5′-nucleotidase by nickel ions [19] or by concanavalin A [23]; or by suppression of alkaline phosphatase by glycerol-2-P, phenylphosphate or L-cysteine, respectively [4, 18, 21]. Coupled methods are now available which use adenosine deaminase to liberate ammonia from adenosine formed by 5′-nucleotidase, thus allowing a continuous assessment of 5′-nucleotidase activity [21, 22].

International reference methods and standards: neither standardization at the international level nor the existence of reference standard materials is known up to now.

Enzyme effectors: 5′-nucleotidase is inhibited by anions (e.g. fluoride, borate, arsenate) and by metal ions (e.g. zinc, cobalt, copper, mercury, nickel) [2, 4]. Manganous ions activate at a concentration of 0.1 mol/l more than at 1 mol/l [24]. Concanavalin A inhibits 5′-nucleotidase activity [23]. Concentrations of adenosine 5′-monophosphate above 3 mmol/l cause an inhibition due to excess of substrate. An end-product inhibition is caused by inorganic phosphate or the dephosphorylated nucleosides, respectively [2, 4].

Assay

Method Design

Principle

(a) $$\text{Adenosine 5'-monophosphate} + H_2O \xrightarrow{\text{5'-nucleotidase}} \text{adenosine} + H_2PO_4$$

(b) $$\text{Adenosine} + H_2O \xrightarrow{\text{adenosine deaminase}} \text{inosine} + NH_4^+$$

(c) $$NH_4^+ + \text{NADH} + \text{2-oxoglutarate} \xrightarrow{\text{glutamate dehydrogenase}} \text{glutamate} + NAD^+ + H_2O$$

The dephosphorylation of the C_5 position of the ribose ring by isodynamic phosphoesterase in the same pH range must be taken into account. The various phosphatases can be distinguished by testing their behaviour towards different substrates (glycerol-2-P [25], phenylphosphate [2, 4]). While 5′-nucleotidase exclusively splits adenosine 5′-monophosphate, the non-specific enzyme activity is directed mainly towards the alternative substrate.

Adenosine formed by 5′-nucleotidase is then converted to inosine and ammonia by adenosine deaminase – adenosine aminohydrolase – (EC 3.5.4.4). The ammonia thus formed reacts with 2-oxoglutarate and NADH in a further coupled reaction catalyzed by glutamate dehydrogenase (EC 1.4.1.2).

The overall reaction can be followed continuously by measuring the decrease in absorbance at 339 nm caused by the conversion of NADH to NAD [21, 22].

To ensure a quantitative conversion of ammonia – i.e. the rate of disappearance of NH_4^+ (reaction c) must exceed that of production (reaction b) – optimized conditions using stabilized glutamate dehydrogenase must be followed [21, 26]. The method is also applicable to automated analyzers.

Optimized conditions for measurement: different pH optima of 5′-nucleotidase from different sources must be taken into account [2, 4]. Most authors have found the pH optimum of 5′-nucleotidase in human serum to be 7.5 [2, 4]. Adenosine 5-monophosphate has the highest rate of hydrolysis of the 5′-nucleotides tested [2]. In the following method L-leucine is used to stabilize glutamate dehydrogenase, and Mn^{2+} to optimize 5′-nucleotidase measurement [21].

Temperature conversion factors: 5′-nucleotidase activity is measured at 30 °C. Though temperature correction factors cannot be applied generally to an individual test, they may help to form an approximate comparison between results obtained at different temperatures.

°C	20	25	30	37
F	1.84	1.35	1.00	0.67

Equipment

A narrow band-width spectrophotometer suitable for measurements at 339 nm, preferably with a thermostatted cuvette holder. Cuvettes with 10 mm light path; water-bath at 30 °C; graduated timer; centrifuge. Pipettes: 0.01 ml, 0.1 ml, 0.2 ml, 10 ml.

Reagents and Solutions

Purity of reagents: the preparations used must be free from 5′-nucleotidase inhibitors (cf. pp. 107, 109, 115). Suitable commercial preparations are available.

Preparation of solutions (for about 5 determinations): all solutions in re-purified water (cf. Vol. II, chapter 2.1.3.2).

1. Buffer/substrate (TEA, 117.5 mmol/l; L-leucine, 23.5 mmol/l; 2-oxoglutarate, 5.87 mmol/l; glycerol 2-phosphate, 58.7 mmol/l; pH 7.6):

   dissolve in 900 ml water 21.82 g triethanolamine/HCl, 3.08 g L-leucine, 987 mg 2-oxoglutarate, sodium salt, 17.97 g glycerol-2-P-$Na_2 \cdot 5H_2O$, 0.198 g $MnSO_4 \cdot H_2O$, adjust to pH 7.6 with HCl, 0.1 mol/l, dilute to 1000 ml.

2. NADH (β-NADH, 41.1 mmol/l):

   dissolve 145.8 mg NADH, disodium salt in 5 ml water.

3. Adenosine deaminase, ADA (400 kU/l):

   use commercial preparation in 50% (v/v) glycerol, potassium phosphate, 10 mmol/l (200 U/mg, 2 mg/ml).

4. Glutamate dehydrogenase, GlDH (1200 kU/l):

   use commercial preparation in 50% (v/v) glycerol (120 U/mg, 10 mg/ml).

5. Adenosine 5′-monophosphate, AMP (23 mmol/l):

   dissolve 9 mg AMP-Na_2 in 1 ml water.

6. Working solution:

   mix 10 ml solution (1), 0.1 ml solution (2), 0.01 ml solution (3) and 0.1 ml solution (4).

Stability of solutions: solutions (1) and (2) are stable at 4 °C for one week. Solution (5) should be prepared daily. The working solution is stable for about 4 h at room temperature.

Procedure

Collection and treatment of specimen: collect blood by venous or arterial puncture. Centrifuge for 10 min at about 3000 *g* in order to obtain serum. Haemolysis should be avoided.

Stability of the enzyme in the sample: haemolysis-free serum may be stored for at least one week at 0–5 °C. Under these conditions, loss of activity after 3 to 6 months ranges between 4 and 8%. At −80 °C, serum 5′-nucleotidase is reported to be stable up to 6 months [18, 27].

Details for measurement in other specimens: since 5′-nucleotidase is a mainly membrane-bound enzyme, total activity in tissue is only extracted quantitatively after complete homogenization. Because of the great variation in pH optima and substrate affinity of 5′-nucleotidase from different sources, these parameters should be determined in preliminary experiments.

Assay conditions: wavelength 339 nm; light path 10 mm; final volume 2.3 ml; 30 °C (thermostatted cuvette holder). Measure against air.

Measurement

Pipette successively into the cuvette:			concentration in the assay mixture	
working solution	(6)	2.0 ml	TEA glycerol-2-P L-leucine 2-oxoglutarate Mn^{2+} NADH GlDH ADA	100 mmol/l 50 mmol/l 20 mmol/l 5 mmol/l 1 mmol/l 0.35 mmol/l 10 kU/l 0.4 kU/l
serum		0.2 ml	volume fraction	0.087
mix thoroughly with plastic spatula and incubate at 30 °C for 30 min,				
AMP solution	(5)	0.1 ml	AMP	1 mmol/l
mix, measure $\Delta A/\Delta t$ after 2 min of temperature equilibration.				

Calculation: use eqn. (k) or (k_1) from "Formulae", Appendix 3. Values for ε are given in Appendix 4. For calculation of the catalytic concentration of the sample the following relationships are valid (t in minutes):

$$b = 1825 \times \Delta A/\Delta t \quad \text{U/l}$$

$$b = 3.04 \times 10^4 \times \Delta A/\Delta t \quad \text{nkat/l}$$

Validation of Method

Precision, accuracy, detection limits and sensitivity: frozen aliquots of pooled sera or commercially available quality control material of known activity should be assayed concurrently to ensure proper performance of the procedure. The method described showed a close correlation with other established procedures for assay of 5′-nucleotidase. The between-run imprecision in the normal range is reported to be 6.5% [21]. With values around 60 U/l the day-to-day coefficient of variation is 4.7%. The lowest practical value to be measured by the method is about 4 U/l.

Sources of error: despite the fact that abnormally high levels of ammonia will consume some of the NADH and thereby reduce the maximum activity measurable, when the ammonia concentration does not exceed 250 μmol/l, normal or moderately elevated activity of 5′-nucleotidase can be assayed correctly. Provided serum ammonia is normal, the maximum activity of 5′-nucleotidase that may be measured correctly by this procedure is about 120 U/l. In sera containing excessive amounts of ammonia, the latter may be removed using a cation-exchange resin without effect on 5′-nucleotidase [21, 28].

Specificity: it should be borne in mind that not only 5′-nucleotidase but also unspecific phosphatases are capable of splitting the substrate used for the assay. The unspecific degradation is minimized by adding glycerol-2-P as an alternative substrate [25].

Reference ranges: with the method described, the activity of 5′-nucleotidase in serum of normal adults is 2 – 10 U/l [21]. With other methods, similar results were achieved (for review cf. [23]). *Belfield & Goldberg* demonstrated an age-dependency of the upper limit of normal, being low up to the age of 20 and doubling with advancing age [29].

References

[1] *M. J. Reis,* La nucléotidase et sa relation avec la désamination des nucléotides dans le coeur et dans le muscle, Bull. Soc. Chim. biol. *16,* 385 – 399 (1934).

[2] *O. Bodansky, M. Schwartz,* 5′-Nucleotidase, Adv. Clin. Chem. *11,* 277 – 328 (1968).

[3] *L. Heppel, R. Hilmoe,* Purification and Properties of 5′-Nucleotidase, J. Biol. Chem. *188,* 665 – 676 (1951).

[4] *D. M. Goldberg,* Biochemical and Clinical Aspects of 5′-Nucleotidase in Gastroenterology, Front. Gastrointest. Res., Vol. *2,* 71 – 108 (1976).

[5] *K. Konopka, M. Gross-Bellard, W. Turski,* The Influence of Detergents on the Activity of 5′-Nucleotidase in Rat Liver Homogenates, Enzyme *13,* 269 – 277 (1972).

[6] *M. Hardonk, A. Jonkers,* 5′-Nucleotidase: VI. Sex Differences in the Localization of Rat Liver 5′-Nucleotidase Activity, Histochemie *32,* 363 – 368 (1972).

[7] *Q. Pletsch, J. Coffey,* Studies on 5′-Nucleotidase of Rat Liver, Biochim. Biophys. Acta *276,* 192 – 205 (1972).

[8] *A. Righetti, M. Kaplan,* Disparate Responses of Serum and Hepatic Alkaline Phosphatase and 5′-Nucleotidase to Bile Duct Obstruction in the Rat, Gastroenterology *62,* 1034 – 1039 (1972).

[9] *U. Gerlach, W. Hiby, L. Paul,* Über den diagnostischen Wert der Bestimmung von Isoenzymen der alkalischen Phosphatase im Serum, Med. Welt *21* (N.F.), 1252 – 1258 (1970).

[10] *N. van Husen, G. Eberhardt, U. Gerlach,* Differentialdiagnostischer Wert der Bestimmung von Gallengangsphosphatase im Serum, Med. Welt *24* (N.F.), 1510 – 1511 (1973).

[11] *N. van Husen, U. Gerlach,* Über die Aktivitätserhöhung der unspezifisch lipoproteidgebundenen alkalischen Phosphatase im menschlichen Serum, Clin. Chim. Acta *53,* 91 – 99 (1974).

[12] *G. Ellis, D. Goldberg, R. Spooner, A. Milford Ward,* Serum Enzyme Tests in Diseases of the Liver and the Biliary Tree, Am. J. Clin. Pathol. *70,* 248 – 258 (1978).

[13] *T. Pollock, J. Mullen, K. Tsou, K. Lo, E. Rosato,* Serum 5′-Nucleotidase as a Predictor of Hepatic Metastases in Gastrointestinal Cancer, Am. J. Surg. *137*, 22 – 25 (1979).
[14] *M. el Kouni, S. Cha,* Isolation and Partial Characterization of a 5′-Nucleotidase Specific for Orotidine-5′-monophosphate, Proc. Natl. Acad. Sci. *79*, 1037 – 1041 (1982).
[15] *H. Maguire, T. Krishnakantha,* Human Placental 5′-Nucleotidase: Purification and Properties, J. Clin. Chem. Clin. Biochem. *17*, 424 (1979).
[16] *M. Hardonk, H. Boer,* 5′-Nucleotidase. III. Determination of 5′-Nucleotidase Isoenzymes in Tissues of Rat and Mouse, Histochemie *12*, 29 – 41 (1968).
[17] *I. Fox,* Human 5′-Nucleotidase: Properties and Regulation, J. Clin. Chem. Clin. Biochem. *14*, 289 (1976).
[18] *J. Beckmann, K. Leybold, L. Weisbecker,* Zur Bestimmung der 5′-Nucleotidase im Serum, J. Clin. Chem. Clin. Biochem. *7*, 18 – 24 (1969).
[19] *D. Campbell,* Determination of 5′-Nucleotidase in Blood Serum, Biochem. J. *82*, 34p (1962).
[20] *M. Gentry, R. Olsson,* A Simple, Specific Radioisotopic Assay for 5′-Nucleotidase, Anal. Biochem. *64*, 624 – 627 (1975).
[21] *C. Arkesteijn,* A Kinetic Method for Serum 5′-Nucleotidase Using Stabilized Glutamate Dehydrogenase, J. Clin. Chem. Clin. Biochem. *14*, 155 – 158 (1976).
[22] *G. Ellis, D. Goldberg,* An Improved Kinetic 5′-Nucleotidase Assay, Anal. Lett. *5*, 65 – 73 (1972).
[23] *E. Zygowicz, F. Sunderman, E. Horak, J. Dooley,* Inhibition by Concanavalin A as the Basis for a Specific Assay of Serum 5′-Nucleotidase Activity, Clin. Chem. *23*, 2311 – 2323 (1977).
[24] *G. Song, O. Bodansky,* Subcellular Localization and Properties of 5′-Nucleotidase in the Rat Liver, J. Biol. Chem. *242*, 694 – 699 (1967).
[25] *A. Belfield, D. Goldberg,* Inhibition of the Nucleotidase Effect of Alkaline Phosphatase by β-Glycerophosphate, Nature *219*, 73 – 76 (1968).
[26] *K. Jung, A. Sokolowski, E. Egger,* An Optimized Assay of Human Serum Glutamate Dehydrogenase Activity, Enzyme *14*, 44 – 54 (1972).
[27] *H. Bartels,* Ein Beitrag zur Bedeutung des Enzyms 5′-Nucleotidase in der Diagnostik innerer Krankheiten, Inaug. Dissertation, Münster 1968.
[28] *A. Ismail, D. Williams,* Scope and Limitations of a Kinetic Assay for Serum 5′-Nucleotidase Activity, Clin. Chim. Acta *55*, 211 – 216 (1974).
[29] *A. Belfield, D. Goldberg,* Normal Ranges and Diagnostic Value of Serum 5′-Nucleotidase and Alkaline Phosphatase Activities in Infancy, Arch. Dis. Child. *46*, 842 – 846 (1971).

1.8 Phosphoprotein Phosphatases

Phosphoprotein phosphohydrolase, EC 3.1.3.16

D. Grahame Hardie and Philip Cohen

General

Protein phosphatases are a heterogeneous group of enzymes of almost ubiquitous occurrence that are involved in reversing the actions of protein kinases (Vol. III, chapter 7.2). Protein phosphatases in the cytoplasm of mammalian tissues have been classified into three types [1 – 3]:

a) a phosphatase termed protein phosphatase 1, which dephosphorylates the β-subunit of phosphorylase kinase and is inactivated by two thermostable proteins, inhibitor 1 and inhibitor 2,

b) phosphatases which dephosphorylate the α-subunit of phosphorylase kinase and are insensitive to inhibitors 1 and 2 (protein phosphatases 2A, 2B, and 2C),

c) phosphatases which only dephosphorylate tyrosine residues. While the methods described in this chapter are also applicable to these protein phosphatases [3], they are not considered further here.

Protein phosphatases also exist in other subcellular fractions and organisms but with the exception of pyruvate dehydrogenase phosphatase in mammalian mitochondria [4, 5] and isocitrate dehydrogenase phosphatase in *E. coli* [6] they have not been purified to homogeneity.

Application of method: in biochemistry.

Enzyme properties relevant in analysis: in general, protein phosphatases are not specific for one substrate, and the dephosphorylation of any protein may be catalyzed by more than one enzyme in tissue extracts. This possibility can be examined by the use of specific inhibitors of type 1 and type 2 protein phosphatases as detailed below. The assay proceeds without a lag phase and is linear with time as long as the extent of dephosphorylation of the protein substrate is low ($< 10-20\%$).

Methods of determination: the catalytic activity is measured in a fixed-time assay by analyzing the release of acid-soluble ^{32}P-radioactivity from ^{32}P-labelled protein substrate. Dephosphorylation can also be measured by a change in function of the protein substrate. However, these procedures are, in general, less sensitive and more laborious than the radioactive assay.

International reference methods and standards: none known.

Enzyme effectors: protein phosphatase 1 is inhibited specifically by inhibitors 1 and 2 [1]. Protein phosphatase 2B has an absolute requirement for calcium ions, is stimulated 8–10 fold by the Ca^{2+}-binding protein calmodulin and completely inhibited by the phenathiazine drug, trifluoperazine [7, 8]. Protein phosphatase 2C has an absolute requirement for magnesium ions [1].

Assay

Method Design

Principle

$$[^{32}P]\text{Phosphoprotein} + n\,H_2O \longrightarrow \text{protein} + n\,[^{32}P]\text{phosphate}$$

The reaction is stopped by adding trichloroacetic acid to inactivate the protein phosphatase and precipitate the $[^{32}P]$protein. The acid-soluble radioactivity is then determined by scintillation counting. A reaction blank is included from which protein phosphatase is omitted. For the assay of type 1 and type 2 protein phosphatases, [^{32}P]phosphorylase *a* or [^{32}P]phosphorylase kinase are generally used as substrates.

Optimized conditions for measurement: maximal activities are generally observed at pH 7.0 – 7.5. Ideally, saturating concentrations of protein substrates should be used (~10 – 20 μmol/l), but this is rarely feasible.

Temperature conversion factors: must be established empirically in each case.

Equipment

Water-bath, stopwatch, micro-centrifuge, scintillation counter, spectrophotometer (for measuring protein substrate concentrations).

Reagents and Solutions

Purity of reagents: the protein substrate must not contain other [^{32}P]proteins, and be essentially free from contaminating protein phosphatases and [^{32}P]phosphate.

Preparation of solutions (for 50 assays): all solutions in re-purified water (cf. Vol. II, chapter 2.1.3.2).

1. Buffer solution (Tris 50 mmol/l, pH 7.0; EDTA 0.1 mmol/l; 2-mercaptoethanol 0.1% v/v):

   dissolve 6 g Tris base and 37 mg EDTA-$Na_2H_2 \cdot 2\,H_2O$ in about 500 ml water, add 1.0 ml 2-mercaptoethanol, adjust to pH 7.0 with HCl, 2 mol/l, and dilute to 1 l with water. The temperature of the solution should be 25 °C when the pH is adjusted.

2. Diluent for protein phosphatases (Tris, 50 mmol/l; EDTA, 0.1 mmol/l; 2-mercaptoethanol, 0.1% (v/v); bovine serum albumin, 1.0 mg/ml; Brij* 35, 0.01% (w/v)):

   add to 10 ml solution (1) 10 mg bovine serum albumin and 1.0 mg Brij 35.

* Abbreviations:
Brij polyoxyethylene lauryl ether
EGTA ethyleneglycol-bis-(2-aminoethyl ether) N,N′-tetraacetic acid

3. [^{32}P]Phosphoprotein ($10^7 - 10^8$ cpm/μmol):

   a) dissolve 3 mg [^{32}P]phosphorylase *a* in 1 ml solution (1), corresp. to 30 μmol/l. Phosphorylase is phosphorylated to 1.0 mol/mol using phosphorylase kinase and [γ-^{32}P]ATP and [^{32}P]phosphorylase *a* prepared as described in [9, 10].

   b) dissolve 2.4 mg [^{32}P]phosphorylase kinase in 1 ml solution (1), corresp. to 7.2 μmol/l. Phosphorylase kinase is phosphorylated to 2 mol/mol using cyclic AMP-dependent protein kinase and [γ-^{32}P]ATP and contains approximately equal amounts of phosphate in the α and β-subunits. The [^{32}P]phosphorylase kinase is prepared as described in [10].

4. Trichloroacetic acid (20%, w/v):

   dissolve 20 g trichloroacetic acid in 50 ml water and dilute to 100 ml.

Stability of solutions: store all solutions at 4°C. The stability of solution (3) will depend on the particular substrate used. If traces of contaminating protein phosphatases are present, the substrate will be slowly dephosphorylated on storage. The [^{32}P]-radioactivity in solution (3) will decay with a half-life of 14 days.

Procedure

Collection and treatment of specimen: the sample can be a purified enzyme or a tissue extract.

Stability of the protein phosphatase in the specimen: this depends on the protein phosphatase under study. The phosphatase should be diluted in solution (3) immediately before use.

Details for measurement in tissues: measurement of protein phosphatase activities in tissue extracts requires attention to the following two problems:

a) the extract is likely to contain more than one protein phosphatase capable of dephosphorylating any particular substrate (see below under "Specificity"),

b) extracts contain both high and low molecular weight substances which are potent inhibitors of protein phosphatases. It is therefore essential to filter the extracts through small columns of Sephadex G50 gel and to assay at as high a dilution as possible [11]. This is facilitated if the specific ^{32}P-radioactivity of the protein substrate is very high ($> 10^8$ cpm/μmol).

Assay conditions: final volume 60 μl; 30°C. The reaction is carried out in a 1.5 ml polypropylene micro-centrifuge tube. Pre-warm assay components to the reaction temperature.

Measurement

Pipette successively into the reaction tube:		sample	blank	concentration in assay mixture	
buffer solution	(1)	20 µl	20 µl	Tris EDTA	50 mmol/l 0.1 mmol/l
diluent solution	(2)	–	20 µl	2-mercaptoethanol	0.1%
sample (in diluent)		20 µl	–	Brij 35 BSA	0.003% 0.3 mg/ml
vortex and pre-incubate for 5 min at 30°,					
[^{32}P]phosphoprotein	(3)	20 µl	20 µl	phosphorylase *a* or phosphorylase kinase	1 mg/ml 0.8 mg/ml
vortex; after a suitable time stop the reaction,					
trichloroacetic acid	(4)	60 µl	60 µl		
vortex; reaction is stopped.					

After standing for 5 min to ensure complete precipitation of [^{32}P]-protein centrifuge the suspension for 2 min in a micro-centrifuge. Transfer a 0.1 ml aliquot of the supernatant to a second micro-centrifuge tube, and add 1.0 ml of an appropriate scintillation fluid. Finally, insert the tube into a scintillation vial and count using a ^{32}P-programme.

Alternatively, the sample can be analyzed by *Cerenkov* counting without added scintillant using a tritium programme. The efficiency of *Cerenkov* counting is approximately half that of standard scintillation counting.

Calculation: the catalytic activity concentration of the sample is

$$b = \frac{C_{\text{sample}} - C_{\text{blank}}}{X \times \Delta t} \times \frac{120}{100} \times 5 \times 10 \quad \text{U/l}$$

in which

C is radioactivity in cpm

X is specific radioactivity of the [^{32}P]protein substrate in cpm/nmol, measured under the same conditions

Δt is incubation time in min

Validation of Method

Precision, accuracy, detection limits and sensitivity: the imprecision of the assay is ± 5% and 0.001 U is easily detectable using a reaction time of 30 min.

Sources of error: the assay is based on the assumption that all the ^{32}P-radioactivity released from the [^{32}P]protein substrate is phosphate. Proteolytic enzymes can release acid-soluble [^{32}P]peptides from [^{32}P]proteins, and this possibility should be eliminated by adding acid ammonium molybdate and extracting into isobutanol toluene (1 : 1), which specifically extracts phosphate-molybdate complexes.

The assays frequently become non-linear when more than 10 – 20% of the [^{32}P]protein is dephosphorylated.

Specificity: the assay described above will measure both protein phosphatase 1 or protein phosphatase 2A, which dephosphorylate both phosphorylase *a* and phosphorylase kinase [1]. However, it is possible to quantitate protein phosphatases 1 and 2A, even in solutions containing a mixture of the two enzymes, by carrying out a further assay in which inhibitor 2 is included at a concentration sufficient to completely inhibit protein phosphatase 1 [11, 12].

Protein phosphatases 2B and 2C are inactive towards phosphorylase *a* and must therefore be measured by the dephosphorylation of other substrates, e.g. phosphorylase kinase [1, 2]. Inhibitor 2 is normally included in the assays to inactivate any contaminating protein phosphatase 1. To estimate protein phosphatase 2B, EDTA is removed from solution (1) and replaced by EGTA/$CaCl_2$, 1.0 mmol/l and 0.85 mmol/l resp. (pH 7.0) to give a free Ca^{2+}-concentration of 3 µmol/l [10, 11]. Calmodulin, 10^{-7} mol/l, is also added to solution 1. The activity is the value obtained in the presence of trifluoperazine, 100 µmol/l, subtracted from the value obtained in the absence of the drug.

In order to estimate protein phosphatase 2C, magnesium chloride, 30 mmol/l, is added to solution (1). The activity is the value obtained in the presence of EDTA, 0.1 mmol/l, subtracted from that measured in the presence of Mg^{2+} [11].

In tissue extracts containing all four phosphatases (1, 2A, 2B and 2C) the following modification is required to measure 2B and 2C [11]:

1. The extracts are incubated for 30 min at 30 °C with EDTA, 10 mmol/l, and NaF, 50 mmol/l. This treatment, which does not affect 2C, is carried out to inactivate protein phosphatase 2A, which would otherwise interfere with the assay. Protein phosphatase 2C is then assayed using [^{32}P]phosphorylase kinase as substrate, after gel filtration in solution (1) to remove EDTA and NaF [11].

2. Protein phosphatase 2B appears to be destroyed very rapidly by Ca^{2+}-dependent proteinases present in tissue extracts [10]. In order not to underestimate the activity, the reaction is therefore initiated with the phosphatase rather than [^{32}P]protein substrate, to avoid pre-incubating the sample with Ca^{2+} prior to the assay. The assay period is reduced to only one minute.

Some of the protein phosphatase 1 in tissue extracts may be present in an inactive form, termed the Mg-ATP-dependent protein phosphatase [13]. Similarly, some of the protein phosphatase 2A also exists as an inactive species, termed protein phosphatase $2A_0$ [14]. Measurements of protein phosphatases 1 and 2A in tissue extracts may therefore underestimate the total potential activities of these enzymes.

Protein phosphatases 1 and 2A are gradually converted during purification and storage to forms that are partly or completely dependent on Mn^{2+} for activity [11]. This can be examined by carrying out assays in the presence of $MnCl_2$, 1 mmol/l. Mn^{2+}, 1 mmol/l, can also substitute for Ca^{2+} and Mg^{2+} in the activation of protein phosphatases 2B and 2C respectively [11, 14].

Reference ranges: the concentrations of protein phosphatases 1, 2A, 2B and 2C in rat and rabbit tissues are given in [11].

References

[1] *T. S. Ingebritsen, P. Cohen,* The Protein Phosphatases Involved in Cellular Regulation 1. Classification and Substrate Specificities. Eur. J. Biochem. *132,* 255 – 261 (1983).

[2] *T. S. Ingebritsen, P. Cohen,* Classification of Protein Phosphatases Involved in Cellular Regulation, Science *221,* 331 – 338 (1983).

[3] *J. G. Foulkes, E. Erikson, R. L. Erikson,* Separation of Multiple Phosphotyrosyl- and Phosphoseryl-protein Phosphatases from Chicken Brain, J. Biol. Chem. *258,* 431 – 438 (1983).

[4] *W. M. Teague, F. H. Pettit, T. L. Wu, S. R. Silberman, L. J. Reed,* Purification and Properties of Pyruvate Dehydrogenase Phosphatase from Bovine Heart and Kidney, Biochemistry *21,* 5585 – 5592 (1982).

[5] *M. L. Pratt, J. F. Maher, T. E. Roche,* Purification of Bovine Kidney and Heart Pyruvate Dehydrogenase Phosphatase on Sepharose Derivatized with the Pyruvate Dehydrogenase Complex, Eur. J. Biochem. *125,* 349 – 355 (1982).

[6] *D. C. LaPorte, D. E. Koshland,* A Protein with Kinase and Phosphatase Activities Involved in the Regulation of the Tricarboxylic Acid Cycle, Nature *300,* 458 – 460 (1982).

[7] *A. A. Stewart, T. S. Ingebritsen, A. Manalan, C. B. Klee, P. Cohen,* Discovery of a Ca^{2+} and Calmodulin-dependent Protein Phosphatase, Probable Identity with Calcineurin (CaM-BP_{80}). FEBS Lett. *137,* 80 – 84 (1982).

[8] *A. A. Stewart, T. S. Ingebritsen, P. Cohen,* The Protein Phosphatases Involved in Cellular Regulation 5. Purification and Properties of a Calcium Ion and Calmodulin Dependent Protein Phosphatase (2B) from Rabbit Skeletal Muscle, Eur. J. Biochem. *132,* 289 – 295 (1983).

[9] *J. F. Antoniw, H. G. Nimmo, S. J. Yeaman, P. Cohen,* Comparison of the Substrate Specificities of Protein Phosphatases Involved in the Regulation of Glycogen Metabolism in Rabbit Skeletal Muscle, Biochem. J. *162,* 423 – 433 (1977).

[10] *A. A. Stewart, B. A. Hemmings, P. Cohen, J. Goris, W. Merlevede,* The MgATP-dependent Protein Phosphatase and Protein Phosphatase-1 have Identical Substrate Specificities, Eur. J. Biochem. *115,* 197 – 205 (1981).

[11] *T. S. Ingebritsen, A. A. Stewart, P. Cohen,* The Protein Phosphatases Involved in Cellular Regulation 6. Measurement of Type-1 and Type-2 Protein Phosphatases in Extracts of Mammalian Tissues; an Assessment of their Physiological Roles, Eur. J. Biochem. 32, 297 – 307 (1983).

[12] *J. G. Foulkes, P. Cohen,* The Regulation of Glycogen Metabolism. Purification and Properties of Protein Phosphatase Inhibitor-2 from Rabbit Skeletal Muscle, Eur. J. Biochem. *105,* 195 – 203 (1980).

[13] *B. A. Hemmings, T. J. Resink, P. Cohen,* Reconstitution of a Mg-ATP-dependent Protein Phosphatase and its Activation through a Phosphorylation Mechanism, FEBS Lett. *150,* 319 – 324 (1982).

[14] *T. S. Ingebritsen, J. G. Foulkes, P. Cohen,* The Protein Phosphatases Involved in Cellular Regulation 2. Glycogen Metabolism, Eur. J. Biochem. *132,* 263 – 274 (1983).

1.9 Cyclic Nucleotide Phosphodiesterase (PDE)

3′: 5′-Cyclic-nucleotide 5′-nucleotidohydrolase, EC 3.1.4.17

W. Joseph Thompson and Samuel J. Strada

General

The hydrolysis of cyclic AMP and cyclic GMP was initially studied in the laboratories of *Sutherland* [1, 2] and *Appleman* [3, 4], respectively. Cyclic AMP and cyclic GMP phosphodiesterases are a family of enzyme forms that occur in various proportions in all eukaryotic and prokaryotic cells. The two best characterized forms hydrolyze both A-3: 5-MP and G-3: 5-MP with preference for the latter: one form is activated by calcium-calmodulin and the other by G-3: 5-MP. Both are localized principally in cytosolic subcellular fractions, but may also be associated with membranes. Other important forms include a predominantly membrane-bound enzyme, specific for A-3: 5-MP, that derives its biological importance from its regulation by and responsiveness to drugs and hormones, and a unique membrane-bound form, which preferentially hydrolyzes G-3: 5-MP. The latter is located in the rod outer segments and thought to be intimately involved in the transduction of light in the retina. Current research suggests, but has not yet proved, that each of these distinct enzymes is a separate gene product. Despite the physical and biological complexities of the PDE system, catalysis of A-3: 5-MP or G-3: 5-MP hydrolysis by any of the enzyme forms can be measured by relatively simple, reproducible and inexpensive methods [4, 6, 7]. A comprehensive monograph describing many aspects of the biochemistry, physiology and pharmacology of the PDE system has recently been published [5].

Application of method: research in biochemistry and pharmacology. Applications to use in clinical laboratories are certainly possible, but are not yet routinely used.

Enzyme properties relevant in analysis: catalytic activities often show complex kinetic properties. Negative and positive co-operativity and allosteric regulation and hysteresis can be seen with both substrates. Measurement of catalysis in crude systems, i.e., systems in which it is not possible to assess the contributions of different enzyme forms, is of limited value unless the cellular system is such that interference by multiple forms is unlikely. Extreme caution must be employed to establish reaction linearity with respect to time and enzyme concentration.

Methods of determination: the catalytic activity of all enzyme forms can be measured by modifications of the original method of *Thompson & Appleman* [4, 6, 7] or by a wide variety of other procedures, each with its own assets and liabilities [7]. Estimates of enzyme mass are not routinely employed but a radioimmunoassay procedure for one enzyme form has been described [8]. The routine and most widely-used procedures to isolate and separate enzyme forms have been reviewed in detail [7].

International reference methods and standards: no standard international unit has been established. Investigators usually rely on enzyme activity values measured at substrate concentrations well above and/or below the K_m of the purified enzyme forms for comparative purposes.

Enzyme effectors: distinct modulators of various PDE include a wide variety of hormones and drugs [5]. Most hormonal responses are observed only in intact cell systems. Pharmacological agents affecting both intact cells and cell-free systems include the methylxanthine and phenothiazine classes of drugs. Many heat-stable and heat-labile peptides or protein activators and inhibitors have been described [5]. The best-characterized effector is the calcium-binding protein, calmodulin. Substrates themselves, e.g. G-3 : 5-MP, can activate A-3 : 5-MP hydrolysis. Calcium ions can affect the enzyme indirectly by binding to calmodulin and thereby influencing its conformation.

Assay

Method Design

Principle

$$(a_1) \quad [^3H]A\text{-}3:5\text{-}MP + H_2O \xrightarrow{PDE} [^3H]AMP$$

$$(a_2) \quad [^3H]G\text{-}3:5\text{-}MP + H_2O \xrightarrow{PDE} [^3H]GMP$$

$$(b_1) \quad [^3H]AMP + H_2O \xrightarrow{5'\text{-nucleotidase}} [^3H]\text{adenosine} + P_i$$

$$(b_2) \quad [^3H]GMP + H_2O \xrightarrow{5'\text{-nucleotidase}} [^3H]\text{guanosine} + P_i\,.$$

Two steps are involved in the radioactive assay of hydrolysis catalyzed by PDE. Activity is measured by complete enzymatic conversion of the reaction products (AMP or GMP) to adenosine or guanosine, respectively. The resulting nucleosides are separated from unreacted substrate by precipitation with an anion-exchange resin or by filtration over the resin in a small column. Both reagent blanks and boiled-tissue blanks should be performed routinely. These should constitute less than 0.4 to 1% of the initiating radioactivity.

Optimized conditions for measurement: PDE(s) have fairly broad pH and temperature optima. pH 8.0 (Tris) is usually used for A-3 : 5-MP hydrolysis and pH 7.4 (TES, Tris or MES) for G-3:5-MP hydrolysis. Some of the enzyme forms are active even below 0°C; it is therefore useful to initiate the reaction by the addition of enzyme protein. With crude or partially purified preparations it is advisable to assay at 30°C rather than 37°C. This minimizes AMP- or GMP-deamination and myokinase-type reactions, and obviates unnecessary ^3H exchange with H_2O which could artefactually raise blank values. To maintain a constant reaction temperature it is advisable to pre-incubate assay tubes for 30 sec prior to the addition of enzyme. Catalysis requires a divalent cation. Magnesium is preferred, but manganese can achieve 50% of the maximal rate. Calcium and several other divalent cations substitute poorly for magnesium. The addition of a sulphydryl reagent (DTT or β-mercaptoethanol) improves both the stability and rate of the reaction for most PDE forms.

Equipment

A good, accurately controlled shaking water-bath. An apparatus is needed to suspend multiple small borosilicate glass columns such as *Pasteur* pipettes (14.6 cm × 0.4 cm) above scintillation vials. We have previously published the specifications for a suitable apparatus, but any standard platform with holes could be used to hold the pipettes.

Reagents and Solutions

Purity of reagents: ^3H-labelled substrates obtained from commercial vendors must be purified, especially when exacting, detailed kinetic determinations are to be made and assay blank values must be as low as possible.

Prepare a *Pasteur* pipette column with a funnel or conical top using 2 ml of a 1:4 slurry of Dowex 1-X2 (4 – 6°) and wash with 5 ml cold water. Place 3 ml water on top of the column and add [^3H]A-3:5-MP or [^3H]G-3:5-MP (1 – 5 mCi). Rinse vial with 1 ml water and add to column. Wash with 20 ml water and 5 ml HCl, 0.01 mol/l, 50% methanol and elute [^3H]G-3:5-MP with 5 ml HCl, 0.5 mol/l, 50% methanol.

Unlabelled cyclic nucleotides should also be purified analogously to the method described above and stored frozen in bacteria-free solutions. Their concentrations are determined spectrophotometrically using absorption coefficients of 1.465 l × $mmol^{-1}$ × mm^{-1} (for A-3:5-MP) and 1.37 l × $mmol^{-1}$ × mm^{-1} for (G-3:5-MP) at 259 nm (pH 7.0). These stock solutions should be adjusted to 1 mmol/l for use as additives in the assay.

Buffers must be of "enzyme" grade.

Preparation of solutions (for about 100 determinations): all solutions are in glass-distilled water and filtered over 0.22 micron filters (cf. also Vol. II, chapter 2.1.3.2).

1a. [^{3}H] A-3 : 5-MP solution (ca. 28.0 Ci/mmol; ca. 100000 cpm/100 μl):

dilute the purified [^{3}H]A-3 : 5-MP above with assay buffer (6) to achieve the indicated radioactivity. The trace quantity will be significant relative to the unlabelled substrate in some situations (e.g. kinetics) and will be the only substrate used in other situations (e.g. limited sample) to achieve the required sensitivity.

1b. [^{3}H] G-3 : 5-MP solution (ca. 8.0 Ci/mmol; ca. 100000 cpm/100 μl):

dilute the purified [^{3}H]G-3 : 5-MP above with assay buffer (6) to achieve the indicated radioactivity. The trace quantity will be significant relative to the unlabelled substrate in some situations (e.g. kinetics) and will be the only substrate used in other situations (e.g. limited sample) to achieve the required sensitivity.

2a. A-3 : 5-MP solution (1 μmol/l):

for routine assays dilute the stock solution above which is determined to be 1 mmol/l by spectral analysis with assay buffer (6) to achieve the indicated concentration. For enzyme kinetics multiple concentrations are diluted to achieve concentrations from 0.1 to 400 μmol/l.

2b. G-3 : 5-MP solution (4 μmol/l):

for routine assays dilute the stock solution above which is determined to be 1 mmol/l by spectral analysis with assay buffer (6) to achieve the indicated concentration. For enzyme kinetics multiple concentrations are diluted to achieve concentrations from 0.1 to 400 μmol/l.

3. 5'-Nucleotidase (1 g/l):

dissolve 12 mg snake venom (*Ophiophagus hannah*) from *Sigma Chemical Co.* in 12 ml water. Other venoms should not be used for this type of assay unless carefully analyzed first for endogenous phosphodiesterase and proteolytic enzyme activities.

4. Tris buffer (2 mol/l; pH 8.0):

dissolve 242 g Tris-HCl in water, adjust pH to 8.0 with HCl, 2 mol/l, and make up to 1 litre.

5. $MgCl_2$ solution (1 mol/l):

dissolve 20.3 g MgCl · 6 H_2O in 100 ml water.

6. Assay buffer (Tris, 40 mmol/l; pH 8.0; $MgCl_2$, 10 mmol/l; 2-mercaptoethanol, 3.75 mmol/l, BSA 125 mg/l):

prepare the assay buffer daily by addition of 1 ml Tris buffer (4), 0.5 ml $MgCl_2$ solution (5) and 30 μl conc. 2-mercaptoethanol, and 6.25 mg bovine serum albumin in a volume of 50 ml.

7. Methanol (100% v/v).

8. Ion-exchange column:

   wash Dowex or A 61 – X2 or X8, 200 – 400 mesh, (*Bio-Rad Laboratories*) with HCl, 0.5 mol/l, NaOH, 0.5 mol/l, HCl, 0.5 mol/l, and then repeatedly with water to pH 5.0. The resin is then allowed to settle for at least 45 min. Add 3 volumes of water to 1 volume of settled resin. Prepare the pipette columns by transferring 1 ml of the resin slurry to each column.

9. Scintillation cocktail:

   use Aquasol (*NEN*) or equivalent.

Stability of solutions: all precautions with reference to ^{3}H-labelled compounds should be observed, especially with [^{3}H]cGMP, since [^{3}H]guanine compounds are inherently more unstable than [^{3}H]adenine compounds. Storage at –20° in 50% ethanol has proved satisfactory for stock solutions which generally need to be purified every 3 – 5 weeks for optimum results.

Radioactive substrate is diluted just before use. Solutions of A-3:5-MP (2a) and G-3:5-MP (2b) are stable for long periods but should be re-purified and standardized periodically. Solutions (4) and (5) are stored at 4°C and can be used for months if growth of micro-organisms is prevented. Assay buffer (6) is discarded after use. Snake venom solution (3) can be stored at 4°C for several weeks but can also be stored frozen indefinitely. Lyophilized venom powder should be stored at –20°C.

Procedure

Collection and treatment of specimen and sample: any tissue source can be used. The preparation is performed according to [7]. Attention must be paid to conditions which might modify activity, especially in crude samples. Potential mechanisms for altering activity values include proteolysis, covalent modification, sulphydryl oxidation, ionic strength, detergents and lipids, and sample dilution.

Stability of the enzyme in the specimen and sample: as a general rule one assumes that PDE(s) are unstable unless shown otherwise. Chelating agents (EDTA and EGTA), fluoride, protease inhibitors such as PMSF, TLCK and leupeptin, DTT, and ethylene glycol (30%) have been used in some preparations to enhance stability. In some tissues hydrolysis of A-3:5-MP and/or G-3:5-MP will markedly increase at 4°C within a few minutes after tissue homogenization. These post-homogenization changes can dramatically influence the results [9, 10].

Assay conditions: pre-incubation 30 s before initiating the reaction; final volume 0.5 ml. 12 × 75 mm glass tubes are useful reaction vessels, with pre-assay additions made on ice. Routine reaction times 5 – 10 min; 30°C; these must be determined for each system being studied. Run a blank with reagents alone or with boiled sample plus reagents.

Measurement

Incubation

Pipette into reaction vessel:			concentration in assay mixture	
$[^3H]$A-3:5-MP solution or $[^3H]$G-3:5-MP solution	(1a) (1b)	0.1 ml	100000 cpm	ca. 10 nmol/l
A-3:5-MP solution or G-3:5-MP solution	(2a) (2b)	0.1 ml	A-3:5-MP G-3:5-MP	0.25 mol/l 1.0 mol/l
assay buffer	(6)	0.1 ml	$MgCl_2$ BSA Tris 2-mercapto-ethanol	10 mmol/l 125 mg/l 40 mmol/l 3.75 mmol/l
mix and pre-incubate for 30 s				
sample		0.1 ml		
mix and incubate for 5 to 10 min; boil 45 s; let cool				
snake venom solution	(3)	0.1 ml		
mix and incubate 10 min at 30°C,				
methanol	(7)	1 ml		

Chromatography and measurement: mix and transfer the entire contents to an ion-exchange column (8) prepared in a *Pasteur* pipette (1 ml of 1:4 resin slurry in each, cf. pp. 129, 131). Elute the reaction mixture into a scintillation vial, add 1 ml methanol and elute columns to dryness. Determine the radioactivity of the combined eluate after the addition of 8 ml scintillation cocktail (9).

Calculation: correct counting rates of samples for blank. Reaction velocity is calculated according to the following equation (cf. "Formulae", Appendix 3):

$$z = \frac{C_{\text{measured}}}{C_{\text{maximum}}} \times \frac{c_{\text{substrate}}}{\Delta t} \quad \text{pmol/min}\,.$$

Velocity is expressed in picomoles per min per assay and consequently specific activity or molecular activity can be determined.

Validation of Method

Precision, accuracy, detection limits and sensitivity: the precision and accuracy of this method are limited only by pipetting ability. The small columns introduce an error of less than 1%. The reproducibility of the assay is limited by factors associated with protein dilution and liquid scintillation-counting errors. The theoretical limits of substrate concentration of this assay method are a function of the specific activity of the labelled cyclic nucleotide. The practical, convenient limit of the method as presented here is ca. 10 nmol/l for A-3 : 5-MP-dependent PDE and 70 nmol/l for G-3 : 5-MP-dependent PDE.

Source of error: the main source of error is the reported problem of non-specific binding of adenosine and guanosine to the Dowex resin. This has been discussed in detail previously [7]; the use of methanol in the assay with the small columns has virtually eliminated this problem. This methodological improvement has also alleviated problems with blank values that can occur with high ionic strength in the enzyme sample (e.g. fractions obtained by high-salt elution from ion-exchange cellulose) or due to magnesium ions. When inhibitors are tested, their effects on nucleotidase activity must be determined first. Also, any reaction that converts the AMP or GMP reaction products to a charged nucleotide will not elute through the resin and can appear as an inhibition (e.g. ATP + AMP $\rightleftarrows$ 2ADP). Conventional thin-layer chromatography or HPLC with Partasil 10 SAX columns [11] can be used to test these possibilities. Alternatively, ^{32}P-cyclic nucleotide substrates can be employed using the methods described by *Nakai & Brooker* [12].

Specificity: a remarkable feature of PDE(s) is their nearly absolute specificity for cyclic nucleotides as substrates. Interference from polynucleotides or mono-, di-, and trinucleotides is negligible.

Reference ranges: are not known, except the reported values and constants for the purified enzyme forms; these vary widely in intrinsic activity [5]. IC_{50} values are often reported for inhibitory drugs, but they have limited comparative value. K_i values are less frequently noted in the literature and should be reported whenever possible.

References

[1] *E. W. Sutherland, T. W. Rall,* Fractionation and Characterization of a Cyclic Adenine Ribonucleotide formed by Tissue Particles, J. Biol. Chem. *232*, 1077 – 1091 (1958).

[2] *R. W. Butcher, E. W. Sutherland,* Adenosine 3′,5′-phosphate in Biological Materials I., J. Biol. Chem. *237*, 1244 – 1250 (1962).

[3] *G. Brooker, L. Thomas, M. M. Appleman,* The Assay of Adenosine 3′5′-Cyclic Monophosphate and Guanosine 3′-5′-Cyclic Monophosphate in Biological Materials by Enzymatic Radioisotopic Displacement, Biochemistry *7*, 4177 – 4181 (1968).
[4] *W. J. Thompson, M. M. Appleman,* Multiple Cyclic Nucleotide Phosphodiesterase Activities in Rat Brain, Biochemistry *10*, 311 – 316 (1971).
[5] *W. J. Thompson, M. L. Pratt, S. J. Strada,* Cyclic Nucleotide Phosphodiesterase, in: *S. J. Strada, W. J. Thompson* (eds.), Advances in Cyclic Nucleotide Research, Vol. *16*, Raven Press, New York 1983.
[6] *W. J. Thompson, G. Brooker, M. M. Appleman,* Assay of Cyclic Nucleotide Phosphodiesterase with Radioactive Substrate, in: *S. P. Colowick, N. O. Kaplan* (eds.), Methods in Enzymology, Vol. *XXXVIII*, Academic Press, New York 1974, pp. 205 – 212.
[7] *W. J. Thompson, W. L. Terasaki, P. M. Epstein, S. J. Strada,* Assay of Cyclic Nucleotide Phosphodiesterase and Resolution of Multiple Molecular Forms of the Enzyme, Adv. Cyclic Nucleotide Res. *10*, 69 – 92 (1979).
[8] *K. Sarada, P. M. Epstein, S. J. Strada, W. J. Thompson,* Analysis of Cyclic Nucleotide Phosphodiesterase(s) by Radioimmunoassay, Arch. Biochem. Biophys. *215*, 183 – 198 (1982).
[9] *P. M. Epstein, W. J. Pledger, E. A. Gardner, G. M. Stancel, W. J. Thompson, S. J. Strada,* Activation of Mammalian Cyclic AMP Phosphodiesterases by Trypsin, Biochim. Biophys. Acta *527*, 442 – 455 (1978).
[10] *S. J. Strada, P. M. Epstein, E. A. Gardner, W. J. Thompson, G. M. Stancel,* Evidence for Convertible Forms of Soluble Uterine Cyclic Nucleotide Phosphodiesterase, Biochim. Biophys. Acta *661*, 12 – 20 (1981).
[11] *D. M. Watterson, B. B. Iverson, L. J. Van Eldik,* Rapid Separation and Quantitation of 3′,5′-Cyclic Nucleotide and 5′-Nucleotides in Phosphodiesterase Reaction Mixtures Using High-Performance Liquid Chromatography, Biochim. Biophys. Methods, *2*, 139 – 146 (1980).
[12] *C. Nakai, G. Brooker,* Assay for Adenylate Cyclase and Cyclic Nucleotide Phosphodiesterase and the Preparation of High Specific Activity ^{32}P-Labelled Substances, Biochim. Biophys. Acta *391*, 222 – 239 (1975).

1.10 Serum Ribonuclease

Ribonucleate 3′-pyrimidino-oligonucleotidohydrolase, EC 3.1.27.5

Timothy P. Corbishley, Philip J. Johnson and Roger Williams

General

Ribonucleases, enzymes which catalyze the depolymerization of RNA, are ubiquitous in their distribution throughout all living organisms. The most extensively studied form is that extracted from bovine pancreatic tissue (EC 3.1.4.22) but, like animal serum ribonucleases, this shows different properties from all forms so far studied in man [1]. Many human tissues, including brain, liver, spleen, kidney, skeletal muscle and leukocytes, show ribonuclease activity with differing substrate specificities and pH buffer-composition optima when assayed *in vitro* [2 – 5]. Normal serum activity is due predominantly to three forms of the enzyme, the source and fate of which are still unclear [6].

Serum ribonucleases also differ in their glycosylation, the extent of which varies from day to day in an individual [7]. Some racial differences in activities of this enzyme have been noted [8], but as there does not appear to be a high degree of genetic polymorphism in human serum ribonucleases [9] these differences may be due to environmental factors. Evidence is now accumulating that human ribonucleases are not all true "isoenzymes" but may be the products of more than one gene, falling into families of post-translationally modified forms [9].

Serum ribonuclease has been proposed as a marker for several human tumours, particularly pancreatic, ovarian and lymphatic cancer [10, 11], and may be involved in the development of neoplasia and in tumour regression. The enzyme has also been implicated in the control of cellular and humoral immunity [12, 13] and as an inhibitor of cellular and viral protein synthesis [14]. Its use in the monitoring of nutritional status, especially nitrogen retention in neonates and infants [15], has been described recently.

Application of method: in biochemical and clinical research.

Enzyme properties relevant in analysis: several isoenzymes have been identified in human serum with different substrate specificities. In normal human serum, three isoenzymes are responsible for most of the catalytic activity; however, in pathological conditions it is possible that others may make a significant contribution. Under the correct assay conditions substrate conversion proceeds linearly with time.

Methods of determination: both colorimetric and turbidimetric assays have been described for the determination of enzyme activity in serum, urine and tissue extracts. In recent years, however, most studies have been based on some form of spectrophotometric assay. The assay of "total" serum activity is performed with RNA as substrate, that of individual isoenzymes requiring synthetic polynucleotide substrates.

International reference method and standards: no standard international reference method presently exists. Bovine pancreatic ribonuclease (EC 3.1.4.22) has been used as a standard preparation by some authors, but as this form of the enzyme exhibits a different substrate specificity and kinetics from any form so far reported in human serum this seems unsatisfactory. However, for standard reference materials cf. Vol. II, chapter 2.3.

Enzyme effectors: several inhibitors have been reported in extracts of various tissues of animals and man [16]; although some of these inhibitors do affect serum forms of the enzyme *in vitro*, the presence of these tissue inhibitors in serum has so far not been described.

Heparin inhibits the activity of most ribonuclease isoenzymes; plasma prepared using this anticoagulant is therefore not suitable for activity determinations.

Polyamines such as spermidine and spermine increase the enzyme activity in low ionic strength buffers [17], and have also been shown to reverse the effects of the tissue inhibitors. There is some confusion in the literature about the effects of buffer composition on enzyme activity. Measurement is dependent on the buffer and substrate combination used. The terms "acid" and "alkaline" ribonuclease are confusing and provide little information about the particular isoenzyme or group of isoenzymes that are being studied, which are better distinguished by their different substrate specificities.

Metals such as copper and zinc in ionic form have been shown to influence the enzyme activity in tissue extracts and serum of animals fed on a metal-deficient diet [18]. However, in normally nourished adult humans it is unlikely that metal ions have any great effect on serum ribonuclease activities. The ratio of substrate concentration to serum dilution is critically important when measuring the activity of this enzyme. An inhibition effect which could be minimized, if not removed, by altering this ratio has been observed using both assay techniques described here [10].

1.10.1 'Total' Serum Ribonuclease

Assays

Method Design

Principle: the endonucleolytic cleavage of RNA or synthetic polynucleotide to produce acid-soluble 3′-phosphomono- and oligonucleotides, ending in cytidine or uridine 5′-phosphates, is monitored spectrophotometrically. The assay is performed by a fixed-time incubation method and both enzyme- and reagent blanks are required.

Optimized conditions for measurement: as mentioned above, the pH optimum for the assay is largely determined by buffer composition. Several methods for precipitation of unreacted substrate have been proposed, including the use of hydrochloric, perchloric or trichloroacetic acids, with or without the addition of uranium or lanthanum salts [19]. Methanol, ethanol, propanol or isopropyl alcohol with or without the addition of sodium chloride or various combinations of alcohols and acids have also been used for this purpose.

Whereas the neutral alcohol techniques do not provide complete precipitation of small polymers of chain length 6 to 30, acidic solutions of rare earths may be too efficient, leading to de-purination and high substrate blanks.

The technique described here utilizes a perchloric acid/lanthanum nitrate solution which, in our hands, gives reproducible results with low blanks.

As ribonuclease isoenzymes differ in their heat stability, this property has been used to a limited extent to distinguish between them. Most "isoenzymes" present in serum have been shown to be stable up to 60°C with the corresponding increase of activity with temperature. As conversion of the substrate proceeds at a convenient rate at 37°C under most assay conditions, there seems no reason to measure the enzyme activity at a temperature above physiological. The optimal time of incubation of the reaction mixture is chosen to retain linearity of conversion of substrate with time and may be reduced when measuring serum specimens of high activity if further dilution of the specimen is not desirable.

The "substrate inhibition effect" referred to previously is observed at low enzyme activities under certain conditions [10]. The substrate concentration/serum dilution ratios used in both assays described have been chosen to produce a linear change of absorbance with time over the whole detection range of the assay.

When RNA is used as substrate, the optimal absorption wavelength of the reaction products is 260 nm; with polycytidylic acid (Poly C) as substrate the best wavelength is 278 nm.

Temperature conversion factors: we are unaware of any published data on this subject.

Equipment

Glass and plastic incubation tubes have been used for this assay procedure. We find disposable polypropylene micro-centrifuge tubes most convenient. Where glassware is used either for assay tubes or storage of reagents, care must be taken to rinse away any detergent which may adversely affect enzyme activity.

A water-bath, stopwatch or alarm timer, micro-centrifuge and ultraviolet spectrophotometer, together with silica or other suitable cuvettes and a range of micropipettes suitable for accurately and precisely measuring volumes in the range 50 µl to 1 ml are also needed for these assays.

Reagents and Solutions

Purity of reagents: commercial preparations of yeast transfer RNA provide a substrate of suitable purity. Samples from different manufacturers do, however, show differing behaviour in the assay, as do different batches from the same manufacturer to a lesser extent. It is therefore a sensible precaution to buy enough substrate from a single batch to complete a study. All other reagents should be of analytical quality.

Preparation of solutions (for about 200 determinations): solutions are prepared using re-purified water (cf. Vol. II, chapter 2.1.3.2).

1. Phosphate buffer (0.1 mol/l, pH 8.0):

   dissolve 13.67 g Na_2HPO_4 and 0.577 g $NaH_2PO_4 \cdot 2\,H_2O$ in 990 ml water, adjust to pH 8.0 with hydrochloric acid or sodium hydroxide if necessary, and make up to 1 litre with water.

2. Substrate solution (tRNA, 10 µmol/l*):

   dissolve 5 mg yeast transfer RNA in 10 ml water.

3. Perchloric acid/lanthanum solution ($HClO_4$, 1.2 mol/l; La^{3+}, 0.022 mol/l):

   dissolve 0.86 g $La(NO_3)_3 \cdot 6\,H_2O$ in 80 ml water, add 7.1 ml of 72% (w/w) $HClO_4$ and make up to 100 ml with water.

Stability of solutions: store all solutions at 4 °C. Serum should be diluted freshly for each assay. If the substrate solution (2) is not prepared freshly every week the substrate blank starts to rise appreciably. The buffer (1) and perchloric acid solutions (3) are stable at 4 °C indefinitely.

Procedure

Collection and treatment of specimens: heparin interferes with ribonuclease activity determinations and although we are unaware of reports of the adverse effects of other anticoagulants we would recommend the use of serum rather than plasma. There are no reports of the use of a tourniquet, time of day, or time after the last meal having any influence on measured enzyme activities. The renal function of subjects under investigation is also important as elevated serum levels have been observed in uraemic patients [20].

Stability of the enzyme in the sample: there is a loss of activity in specimens stored frozen at either −20 °C or −70 °C, and measurement of activity on fresh specimens is preferable. Where this is not possible, samples which have only been frozen and thawed once, even if they have been in storage for many months, are preferable to those that have been repeatedly thawed and re-frozen.

Details of measurement in tissue: this method can be applied to tissue extracts. However, a very different pattern of isoenzymes is present in some tissues compared with that found in serum, which may require a different buffer and substrate condition for optimal activity measurement.

---

* assuming an approximate molecular weight of 50000.

Assay conditions: wavelength 260 nm; light path 10 mm; 37 °C. Incubation volume 250 µl, final volume of measurement 2.0 ml. Use serum samples diluted approximately 1 + 9 with water (further dilution may be necessary for samples of high activity); keep cold but pre-warm immediately before measurement. Perchloric acid/lanthanum solution is used ice-cold.

Run a reagent blank with water instead of serum, and a sample blank with water instead of substrate solution (2), simultanously.

Measurement

Pipette successively into a micro-centrifuge tube:			concentration in assay mixture	
buffer substrate solution	(1) (2)	150 µl 50 µl	phosphate RNA	0.06 mol/l 2 µmol/l
mix and allow to warm to 37 °C				
sample		50 µl	volume fraction	0.2
mix and incubate for 30 min				
perchloric acid/lanthanum solution	(3)	250 µl	$HClO_4$ La^{3+}	0.6 mol/l 0.011 mol/l
mix, let stand on ice for 20 min, spin at 10000 *g* for 10 min, take 400 µl aliquot of supernatant, dilute to 2 ml with water, read absorbance at 260 nm.				

If the measured absorbance is greater than 0.6 then the linear region of the change of absorbance with time plot has been exceeded and the sample must be re-assayed either at greater dilution or by incubating for a shorter period of time.

Calculation: when using RNA as substrate the most commonly used unit is that of absorbance units per ml serum per minute. Correct measured absorbance for blanks. The calculation formula for the catalytic concentration of the enzyme in the sample is:

$$b = (A_t - A_b) \times 1.67 \times F \quad \text{arbitr. units/ml}$$

where is

A_t absorbance of test solution at 260 nm after 30 min incubation
A_b absorbance of blank (sum of serum and substrate blanks)
F dilution factor for serum

1.10.2 Cytidine Specific Ribonuclease

The assay procedure is similar to that described above for "total" ribonuclease activity.

Reagents and Solutions

Purity of reagents: preparations of the synthetic polymer of cytidylic acid (poly C) used in this assay from several different manufacturers gave markedly different results. The batch-to-batch variability of poly C from any single source is, however, low. All other reagents should be of analytical quality.

Preparation of solutions (for about 200 determinations): re-purified water is used to make all solutions (cf. Vol. II, chapter 2.1.3.2).

1. Phosphate/borate buffer (phosphate, 0.1 mol/l, pH 6.5):

   dissolve 3.55 g Na_2HPO_4 in 200 ml water. The pH of this solution is then adjusted to 6.5 by addition of crystalline boric acid. This solution is then made up to 250 ml with water.

2. Polycytidylic acid solution (800 mg/l):

   dissolve 8 mg poly C in 10 ml water.

3. Perchloric acid/lanthanum solution ($HClO_4$, 1.2 mol/l, lanthanum nitrate, 0.022 mol/l):

   as described in the "total" activity method.

Stability of solutions: the poly C substrate solution appears to be stable for up to 1 week at 4°C. However, we prefer to prepare this solution freshly before each assay.

Procedure

Collection and treatment of specimen: fresh serum is preferable for this assay.

Stability of enzyme in the sample: cf. chapter 1.10.1.

Details for measurement in tissue: this assay can be used for activity measurements in tissue extracts. Together with assays based on polyuridylic, polyadenylic and polyguanylic acids and cyclic cytidine 2′3′-monophosphate it has been used to characterize the substrate specificity of ribonuclease isoenzymes extracted from various organs.

Assay conditions: the assay conditions using this substrate differ only in two respects from those previously described for use with transfer RNA: the reaction mixture is incubated for 15 minutes at 37 °C and the optimal wavelength for measurement of absorbance of the reaction products is 278 nm.

Measurement

Pipette successively into a micro-centrifuge tube:			concentration in assay mixture	
phosphate/borate buffer poly C solution	(1) (2)	150 µl 50 µl	phosphate poly C	0.06 mol/l 160 mg/l
mix and warm to 37 °C				
sample		50 µl	volume fraction	0.2
mix and incubate for 15 min				
$HClO_4/La^{3+}$ solution	(3)	250 µl	$HClO_4$ La^{3+}	0.6 mol/l 0.011 mol/l
mix, let stand on ice for 20 min, spin at 10000 *g* for 10 min, take 400 µl aliquot of supernatant, dilute to 2 ml with water, read absorbance at 278 nm.				

If the measured absorbance is greater than 0.25 then the linear region of the change of absorbance with time plot has been exceeded and the sample must be re-assayed either at greater dilution or by incubating for a shorter period of time.

Calculation: a linear correlation of absorbance and the amount of substrate hydrolyzed was established. Substrate hydrolyzed (µg) is 85.9 × net absorbance change − 2.9 ($r = 0.994$, $p < 0.01$, $n = 90$).

Correct measured absorbance for blanks. The catalytic concentration of the enzyme in the sample is (t in minutes):

$$b = \frac{1000 \times \mathrm{F}}{M_r} \times (85.9 \times \Delta A - 2.9)/\Delta t \qquad \mathrm{U/l}$$

in terms of mononucleotide units liberated

where

F final serum dilution 1 : 500

M_r mean molecular weight of a single mononucleotide unit of poly C, 417 in our preparation.

Validation of Methods

Precision, accuracy, detection limits and sensitivity: "Total ribonuclease activity" (transfer RNA substrate). This assay gives an intra-assay variability of 4% and inter-assay of 6% on replicate analysis. We are unable to give any information about the accuracy of the methods described as we have not had access to any reference material.

With the stated serum dilution and incubation time the detection limits of the assay are 0.007 – 0.0133 absorbance units/min. The lower detection limit of the assay is approximately one order of magnitude lower than the enzyme activity observed in normal serum.

Cytidylic acid specific assay: inter- and intra-assay coefficients of variability for this assay are both approximately 6.5%.

The detection limits with the incubation time and serum dilutions described above are 0.37 to 1.5 μmol poly C hydrolyzed per ml serum per minute.

Sources of error: the main source of error in both methods is the timing of the incubation step of the procedure. The magnitude of this error increases with decreasing incubation time, especially below 5 minutes.

Specificity: several forms of ribonuclease with different substrate specificities have been described in tissue extracts of many organs but few of these are detectable in normal human serum [1 – 4]. The methods described have been optimized to measure the three forms found in normal serum. It is possible that isoenzymes of different specificity, requiring different pH buffer composition combinations may be present in the serum of patients with some pathological conditions which would not be detected by the assay systems described here.

Reference ranges: due to the lack of standardization of methods used by most investigators and the very large discrepancy in quoted "normal ranges" from authors using apparently similar techniques, no reference values exist for serum ribonuclease activity. There is a consensus that the mean serum activity of the enzyme does increase with age. Neonates have enzyme activities higher than those of the adult mean value, whereas infants of one to four years have activities lower than this value. Above this age ribonuclease activity increases linearly with age. When performing any study on a particular condition it is, therefore, important to use an age matched control group of subjects.

References

[1] *T. Umeda, T. Mariyama, H. Oura, K. Tsakala,* Rat Serum Ribonuclease, Biochim. Biophys. Acta *171*, 260 – 264 (1969).

[2] *R. Delaney,* Chemical, Physical and Enzymatic Properties of Several Human Ribonucleases, Biochemistry *2*, 438 – 444 (1963).

[3] *J. Sznajd, J. W. Naskolski,* Ribonuclease from Human Granulocytes, Biochim. Biophys. Acta *302*, 282 – 292 (1973).

[4] *J. S. Roth,* Some Observations on the Assay and Properties of Ribonucleases in Normal and Tumour Tissue, Methods Cancer Res. *3*, 153 – 242 (1967).

[5] *J. J. Frank, C. C. Levy,* Properties of a Human Liver Ribonuclease, J. Biol. Chem. *251*, 5745 – 5751 (1976).

[6] *E. A. Neuwelt, M. S. Bogaski, J. J. Frank, K. Proctor-Appich, C. C. Levy,* Possible Sites of Origin of Human Plasma Ribonucleases as Evidenced by Isolation and Partial Characterisation of Ribonucleases from Several Human Tissues, Cancer Res. *38*, 88 – 93 (1978).

[7] *J. M. Thomas, M. E. Hodes,* Isoenzymes of Human Urine Ribonuclease Demonstrated by Isoelectric Focusing, Proc. Indiana. Acad. Sci. *88*, 153 – 159 (1979).

[8] *J. G. Bissenden, P. H. Scott, J. Hallum, H. M. Mansfield, P. Scott, B. A. Wharton,* Anthropometric and Biochemical Changes During Pregnancy in Asian and European Mothers Having Well Grown Babies, Brit. J. Obstet Gynaecol. *88*, 992 – 998 (1981).

[9] *J. M. Thomas, M. E. Hodes,* Isoenzymes of Ribonuclease in Human Serum and Urine. I. Methodology and a Survey of a Control Population, Clin. Chim. Acta *111*, 185 – 197 (1981).

[10] *T. P. Corbishley, B. Greenway, P. J. Johnson, R. Williams,* Serum Ribonuclease in the Diagnosis of Pancreatic Carcinoma and in Monitoring Chemotherapy, Clin. Chim. Acta *124*, 225 – 233 (1982).

[11] *R. H. Kottel, S. O. Hoch, R. G. Parsons, I. A. Hoch,* Serum Ribonuclease Activity in Cancer Patients, Br. J. Cancer *38*, 280 – 286 (1978).

[12] *R. C. Davis, S. R. Cooperband, J. A. Mannick,* The Immunosuppressive Effects of Ribonuclease Complex in vivo and in vitro, J. Reticuloendothel. Soc. *7*, 43 – 52 (1970).

[13] *J. F. Mowbray, J. Scholand,* Inhibition of Antibody Production by Ribonucleases, Immunology *11*, 421 – 426 (1966).

[14] *C. Baglioni,* Interferon Induced Enzymatic Activities and their Role in the Antiviral State, Cell *17*, 255 – 264 (1979).

[15] *P. H. Scott, H. M. Berger, C. Kenwood, P. Scott, B. A. Wharton,* Plasma Alkaline Ribonuclease (EC 3.1.4.22) and Nitrogen Retention in Low Birth Weight Infants, Br. J. Nutr. *40*, 459 – 464 (1978).

[16] *P. Blackburn, G. Wilson, S. Moore,* Placental RNAse Inhibitor, J. Biol. Chem. *252*, 5904 – 5910 (1977).

[17] *M. Schukler, P. B. Jewett, C. C. Levy,* The Effects of Polyamines on a Residue Specific Human Plasma Ribonuclease, J. Biol. Chem. *250*, 2206 – 2212 (1975).

[18] *A. S. Prasad, D. Oberleas,* Ribonuclease and Deoxyribonuclease Activities in Zinc-deficient Tissue, J. Lab. Clin. Med. *82*, 461 – 466 (1973).

[19] *W. E. Razzell,* The Precipitation of Polyribonucleotides with Magnesium Salts and Ethanol, J. Biol. Chem. *238*, 3053 – 3057 (1963).

[20] *M. Rabinovitch, B. Lieberman, N. Fausto,* Plasma Ribonuclease in Human Uremia, J. Lab. Clin. Med. *53*, 563 – 568 (1959).

2 Glycosidases

2.1 α-Amylase

1,4-α-D-glucan glucanohydrolase, EC 3.2.1.1

2.1.1 UV-method with Maltotetraose

Kenneth J. Pierre and Ker-Kong Tung

General

α-Amylase (1,4 α-D-glucan glucanohydrolase, EC 3.2.1.1) catalyzes the hydrolysis of α-1,4 glucan linkages to produce maltose and larger oligosaccharides. As a class of enzymes, the amylases are found in plants, animals and micro-organisms. The enzymes from porcine pancreas [1], human pancreas [2], human saliva [3], *B. subtilis* [4], *A. oryzae* [5] and barley [6] have been purified to homogeneity and studied extensively. Although α-amylases are stable and easy to purify, the measurement of activities is complicated by the variety of reaction mechanisms, the complex mixture of products, and by the lack of substrates with defined chemical structures.

In human beings, the enzyme is produced and stored in the salivary glands and the pancreas. It is secreted into the digestive tract, eventually into the blood stream and is excreted in the urine. Enzyme levels in serum and urine are low and relatively constant in normal individuals, becoming elevated during certain disease states such as acute pancreatitis and salivary lesions.

Application of method: in biochemistry, the food industry and clinical chemistry.

Enzyme properties relevant in analysis: the catalytic activity of α-amylase in serum and urine is attributable to several isoenzymes from the pancreas and parotid glands. Substrate hydrolysis proceeds linearly with time after a 5-minute lag phase.

Methods of determination: measurement of α-amylase activity by saccharogenic, iodometric, and dye-glucan methods, and by an enzymatic method with α-glucosidase, has been described previously [7]. Methods employing 4-nitrophenyl malto-oligosaccharides and α-glucosidase have been reported elsewhere [8, 9]. The method described here is a well-tested continuous-monitoring assay with maltotetraose as a defined substrate, and maltose phosphorylase, MP (EC 2.4.1.8), β-phosphoglucomutase, β-PGM (EC 5.3.1.9) and glucose-6-phosphate dehydrogenase, G6P-DH (EC 1.1.1.49) as the coupling enzymes [10]. This system provides a simple, precise and easily adaptable method for α-amylase measurement. Estimation of enzyme mass or active-site titration is not in use. Recently a wheat germ inhibitor was coupled with this system to assay the pancreatic isoenzyme of α-amylase in serum [11].

International reference methods and standards: neither reference methods nor materials have been adopted at the international level; however, cf. p. 153 and Vol. II, chapter 2.3.

Enzyme effectors: calcium and chloride ions are required for enzyme activity. Specific inhibitors other than chelating agents have been isolated from wheat germ [12] and beans [13].

Assay

Method Design

Principle: the enzymatic reaction sequence employed in this assay of α-amylase is as follows.

(a) $$\text{Maltotetraose} + H_2O \xrightarrow{\alpha\text{-amylase}} 2\ \text{maltose}$$

(b) $$\text{Maltose} + P_i \xrightarrow{\text{MP}} \text{glucose} + \beta\text{-glucose-1-P}$$

(c) $$\beta\text{-Glucose-1-P} \xrightarrow[\text{G-1,6-}P_2]{\beta\text{-PGM}} \text{glucose-6-P}$$

(d) $$\text{Glucose-6-P} + NAD^+ \xrightarrow{\text{G6P-DH}} \text{gluconate-6-P} + NADH + H^+.$$

Maltotetraose is hydrolyzed by α-amylase to yield two moles of maltose per mole of substrate. Maltose phosphorylase catalyzes the phosphorolysis of each mole of maltose to one mole each of glucose and β-glucose 1-phosphate. β-Phosphoglucomutase converts β-glucose 1-phosphate to glucose 6-phosphate which is then oxidized to gluconate 6-phosphate with the concomitant reduction of nicotinamide-adenine dinucleotide to NADH in the reaction catalyzed by glucose 6-phosphate dehydrogenase from *Leuconostoc*. The rate of production of NADH is followed by measuring the increase in absorbance at 339 nm which is directly related to the α-amylase activity in the specimen.

Optimized conditions for measurement: maltotetraose instead of starch is used as the substrate for α-amylase measurement because of its high purity, defined structure and simple product pattern. The assay value with maltotetraose is proportional to enzyme concentration. The optimum concentration of maltotetraose is 7.3 mmol/l for both human and porcine α-amylases and the optimum pH is 6.6 centred on a plateau between pH 6.3 and pH 6.9. Optimum concentrations of cofactors and coupling enzymes in 1 litre reaction mixture are: 2.5 mmol NAD, 0.5 mmol glucose 1,6-bisphosphate, 6000 U MP, 1000 U PGM, 6000 U G6P-DH.

Temperature conversion factors: we have found the activity at 37 °C to be 1.46 times that measured at 30 °C.

Equipment

Spectrophotometer or spectral-line photometer suitable for accurate measurements at 339 nm, Hg 334 or Hg 365 nm. Water-bath and stopwatch.

Reagents and Solutions

Purity of reagents: the enzyme preparations must be free from α-amylase. Contamination with 6-phosphogluconate dehydrogenase (EC 1.1.1.44), alcohol dehydrogenase (EC 1.1.1.1) and "NADH oxidase" should be less than 0.02% of the β-PGM activity. Maltotetraose must be free from maltose. Contamination of maltotetraose with glucose should be less than 3% and less than 2% of larger malto-oligosaccharides should be present.

Preparation of solutions (for about 500 determinations): all solutions in re-purified water (cf. Vol. II, chapter 2.1.3.2).

1. Phosphate buffer (25 mmol/l, pH 6.6):

   dissolve 1.63 g K_2HPO_4 and 2.13 g KH_2PO_4 in about 800 ml water. If necessary adjust to pH 6.6 with KOH, 1 mol/l, and dilute to 1000 ml with water.

2. Enzyme solutions:

   a) dilute a sufficient quantity of MP in 10 ml phosphate buffer (1) to give a final concentration of 300 U/ml;

   b) dilute a sufficient quantity of β-PGM in 10 ml phosphate buffer (1) to give a final concentration of 50 U/ml;

   c) dilute a sufficient quantity of G6P-DH in 10 ml phosphate buffer (1) to give a final concentration of 300 U/ml.

3. Assay solution:

   dissolve 500 mg maltotetraose, 180 mg NAD, free acid, 20 mg G-1,6-P_2-$Na_2 \cdot H_2O$, in 75 ml phosphate buffer (1) and adjust to pH 6.6 with KOH, 1 mol/l. Add 2 ml each of the MP, β-PGM and G6P-DH solutions and adjust the volume to 100 ml with phosphate buffer (1).

Stability of solutions: the phosphate buffer (1) is stable for at least 1 month at room temperature if microbial contamination is avoided. The enzyme solution is stable for 1 week at 0 °C to 4 °C. The final assay solution is stable for 24 h at 0 °C to 4 °C and 8 h at room temperature.

Procedure

Collection and treatment of specimen: blood can be collected either with a vacuum tube (Vacutainer) or a clean, dry needle and syringe. After the blood has clotted, the tube is centrifuged and the serum removed for assay. Urine samples can be collected in a clean container.

Details for measurement in urine: measurement as for serum specimens, except that 0.025 ml is used instead 0.05 ml.

Stability of enzyme in specimen or sample: the stability of amylase in serum or urine is reported to be one week at room temperature and several weeks at 4 °C if kept free from microbial contamination [14].

Assay conditions: wavelength 339 nm; light path 10 mm; 30 °C; final volume 1.05 ml for serum and 1.025 ml for urine; read against water. Equilibrate the assay solution at 30 °C before the start of the assay.

Measurement (for serum)

Pipette successively into cuvettes:			concentration in assay mixture	
assay solution	(3)	1.00 ml	phosphate	25 mmol/l
serum		0.05 ml	maltotetraose	7.3 mmol/l
mix, start stopwatch and read absorbance at exactly 5 min and 8 min or monitor the reaction on a recorder.			NAD	2.5 mmol/l
			$MgCl_2$	4.0 mmol/l
			G-1,6-P_2	0.5 mmol/l
			MP	6000 U/l
			β-PGM	1000 U/l
			G6P-DH	6000 U/l
			volume fraction	0.048

With absorbance changes per min greater than 0.2, dilute the sample ten-fold with phosphate buffer (1) and repeat the assay.

Calculation: subtract the 5-min absorbance reading from the 8-min reading to give ΔA for 3 min reaction time. Calculate catalytic activity concentration in the sample using eqn. (k) or (k_1), cf. "Formulae", Appendix 3. $\varepsilon = 0.63\ l \times mmol^{-1} \times mm^{-1}$.

For serum assay:

$$b = \frac{\mathrm{V} \times \Delta A \times 1000}{\mathrm{v} \times \Delta t \times 0.63 \times 10} = 3333 \times \Delta A/\Delta t \qquad \mathrm{U^*/l}$$

For urine assay:

$$b = \frac{\mathrm{V} \times \Delta A \times 1000}{\mathrm{v} \times \Delta t \times 0.63 \times 10} = 6508 \times \Delta A/\Delta t \qquad \mathrm{U^*/l}$$

Validation of Method

Precision, accuracy, detection limits and sensitivity: the day-to-day coefficients of variation for serum pools over a thirty-day period were 3.6%, 3.1% and 2.7% with mean values of 71, 101 and 139 U/l, respectively. This method was compared with the saccharogenic method of *Henry & Chiamori* [15] for analysis of both urine and serum samples. One hundred and eight serum samples and sixty urine samples were analyzed by both methods at 37 °C for the enzymatic method and 40 °C for the saccharogenic method. Correlation coefficients of 0.96 and 0.97 were obtained for serum and urine samples, respectively. The limits of detection and dynamic range of the assay are determined by the instrument used rather than by the method. Absorbance changes between 0.003 to 0.200 per minute can be measured accurately on a Beckman DU spectrophotometer. The practical sensitivity limit is $\Delta A/\Delta t = 0.003/\mathrm{min}$.

Sources of error: contamination with salivary amylase will give erroneous results; therefore, pipetting by mouth should be avoided.

Because the assay method is based on the measurement of maltose production, specimens with high levels of maltose will give erroneously high results.

A list of drugs and other substances which affect the level of amylase or may interfere with its determination has been prepared by *Young et al.* [16]. Various substances were tested for interference with this method. No measurable effect on the activity of α-amylase was observed when the following compounds were added to serum at the levels (mg/l) indicated: ascorbic acid (250), bilirubin (100), creatinine (1000), ethanol (2000), pyruvate (25), glucose (5000), glutathione (100), and heparin (200).

---

* One unit of α-amylase is defined as the amount of enzyme which catalyzes the production of 1 µmole of NADH per minute under the conditions of the assay.
Note by the editor: definition of one international unit is 1 µmole substrate conversion per minute. Values obtained with the above mentioned formulae may be converted to international units by multiplying by the factor 0.5.

Specificity: this assay method is specific for the rate of maltose production. Thus, the presence of reducing substances, glucose and other carbohydrates (except maltose) has no effect on the assay values.

Reference ranges: in human serum the reference values are 20 to 110 U/l at 37 °C, 18 to 75 U/l at 30 °C and 16 to 65 U/l at 25 °C. In human urine the reference values are 1 to 17 U/l at 37 °C.

References

[1] *J. F. Robyt, D. French,* The Action Pattern of Porcine Pancreatic α-Amylase in Relationship to the Substrate Binding Site of the Enzyme, J. Biol. Chem. *245*, 3917 – 3927 (1970).

[2] *E. H. Fischer, F. Duckert, P. Bernfeld,* Isolement et cristallisation de l'α-amylase de pancreas humain, Helv. Chim. Acta *33*, 1060 – 1064 (1960).

[3] *H. Mutzbauer, G. V. Schulz,* Die Bestimmung der molekularen Konstanten von α-Amylase aus Humanspeichel, Biochim. Biophys. Acta *102*, 526 – 532 (1965).

[4] *J. F. Robyt, D. French,* Action Pattern and Specificity of an Amylase from Bacillus subtilis, Arch. Biochem. Biophys. *100*, 451 – 467 (1963).

[5] *J. D. Allen, J. A. Thoma,* Multimolecular Substrate Reactions Catalyzed by Carbohydrases. Aspergillus oryzae α-Amylase Degradation of Maltooligosaccharides, Biochemistry *17*, 2338 – 2344 (1978).

[6] *S. Schwimmer, A. K. Balls,* Isolation and Properties of Crystalline α-Amylase from Germinated Barley, J. Biol. Chem. *179*, 1063 – 1074 (1949).

[7] *W. Rick, H. P. Stegbauer,* α-Amylase, Measurement of Reducing Groups, in: *H. U. Bergmeyer* (ed.), Methods of Enzymatic Analysis, 2nd edit., Verlag Chemie, Weinheim, and Academic Press, New York 1974, pp. 885 – 890.

[8] *E. Rauscher, S. v. Buelow, E.-O. Haegele,* Optimized Determination of α-Amylase in Serum and Urine using p-Nitrophenyl-α,D-maltoheptaoside as Substrate, Fresenius Z. Anal. Chem. *311*, 454 (1982).

[9] *R. McCroskey, T. Chang, H. David, E. Winn,* p-Nitrophenylglycosides as Substrates for Measurement of Amylase in Serum and Urine, Clin. Chem. *28*, 1787 – 1791 (1982).

[10] *K. T. Whitlow, N. Gochman, R. L. Forrester, L. J. Wataji,* Maltotetraose as a Substrate for Enzyme-Coupled Assay of Amylase Activity in Serum and Urine, Clin. Chem. *25*, 481 – 483 (1979).

[11] *W. Y. Huang, N. W. Tietz,* Determinations of Amylase Isoenzymes in Serum by Use of a Selective Inhibitor, Clin. Chem. *28*, 1525 – 1527 (1982).

[12] *M. D. O'Donnell, K. F. McGeeney,* Purification and Properties of an α-Amylase Inhibitor from Wheat, Biochim. Biophys. Acta *422*, 159 – 169 (1976).

[13] *J. J. Marshall, C. M. Lauda,* Purification and Properties of Phaseolamin, and Inhibitor of α-Amylase, from the Kidney Bean, Phaseolus vulgaris, J. Biol. Chem. *250*, 8030 – 8037 (1975).

[14] *R. J. Henry, D. C. Cannon, J. W. Winkelmann,* Clinical Chemistry – Principles and Technics 2nd Edit., Harper & Row, Hagerstown, MD. 1974; p. 948.

[15] *R. J. Henry, N. Chiamori,* Study of the Saccharogenic Method for the Determination of Serum and Urine Amylase, Clin. Chem. *5*, 434 – 452 (1960).

[16] *D. S. Young, D. W. Thomas, R. B. Friedman, L. C. Pestaner,* Effects of Drugs on Clinical Laboratory Tests, Clin. Chem. *18*, 1041 – 1042 (1972).

2.1.2 UV-method with Maltoheptaose

Elli Rauscher

General

α-Amylases are present in all kinds of living organisms. They occur as endoamylases in animal tissues and as exoamylases (β-amylase) in plants and micro-organisms. In mammals α-amylases are present mainly in pancreatic and salivary glands and, in lesser concentration, in other organs also. In acute pancreatitis the pancreatic type increases quickly to a high level in serum and is eliminated in the urine because of its relatively low molecular weight. In inflammation of the salivary gland the level of the salivary type is markedly increased. Low concentrations of both isoenzymes are present in the serum of healthy persons in almost equal amounts [1 – 5].

Exoamylases attack saccharides from the non-reducing ends of the chains in a regular manner, e.g. β-amylase produces maltose. Endoamylases hydrolyze α-1 → 4 glucan bonds of high molecular-weight substrates such as starch, amylose or glycogen almost arbitrarily and produce saccharides of shorter chain length [4], but hydrolyze oligosaccharides to defined reaction products [6].

Application of method: in biochemistry and clinical chemistry.

Enzyme properties relevant in analysis: the catalytic activity in serum is due to the existence of isoenzymes, mainly of salivary and pancreatic type. The existence of conglomerates of amylase and immunoglobulins (macroamylasaemia) has been described [7, 8]. Substrate conversion proceeds with a lag phase of several minutes.

Methods of determination: the catalytic activity can be measured by several techniques using poly- or oligosaccharides containing α-1 → 4 glucan bonds. For a long time mainly saccharogenic and iodine-staining (amyloclastic) methods were used, with starch or amylose as substrates [4]; then colorimetric assays with dye-labelled insoluble polysaccharides (cf. p. 161) were introduced. Turbidimetric [9], nephelometric [10, 11], and polarimetric [12] methods, and a UV method [13], all using high molecular-weight substrates, have been described. In recent years methods using low molecular-weight defined substrates are preferred and recommended [14] because the purity of the substrates and the reaction patterns can be determined exactly. Several reaction products may be measured, e.g. glucose (cf. p. 153), glucose 1-phosphate (cf. p. 147) or 4-nitrophenol (cf. p. 157) in the presence of suitable auxiliary enzymes. Electrophoretic and chromatographic methods, or the addition of an inhibitor (e.g. from wheat germ) with differential inhibition of human salivary and pancreatic amylases, may be used for determination of specific isoenzymes (cf. p. 167). Estimation of enzyme mass or active-site titration is not in use.

International reference methods and standards: neither standardization at the international level nor the existence of reference standard materials is known so far [15]. However, there are strong efforts in the pharmaceutical field to achieve a standardized method specially for the determination of amylase in pancreas preparations. A recommended procedure is described in the 4th report of the International Commission for Standardization of Pharmaceutical Enzymes published in Journal Mondial de Pharmacie *3*, 337 – 354 (1968), cf. also Vol. II, chapter 2.3.

Enzyme effectors: chloride activates α-amylase when added in concentrations of 10 to 100 mmol/l to the assay, whereas it inhibits the enzyme when added in higher concentration.

Assay

Method Design

Principle (simplified)

(a) $$\text{Maltoheptaose} + H_2O \xrightarrow{\alpha\text{-amylase}} \text{maltotetraose} + \text{maltotriose}$$

(b) $$\text{Maltotriose} + 2\,H_2O \xrightarrow{\alpha\text{-glucosidase}} 3\ \text{glucose}$$

(c) $$3\ \text{Glucose} + 3\ \text{ATP} \xrightarrow{\text{hexokinase}} 3\ \text{glucose-6-P} + 3\ \text{ADP}$$

(d) $$3\ \text{Glucose-6-P} + 3\,NAD^+ \xrightarrow{\text{G6P-DH*}} 3\ \text{gluconate-6-P} + 3\,NADH + 3\,H^+.$$

Maltoheptaose is split by α-amylase mainly to maltotetraose and maltotriose and to a negligible extent to maltopentaose and maltose [6].

The amount of maltoheptaose reacting per unit time, measured by the increase in absorbance due to the reaction scheme given above, is therefore a measure of the catalytic activity of α-amylase. α-Glucosidase hydrolyzes the substrate slowly to glucose and therefore the reagent blank has to be determined and subtracted from the results for the samples.

The method may be adapted to automated equipment.

Optimized conditions for measurement: whereas the pH optima of the isoenzymes are very similar, due to the very small difference in the molecular structures of human pancreatic and saliva amylases, the substrate specificity shows some differences [16]. These effects are still under study and discussion. In the following method maltoheptaose, 10 mmol/l, and buffer, pH 7.0, containing NaCl, 50 mmol/l, are used.

Temperature conversion factors: the following factors were found for the measurement of α-amylase in human sera at 30° or 37° relative to the value measured at 25 °C:

°C	25	30	37
F	1.00	0.77	0.61

* D-Glucose-6-phosphate : $NADP^+$ 1-oxidoreductase, EC 1.1.1.49.

Equipment

Spectrophotometer or spectral-line photometer capable of exact measurement at 339 nm, Hg 334 nm or Hg 365 nm with a thermostatted cuvette holder; water-bath; recorder or stopwatch.

Reagents and Solutions

Purity of reagents: maltoheptaose must be free from glucose, maltose, maltotriose, maltotetraose and β-cyclodextrin. Contamination with maltopentaose should be $<1\%$ and with maltohexaose $<5\%$, determined by high-pressure liquid chromatography in relation to maltoheptaose. α-Glucosidase, hexokinase and G6P-DH should contain no amylase, $<0.001\%$ NADH-oxidizing contaminants and $<0.001\%$ 6-PGDH*. All enzymes must be free from ammonium ions. Suitable commercial preparations are available.

Preparation of solutions (for about 25 determinations): all solutions in re-purified water (cf. Vol. II, chapter 2.1.3.2).

1. Buffer/coenzyme/auxiliary enzymes (phosphate, 55.5 mmol/l; pH 7.0; chloride, 55.5 mmol/l; Mg^{2+}, 2.22 mmol/l; ATP, 1.50 mmol/l; NAD, 2.22 mmol/l; hexokinase, $\geqq$ 2.2 U/ml; G6P-DH, $\geqq$ 2.2 U/ml; α-glucosidase, $\geqq$ 11 U/ml):

   dissolve 326 mg $Na_2HPO_4 \cdot 2\,H_2O$ + 130 mg $NaH_2PO_4 \cdot H_2O$ + 162 mg NaCl + 27 mg $MgSO_4 \cdot 7\,H_2O$ + 45 mg $ATPNa_2H_2 \cdot 3\,H_2O$ + 74 mg NAD, free acid, in 40 ml water, then add 110 U hexokinase from yeast, lyophilized, and 110 U G6P-DH from *Leuconostoc mesenteroides,* lyophilized, and 550 U α-glucosidase from yeast, lyophilized, mix gently and dilute to 50 ml with water.

2. Maltoheptaose solution (111 mmol/l):

   dissolve 640 mg maltoheptaose in 5 ml water.

Stability of solutions: store all solutions at 0 – 4 °C. Solutions (1) and (2) are stable for two weeks, provided that no microbial contamination occurs.

Procedure

Collection and treatment of specimen: collect blood from the vein without stasis. Addition of heparin, 0.2 mg/ml, is acceptable but other anticoagulants cannot be used. Centrifuge for 10 min at about 3000 *g* in order to obtain serum or plasma.

Collect urine without any additive. Use only fresh urine. Dilute urine 1 + 2 (v/v) with sodium chloride solution 0.154 mol/l.

---

* 6-Phospho-D-gluconate : $NAD(P)^+$ 2-oxidoreductase, EC 1.1.1.43 or 1.1.1.44.

Stability of the enzyme in the sample: the enzyme is stable in human serum between 0 °C and 20 °C for one week at least. In urine the enzyme activity decreases quickly, even at −20 °C.

Measurements in other samples: pancreatic juice or duodenal content or saliva have to be diluted 1 + 100 to 1 + 1000 (v/v) with a solution of sodium chloride, 0.154 mol/l.

Assay conditions: wavelength 339 nm or Hg 365 nm, Hg 334 nm; light path 10 mm; final volume 2.22 ml; 25 °C (thermostatted cuvette holder). Measure against air. Before starting the assay, adjust temperature of solutions to 25 °C.

Measurement

Pipette successively into cuvettes:		blank	sample	concentration in assay mixture	
reagent solution	(1)	2.00 ml	2.00 ml	phosphate chloride Mg^{2+} NAD ATP hexokinase G6P-DH α-glucosidase	50 mmol/l 50 mmol/l 2.0 mmol/l 2.0 mmol/l 1.3 mmol/l 2 U/ml 2 U/ml 10 U/ml
substrate	(2)	0.20 ml	0.20 ml	maltoheptaose	10 mmol/l
water		0.02 ml	–		
sample (e.g. serum)		–	0.02 ml	volume fraction	0.009
mix thoroughly with a plastic spatula, wait for 10 min (lag phase), read absorbance and start stopwatch. Repeat the reading after exactly 1, 2 and 3 min, or monitor the reaction on a recorder.					

The values $\Delta A/\Delta t$ must be <0.080/min at Hg 365 nm (<0.160/min at Hg 334 nm or 339 nm). Otherwise, dilute the sample 10-fold with sodium chloride solution, 0.154 mol/l, or measure at shorter intervals.

The absorbance at the beginning of reading should not be higher than 1.5. Otherwise, dilute the sample as described above.

When measuring at 30 °C the reading may begin 8 min after adding the sample because of the shorter lag phase. Measuring at 37 °C is not recommended.
The reagent blank has to be determined once for the reagent in use.

Calculation: use calculation formula (k) or (k_1), resp. (cf. "Formulae", Appendix 3), but divide by 3 since one molecule of substrate gives rise to three molecules of glucose and thus to three molecules of NADH ($\nu_i = 3$). For values of ε cf. Appendix 4. Subtract the value of the reagent blank from the value of each sample.

The following relations are valid for the catalytic activity concentration in the sample:

	Hg 334 nm	339 nm	Hg 365 nm	
$b =$	$5987 \times \Delta A/\Delta t$	$5873 \times \Delta A/\Delta t$	$10882 \times \Delta A/\Delta t$	U/l
$b =$	$9.98 \times 10^4 \times \Delta A/\Delta t$	$9.79 \times 10^4 \times \Delta A/\Delta t$	$18.1 \times 10^4 \times \Delta A/\Delta t$	nkat/l

Validation of Method

Precision, accuracy, detection limits and sensitivity: for activities of about 100 U/l in serum, coefficients of variation in the range of 5 – 10% were found within run and 7 – 12% from day to day for manual performance. Data about accuracy are not available since standard reference material is not established yet. For measurements at Hg 365 nm with sufficiently sensitive photometers the lowest possible value which can be measured is approximately 20 U/l, with an imprecision of $< \pm 20\%$. Sensitivity is found to be $\Delta A/\Delta t = 0.005/10$ min (*Eppendorf* photometer).

Sources of error: influences of therapeutics on the measured activity are unknown. Marked interference results from contamination of the reagents by saliva or sweat, as they contain high activities of α-amylase. Therefore, solutions may not be pipetted by mouth and contact with skin must be avoided. Anticoagulants such as oxalate, citrate, fluoride or EDTA inhibit the reaction and cannot be used. Ammonium ions are also inhibitors. Endogenous glucose reacts during the pre-incubation time and does not interfere with the measurement of amylase activity.

Specificity: the amylase activity is measured specifically. The presence of α-glucosidase in the sample does not interfere.

Reference ranges: in human serum, 30 – 100 U/l at 25 °C, 40 – 130 U/l at 30 °C; in urine, 500 U/l at 25 °C [17]. Sex dependency was not demonstrable.

References

[1] *E. H. Fischer, E. A. Stein,* α-Amylase, in: *P. D. Boyer, H. Lardy, K. Myrbäck* (eds.), The Enzymes, Vol. *4*, Academic Press, New York 1960, pp. 313 – 343.

[2] *D. French,* β-Amylases, in: *P. D. Boyer, H. Lardy, K. Myrbäck* (eds.), The Enzymes, Vol. *4*, Academic Press, New York 1960, pp. 345 – 368.

[3] *J. A. Thoma, J. E. Spradlin, St. Dygert,* Plant and Animal Amylases, in: *P. D. Boyer* (ed.), The Enzymes, Vol. *V*, Academic Press, New York 1971, pp. 115 – 189.

[4] *J. F. Robyt, W. J. Whelan,* Chapters 13, 14 and 15 in: *J. A. Radley* (ed.), Starch and its Derivatives, 4th edit., Chapman & Hall, London 1968, pp. 423 – 497.

[5] *W. B. Salt, St. Schenker,* Amylase – its Clinical Significance: a Review of the Literature, Medicine *55*, 269 – 289 (1976).

[6] *E. O. Haegele, E. Schaich, E. Rauscher, P. Lehmann, M. Graßl,* Action Pattern of Human Pancreatic α-Amylase on Maltoheptaose, a Substrate for Determining α-Amylase in Serum, J. Chromatogr. *223*, 69 – 84 (1981).
[7] *L. Fridhandler, J. E. Berk,* Macroamylasemia, in: *O. Bodansky, A. L. Latner* (ed.), Advances in Clinical Chemistry, Vol. *20*, Academic Press, New York 1978, pp. 267 – 286.
[8] *G. H. K. Dürr,* Nachweis und Häufigkeit der Makroamylasämie, Laboratoriumsmedizin *6*, 1 – 9 (1982).
[9] *A. G. Ware, C. B. Walberg, R. E. Sterling,* Turbidimetric Measurement of Amylase: Standardisation and Control with Stable Serum, in: *D. Seligson* (ed.), Standard Methods of Clinical Chemistry, Vol. *4*, Academic Press, New York 1963, pp. 15 – 21.
[10] *J. R. Shipe, J. Savory,* Kinetic Nephelometric Procedure for Measurement of Amylase Activity in Serum, Clin. Chem. *18*, 1323 – 1325 (1972).
[11] *L. Zinterhofer, S. Wardlaw, P. Jatlow, D. Seligson,* Nephelometric Determination of Pancreatic Enzymes I. Amylase, Clin. Chim. Acta *43*, 5 – 12 (1973).
[12] *W. Hönig, E. Moshudis, K. Oette,* Die polarimetrische Messung der α-Amylase-Aktivität, J. Clin. Chem. Clin. Biochem. *19*, 1057 – 1061 (1981).
[13] *H. W. Schiwara,* Ein UV-Test zur Messung der α-Amylase-Aktivität in Serum und Urin, Z. klin. Chem. u. klin. Biochem. *10*, 12 – 16 (1972).
[14] *K. Lorentz,* α-Amylase Assay: Current State and Future Development, J. Clin. Chem. Clin. Biochem. *17*, 499 – 504 (1979).
[15] *J. P. Bretaudiere, R. Rej, P. Drake, A. Vassault, M. Bailly,* Suitability of Control Materials for Determination of α-Amylase Activity, Clin. Chem. *27*, 806 – 815 (1981).
[16] *V. W. Lee, Ch. Willis,* Activity of Human and Honhuman Amylases on Different Substrates Used in Enzymatic Assay Methods – A Pitfall in Interlaboratory Quality Control, Am. Soc. Clin. Pathol. *77*, 290 – 296 (1982).
[17] *H. D. Weißhaar, H. Sudhoff, P. U. Koller, K. D. Willamowski,* Referenzwerte für eine neue α-Amylase-Bestimmung im Serum und Urin, Dtsch. med. Wochenschr. *106*, 936 – 939 (1981).

2.1.3 Colorimetric Method

Elli Rauscher

Assay

Method Design

Principle (simplified):

$$\text{(a)}\quad 3\ \ \text{4-Nitrophenyl maltoheptaoside} \xrightarrow[H_2O]{\alpha\text{-amylase}} \text{2 maltotriose} + \text{maltotetraose} +$$

$$2\ \ \text{4-nitrophenyl maltotetraoside} + \text{4-nitrophenyl maltotrioside}$$

$$\text{(b)}\quad \text{4-Nitrophenyl maltotrioside} + 3\,H_2O \xrightarrow{\alpha\text{-glucosidase}} \text{3 glucose} + \text{4-nitrophenol}.$$

The substrate is split by α-amylase mainly in the above way, with only a small amount being converted to 4-nitrophenyl maltopentaoside and maltose [1].

The amount of 4-nitrophenyl maltoheptaoside reacting per unit time, measured by the increase in absorbance due to the formation of 4-nitrophenol, is a measure of the catalytic activity of α-amylase. An assay blank is not necessary for routine analysis. The method described below can be adapted to automated equipment.

Optimized conditions for measurement: the pH optimum of α-amylase is 6.9 – 7.1. The dissociation of 4-nitrophenol and its colour intensity are dependent on the pH and temperature. Therefore, the values of ε were determined for the Hg 405 nm line under the assay conditions at the measuring temperatures most commonly used, and were found to be 0.90 at 25 °C, 0.95 at 30 °C and 1.06 at 37 °C ($l \times mmol^{-1} \times mm^{-1}$). In the method described here [2, 3], 4-nitrophenyl maltoheptaoside, 5 mmol/l, was chosen as substrate with phosphate buffer, pH 7.1, containing NaCl, 50 mmol/l. The use of 4-nitrophenylated oligosaccharides of shorter chain length, with somewhat varying assay conditions, is also described by other authors [4 – 8].

Temperature conversion factors: the following factors were found for measurements in human sera at 30 °C and 37 °C, respectively, relative to the values measured at 25 °C:

°C	25	30	37
F	1.00	0.74	0.56

Equipment

Spectrophotometer or spectral-line photometer capable of exact measurement at Hg 405 nm or between 400 and 415 nm with a thermostatted cuvette holder; water-bath; recorder or stopwatch.

Reagents and Solutions

Purity of reagents: 4-nitrophenyl maltoheptaoside must be free from 4-nitrophenol, 4-nitrophenylglucoside, -maltoside and -maltotrioside, and should contain less than 5% of other nitrophenylated oligosaccharides or unsubstituted oligosaccharides. α-Glucosidase must be free from α-amylase and from ammonium ions.

Preparation of solutions (for about 25 determinations): all solutions in re-purified water (cf. Vol. II, chapter 2.1.3.2).

1. Buffer/auxiliary enzyme solution (phosphate buffer, 115 mmol/l; pH 7.1; chloride, 57.5 mmol/l; α-glucosidase, 35 U/ml):

   dissolve 730 mg $Na_2HPO_4 \cdot 2\,H_2O$ + 228 mg $NaH_2PO_4 \cdot H_2O$ + 168 mg NaCl in 40 ml water, then add 1750 U α-glucosidase from yeast, lyophilized, mix gently and dilute to 50 ml with water.

2. 4-Nitrophenyl maltoheptaoside solution (58 mmol/l):

   dissolve 370 mg 4-nitrophenyl maltoheptaoside in 5 ml water.

Stability of solutions: store all solutions at 0 – 4 °C. Solutions (1) and (2) are stable for two weeks, provided that no microbial contamination occurs.

Procedure

Collection and treatment of specimen: collect blood from the vein without stasis. Addition of heparin, 0.2 mg/ml, is permissible, whereas other anticoagulants cannot be used. Centrifuge for 10 min at about 3000 *g* in order to obtain serum or plasma. Collect urine without any additive. Use only fresh urine. Dilute urine 1 + 2 (v/v) with sodium chloride solution, 0.154 mol/l.

Stability of the enzyme in the sample: the enzyme is stable in human serum kept between 0 °C and 20 °C for at least one week. In urine the enzyme activity decreases quickly, even at – 20 °C.

Measurements in other samples: pancreatic juice or duodenal content or saliva have to be diluted 1 : 100 to 1 : 1000 (v/v) with a solution of sodium chloride, 0.154 mol/l.

Assay conditions: wavelength Hg 405 nm (400 – 415 nm); light path 10 mm; final volume 2.30 ml; 30 °C (thermostatted cuvette holder). Measure against air. Before starting the assay, adjust temperature of solutions to 30 °C.

Measurement

Pipette successively into the cuvette:			concentration in assay mixture	
reagent solution	(1)	2.00 ml	phosphate chloride α-glucosidase	100 mmol/l 50 mmol/l 30 U/ml
substrate solution	(2)	0.20 ml	4-nitrophenyl malto-heptaoside	5 mmol/l
sample (serum or other sample solution)		0.10 ml	volume fraction	0.0435
mix thoroughly with a plastic spatula. Wait for 3 min (lag phase), read absorbance and start stopwatch. Repeat the reading after exactly 1, 2 and 3 min or monitor the reaction on a recorder.				

The values $\Delta A/\Delta t$ must be < 0.15 at Hg 405 nm. Otherwise, dilute the sample 10-fold with sodium chloride solution, 0.154 mol/l, or measure at shorter intervals.

Calculation: use calculation formula (k) or (k_1), respectively (cf. "Formulae", Appendix 3) but multiply by 3, since only one molecule of 4-nitrophenol is formed from three molecules of substrate. The value of ε for 4-nitrophenol is dependent on temperature and on the assay conditions [2]; the following relations are valid for measurements at Hg 405 nm (Δt in minutes):

t	25 °C	30 °C	37 °C	
ε	0.90	0.95	1.06	($l \times mmol^{-1} \times mm^{-1}$)
$b =$	$7667 \times \Delta A/\Delta t$	$7263 \times \Delta A/\Delta t$	$6509 \times \Delta A/\Delta t$	U/l
$b =$	$1.28 \times 10^5 \times \Delta A/\Delta t$	$1.21 \times 10^5 \times \Delta A/\Delta t$	$1.08 \times 10^5 \times \Delta A/\Delta t$	nkat/l

If measurements are made at another wavelength, the appropriate value of ε for 4-nitrophenol has to be determined by calibrating with a standard solution of 4-nitrophenol under the conditions of the assay.

Validation of Method

Precision, accuracy, detection limits and sensitivity: for activities of about 100 U/l in serum, relative standard deviations in the range of 2 – 4% were found within run and from 3 – 5% from day to day, for manual performance. Data on accuracy are not available since standard reference material has not been established yet. For measurements at Hg 405 nm with sufficiently sensitive photometers, the lowest value which can be measured is approximately 20 U/l with an imprecision of $< \pm 20\%$. Sensitivity is found to be $\Delta A/\Delta t = 0.005/10$ min (*Eppendorf* photometer).

Sources of error: most commonly-used therapeutics do not influence the measured activity when added to serum in therapeutic doses (*G. Staber,* unpublished). Contamination of the reagents by saliva or sweat causes a marked interference, as they contain high activities of α-amylase. Therefore, solutions may not be pipetted by mouth and contact with skin must be avoided. Anticoagulants such as oxalate, citrate, fluoride or EDTA inhibit the reaction and cannot be used. Ammonium ions are also inhibitors. Tris(hydroxymethyl)aminomethane inhibits the auxiliary enzyme α-glucosidase.

Specificity: amylase activity is measured specifically. The presence of α-glucosidase in the sample does not interfere.

Reference ranges: in preliminary studies activities found in human serum were up to 120 U/l at 25 °C, 150 U/l at 30 °C and 200 U/l at 37 °C, and in urine up to 500 U/l at 25 °C, 650 U/l at 30 °C and 900 U/l at 37 °C (*G. Staber,* unpublished data).

References

[1] *E. O. Haegele, E. Schaich, E. Rauscher, P. Lehmann, H. Bürk, A. W. Wahlefeld,* Mechanism of Human Pancreatic and Salivary α-Amylase on α-4-Nitrophenyl-maltoheptaoside Substrate, Clin. Chem. *28*, 2201 – 2205 (1982).
[2] *E. Rauscher, S. v. Buelow, E. O. Haegele,* Optimized Determination of α-Amylase in Serum and Urine using p-Nitrophenyl-α,D-maltoheptaoside as Substrate, Fresenius Z. Anal. Chem. *311*, 454 (1982).
[3] *G. Staber, G. Möller, E. Rauscher,* Evaluation of a Kinetic Colour-Test for the Determination of α-Amylase, LAB *IX*, N. 2, Abstract B 9 (1982).
[4] *R. A. Kaufman, N. W. Tietz,* Recent Advances in Measurement of Amylase Activity – A Comparative Study. Clin. Chem. *26*, 846 – 853 (1980).
[5] *E. Munz,* Erfahrungen mit einem Farbtest zur Bestimmung von α-Amylase, Fresenius Z. Anal. Chem. *311*, 453 (1982).
[6] *J. Fenton, R. Foery, L. Piatt, K. Geschwindt,* A New Chromogenic Amylase Method Compared with Two Established Methods, Clin. Chem. *28*, 704 – 706 (1982).
[7] *H. David,* Hydrolysis by Human α-Amylase of p-Nitrophenyloligosaccharides Containing Four to Seven Glucose Units, Clin. Chem. *28*, 1485 – 1489 (1982).
[8] *K. Wallenfels, G. Laule, B. Meltzer,* Action Pattern of Human Pancreatic and Salivary α-Amylase on 1,4-α-D-Nitrophenylmaltooligosaccharides, J. Clin. Chem. Clin. Biochem. *20*, 581 – 586 (1982).

2.1.4 Determination with Coloured Insoluble Substrates

August Wilhelm Wahlefeld

General

Application of method: in biochemistry, clinical chemistry and food chemistry [1].

Enzyme properties relevant in analysis: the catalytic activity of α-amylase in serum or urine is due to at least two distinct isoenzymes, α-amylase from pancreas and from saliva. The two isoenzymes show different electrophoretic mobility (for details see [2]) and different inhibition characteristics when using α-amylase inhibitors from plant seeds. *O'Donnel & McGeeney* [3] showed that the lectin type of inhibitor, isolated from wheat germ, inhibits pancreatic α-amylase activity up to 20% whereas salivary α-amylase activity is inhibited up to 80%, using human serum as samples and an assay procedure with coloured insoluble substrates as described below.

Methods of determination: in 1967, *H. Rinderknecht et al.* [4] described an entirely novel substrate for the determination of the catalytic activity of α-amylase. Insoluble starch is stained under alkaline conditions with the reactive dye Remazol Brilliant Blue R (manufacturer: *Farbwerke Hoechst AG,* F. R. Germany). The result is a covalently-labelled, insoluble substrate for the determination of α- or β-amylase activities. After incubation under defined conditions, the cleavage of the coloured substrate is stopped with the aid of a suitable precipitant, with simultaneous precipitation of the insoluble and excess components of the substrate. The soluble coloured cleavage products are determined photometrically in the supernatant.

Numerous modifications of methods based on this principle have been published. The procedure has been well accepted in routine clinical chemical laboratories for practical reasons, although an only two-point-kinetic approach is possible and the procedure is not applicable to automated instruments.

A survey of different modifications is given in the following table:

Dyed amylopectin	Dyed starch	Dyed amylose
Babson et al. [5] Reactone Red 2B	*Ceska et al.* [7] Cibachron Blue F 3 G-A	*Klein et al.* [9, 10] Cibachron Blue F 3 G-A
Sax et al. [6] Procion Brilliant Red M-2 BS	*Hall et al.* [8] Remazol brilliant Blue R	

Additional dyes, given in this table, are manufactured by *Ciba AG,* Switzerland (for Cibachron Blue F 3 G-A and Reactone Red 2 B) and *ICI America* (for Procion Brilliant Red M-2 BS).

The use of dyed amylose has been investigated in more detail: *Ewen* [11] described parameters influencing the synthesis and optimal conditions. More recently, *Klein* [12] investigated the mechanism of the hydrolysis of this substrate by amylase.

However, the method for determination of α-amylase activity using the dyed starch substrate developed by *Ceska et al.* [13, 14] is described below because of its broad acceptance in routine clinical chemistry laboratories.

A comparative evaluation of various methods with insoluble substrates, including the method according to *Ceska,* was published by *Rosalki* in 1973 [15].

International reference methods and standards: at the present time, no international reference method has been achieved and no standard reference material has been accepted.

Enzyme effectors: α-amylase catalytic activity depends on the presence of calcium ions as well as monovalent anions, e.g. chloride; bromide is a less effective activator. Metal chelating agents are therefore inhibiting; however, higher concentrations of chloride (100 mmol/l) or phosphate anions are also inhibitory.

Assay

Method Design

Principle

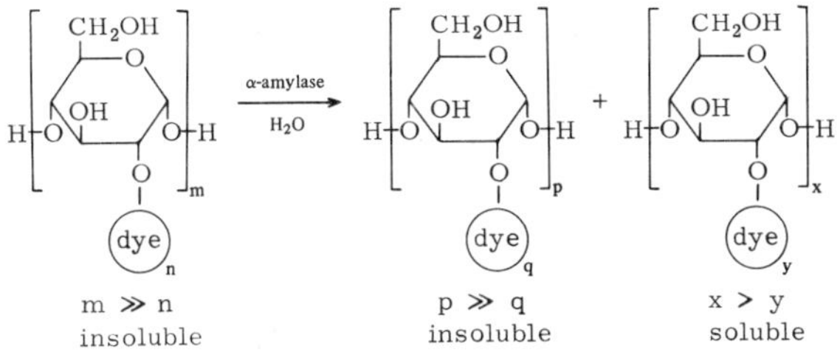

α-amylase, H_2O

m ≫ n insoluble → p ≫ q insoluble + x > y soluble

Soluble starch is stained with the dye Cibachron Blue F 3 G-A and made insoluble by cross-linking with 1.4-butanediol glycid ether. By this procedure the substrate becomes capable of swelling in buffer solutions. The reaction product has been called "Blue Starch Polymer". The action of α-amylase results in soluble coloured cleavage products. After the reaction has been stopped, the insoluble products are separated by centrifugation: the soluble products can then be determined photometrically in the supernatant. The α-amylase activity is determined by comparison with a standard. The reaction is irreversible. A reagent blank is necessary for each series of measurements. Application to automated instruments is not possible.

Optimized conditions for measurement: substrate saturation of the α-amylase of serum is reached above 30 mg of Blue Starch Polymer per ml. However, when considering length of incubation time at 30 °C (or 37 °C) and the photometrically measurable range (i.e. absorbance below 2.0), 10 mg substrate per ml will also provide zero-order kinetics for up to 15 min at 30 °C with catalytic activity concentrations of α-amylase up to 2000 U/l at 30 °C.

The pH optimum is relatively sharp at 7.0; the dependence of activity on the concentration of phosphate and of sodium chloride is less pronounced. Therefore, a sodium phosphate buffer, 20 mmol/l, pH 7.0, containing sodium chloride, 50 mmol/l is suitable.

The reaction mixture should not be incubated for more than 30 min. An incubation period of 30 min may be necessary, for example to measure sub-normal activities more precisely. However, at intermediate activities the reaction rate frequently increases with longer incubation times, since the substrates with decreasing chain-length formed during the reaction react more rapidly.

The enzyme reaction is completely stopped by adding alkaline reagents.

Temperature conversion factors: *Soininen et al.* [16] determined the dependence of this assay on temperature to be 4.73 ± 0.59 (1 SD) % per °C between 30 °C and 37 °C. Calculated factors are therefore

measured	factor for conversion to	
°C	30 °C	37 °C
30	1.0	1.49
37	0.67	1.0

Equipment

Spectrophotometer or spectral-line photometer capable of exact measurement at 620 nm, or Hg 623 nm, Hg 578 nm; thermostatted cuvette holder. The water-bath for incubation of assays must keep the temperature constant within ± 0.1 °C. Stopwatch; magnetic stirrer; vortex mixer.

Reagents and Solutions

Purity of reagents: only substances of A. R. quality should be used. The dyed substrate, Blue Starch Polymer, was obtained from *Pharmacia*, Uppsala, Sweden. Bovine serum albumin should be of the highest purity available. α-Amylase from hog pancreas can be used as a standard; it should be a crystalline preparation.

Preparation of solutions: (for about 25 determinations): all solutions in re-purified water (cf. Vol. II, chapter 2.1.3.2).

1. Phosphate solution (phosphate buffer, 20 mmol/l, pH 7.0; sodium chloride, 50 mmol/l; sodium azide, 0.2 mg/ml):

   dissolve 3.58 g $Na_2HPO_4 \cdot 12\ H_2O$ in 500 ml water, adjust the pH accurately to 7.0 by adding a solution of 1.38 g $NaH_2PO_4 \cdot H_2O$ in 500 ml water. To 100 ml of this phosphate buffer, pH 7.0, add 292.2 mg sodium chloride and 20 mg of sodium azide for preservation.

2. Substrate/buffer solution (phosphate buffer, 20 mmol/l, pH 7.0; sodium chloride, 50 mmol/l; sodium azide, 0.2 mg/ml; Blue Starch Polymer, 10 mg/ml):

   shortly before use, suspend 1.0 g Blue Starch Polymer in 100 ml phosphate solution (1). On pipetting, stir well with a magnetic stirrer.

3. Sodium hydroxide solution (sodium hydroxide 0.5 mol/l):

   dissolve 20 g sodium hydroxide in water and make up to 1.0 l.

4. α-Amylase standard solution (approx. 1000 U/l):

   dissolve 1 mg α-amylase from hog pancreas, crystalline suspension in ammonium sulphate solution, 3.2 mol/l, in 10 ml of a solution of 6 g bovine serum albumin in 100 ml phosphate solution (1). (Stock solution; stable for at least 4 weeks at 4 °C). Make a dilution of 1:100 of this stock solution with the same 6% bovine serum albumin solution as required (concentration of α-amylase approx. 1000 U/l). Determine the exact α-amylase activity in this solution by the saccharogenic method (cf. p. 446 in ref. [14]).

Stability of solutions: store all solutions at 0 – 4 °C. The phosphate solution (1) and sodium hydroxide solution (3) are stable as long as no microbial contamination occurs. Prepare substrate/buffer solution (2) daily. α-Amylase standard solution is stable for 5 days.

Procedure

Collection and treatment of specimen: collect blood from vein without stasis. Obtain serum in the usual manner; only heparin plasma (0.2 mg heparin/ml) can be used. Urine samples should be stabilized [16] by dilution 1:1 with a solution of 500 mg bovine serum albumin in 100 ml phosphate solution (1).

Stability of the enzyme in the sample: the enzyme in serum and in stabilized urine is stable for at least 1 week at 20 – 25 °C, or for 2 weeks at +4 °C. However, it should be mentioned that, during storage, α-amylase may be transformed into more anionic forms by non-enzymatic deamidation without a significant change in total catalytic activity [17].

Assay conditions: wavelength 620 nm or Hg 578 nm; light path 10 mm; final volume 5.2 ml; 30 °C (or 37 °C). Measure against reagent blank. Before starting the assay, adjust temperature of substrate/buffer solution (2) to 30 °C (or 37 °C).

Measurement

<table>
<tr><th>Pipette into test tubes:</th><th>sample or standard</th><th>blank</th><th colspan="2">concentration in assay mixture</th></tr>
<tr><td>substrate/buffer solution (2)</td><td>4.0 ml</td><td>4.0 ml</td><td>phosphate
NaCl
NaN_3
Blue Starch Polymer</td><td>20 mmol/l
50 mmol/l
0.2 mg/ml
10 mg/ml</td></tr>
<tr><td colspan="3">maintain for 5 min in thermostatic water-bath at 30 °C (or 37 °C)</td><td colspan="2"></td></tr>
<tr><td>sample (serum or standard)
water</td><td>0.2 ml
–</td><td>–
0.2 ml</td><td colspan="2"></td></tr>
<tr><td colspan="3">mix quickly (vortex mixer), incubate exactly for 15 min at 30 °C (or 37 °C)</td><td colspan="2"></td></tr>
<tr><td>sodium hydroxide solution (3)</td><td>1.0</td><td>1.0</td><td>NaOH</td><td>98 mmol/l</td></tr>
<tr><td colspan="3">mix quickly (vortex mixer; reaction is stopped). Centrifuge hard or filter. Measure absorption of sample (A_{sample}) and standard ($A_{standard}$) against blank in the filtrates or supernatants.</td><td colspan="2"></td></tr>
</table>

When urine is used as sample, pipette 0.1 ml; also, 0.1 ml of the standard is pipetted.
When the measured values exceed $\Delta A > 1.5$ at Hg 578 nm, the serum sample has to be diluted 5-fold with phosphate solution (1), or urine samples with the phosphate solution (1), containing bovine serum albumin, 2 mg/ml.

Calculation: 1 U denotes the formation of one reducing group per min at 30 °C (cf. ref. [14]). For calculation of the catalytic concentration of the sample use eqn. (m_1), cf. "Formulae", Appendix 3:

$$b = \frac{A_{\text{sample}}}{A_{\text{standard}}} \times b_{\text{standard}} \qquad \text{U/l, nkat/l.}$$

Validation of Method

Precision, accuracy, detection limits and sensitivity: with an α-amylase activity in serum of 321 U/l, the imprecision within series is 1.1%, and from day to day less than 5%. Data about accuracy cannot be given, since reference methods or standard reference material are not established yet.
For measurements at Hg 578 nm the lowest possible value which can be measured is approximately 20 U/l; for example 34 U/l were measured with an imprecision of 12.2%.

Sources of error: the following interferences in the assay technique should be stressed: if the precipitate is not centrifuged down very efficiently, coloured particles are transferred into the cuvette with the supernatant and cause erroneous absorption values. Therefore, filtration is recommended.
Contamination with α-amylase from saliva and/or sweat from the hands should be avoided by all means. Detergents can interfere, as can citrate- or EDTA-treated plasma. Effects of drugs and other therapeutic measures have been reported in single cases, except for corticotropin and ethanol [18].

Reference ranges: in human serum, 80 – 300 U/l at 37 °C is reported as the reference range, with a slight sex dependence in the upper normal range: males 282 U/l; females 226 U/l [14]. The same authors report 100 – 1500 U/l at 37 °C for urine samples [14].
The values obtained by this and by other methods are not strictly comparable.

References

[1] *P. R. Mathewson, Y. Pomeranz,* Modified Chromogenic α-Amylase Assay for Sprouted Wheat, J. Assoc. Anal. Chem. *62*, 198 – 200 (1979).
[2] *B. K. Gillard,* Quantitative Gel-Electrophoretic Determination of Serum Amylase Isoenzyme Distributions, Clin. Chem. *25*, 1919 – 1923 (1979).

[3] *M. D. O'Donnel, K. F. McGeeney,* Purification and Properties of an α-Amylase Inhibitor from Wheat, Biochim. Biophys. Acta *422*, 159–169 (1976).
[4] *H. Rinderknecht, P. Wilding, B. J. Haverback,* A New Method for the Determination of α-Amylase, Experientia (Basel) *23*, 805 (1967).
[5] *A. L. Babson, S. A. Tenney, R. E. Megraw,* New Amylase Substrate and Assay Procedure, Clin. Chem. *16*, 39–43 (1970).
[6] *S. M. Sax, A. B.Bridgewater, J. J. Moore,* Determination of Serum and Urine Amylase with Use of Procion Brilliant Red M-2BS Amylopectin, Clin. Chem. *17*, 311–315 (1971).
[7] *M. Ceska, E. Hultman, B. G. A. Ingelman,* A New Method for Determination of α-Amylase, Experientia (Basel) *25*, 555–556 (1969).
[8] *F. F. Hall, T. W. Culp, T. Hayakawa, C. R. Ratliff, N. C. Hightower,* An Improved Amylase Assay Using a New Starch Derivative, Am. J. Clin. Pathol. *53*, 627–634 (1970).
[9] *B. Klein, J. A. Foreman, R. L. Searcy,* The Synthesis and Utilization of Cibachron Blue-Amylose: A New Chromogenic Substrate for Determination of Amylase Activity, Anal. Biochem. *31*, 412–425 (1969).
[10] *B. Klein, J. A. Foreman, R. L. Searcy,* New Chromogenic Substrate for Determination of Serum Amylase Activity, Clin. Chem. *16*, 32–38 (1970).
[11] *L. M. Ewen,* Synthesis of Cibachron Blue F 3 G-A-Amylase with Increased Sensitivity for Determination of Amylase Activity, Clin. Chim. Acta *47*, 233–245 (1973).
[12] *B. Klein, J. A. Foreman,* Amylosis of a Chromogenic Substrate, Cibachron Blue F 3 G-A-Amylose, Kinetics and Mechanism. Clin. Chem. *26*, 250–258 (1980).
[13] *M. Ceska, K. Birath, B. Brown,* A New and Rapid Method for the Clinical Determination of α-Amylase Activities in Human Serum and Urine. Optimal Conditions, Clin. Chim. Acta *26*, 437–444 (1969).
[14] *M. Ceska, B. Brown, K. Birath,* Ranges of α-Amylase Activities in Human Serum and Urine and Correlations with Some Other α-Amylase Methods, Clin. Chim. Acta *26*, 445–453 (1969).
[15] *S. B. Rosalki, D. Tarlow,* Amylase Determination Using Insoluble Substrates, Ann.Clin. Biochem. *10*, 47–52 (1973).
[16] *K. Soininen, M. Härkäonen, M. Ceska, H. Adlerkreutz,* Comparison between a New Chromogenic α-Amylase Test (Phadebas) and the Wohlgemuth Amyloclastic Method in Urine, Scand. J. Clin. Lab. Invest. *30*, 291–297 (1972).
[17] *K. Lorentz, B. Flatter,* Studies on Isoamylase Formation in Biological Fluids, Enzyme *24*, 163–168 (1979).
[18] *T. W. Challis, L. C. Reid, J. W. Hinton,* Study of Some Factors which Influence the Level of Serum Amylase in Dogs and Humans, Gastroenterology *33*, 818 (1957).

2.1.5 Isoenzymes of α-Amylase

Ying Foo and Sidney B. Rosalki

General

In man, the pancreas and salivary glands are the major producers of amylase and the source of the amylase activity of plasma, serum and urine. Human amylase exists in multiple molecular forms or isoamylases of pancreatic (P-) and salivary (S-) type.

These are true isoenzymes since the amylases secreted by the exocrine pancreas and by the salivary glands are coded by separate gene loci which are closely linked on the same chromosome – chromosome 1 [1]. Allelic variation occurs at each gene locus, giving rise to genetic polymorphism with P- and S-subfractions [2]. In addition, isoamylases undergo post-translational modification by addition or removal of carbohydrate residues and by deamidation of asparagine and glutamine moieties to produce additional multiple forms [3].

Application of method: serum isoamylase determination improves the diagnostic specificity of total amylase determination by indicating the tissue of origin of the serum amylase. When serum total amylase activity is increased (hyperamylasaemia), separation of serum isoamylases into those of pancreatic and salivary type can reveal or exclude involvement of the pancreas. Serum pancreatic isoamylase activity is elevated in pancreatic inflammation, as in acute pancreatitis. Serum salivary isoamylase activity is increased in disease of the salivary glands, such as mumps and parotitis. Isoamylase of salivary type is also increased in serum in a variety of other diseases which can result in elevated total amylase activity; for example, diabetic keto-acidosis, inflammatory and neoplastic lung disease and ruptured ectopic pregnancy [4].

In macroamylasaemia, isoamylases of abnormally high molecular weight are present in plasma, usually as a result of binding to immunoglobulins to form isoenzyme-immunoglobulin complexes [5]. Macroamylasaemia may result in the finding of unexpectedly high total amylase activity accompanied by isoamylase of abnormal electrophoretic mobility, but appears to be without any specific disease association. In the presence of hyperamylasaemia, it is important to distinguish macroamylasaemia from pancreatic disease.

When pancreatic exocrine function is impaired, as in chronic pancreatitis and cystic fibrosis, serum pancreatic isoamylase activity becomes markedly diminished. This may occur despite total amylase activity remaining within the usual reference range.

Enzyme properties relevant in analysis: amylase activity in serum is a composite of the action of pancreatic and salivary isoamylases and their subfractions. Although isoamylases do not differ appreciably in their biochemical and immunological characteristics, they can be distinguished by their differences in certain physicochemical properties. These include net molecular charge, ionization properties, rate of reaction with different substrates and response to specific inhibitors.

Methods of determination: since the discovery of amylase heterogeneity, many methods have been used to demonstrate and quantitate isoamylase activity in serum and other biological fluids. These include electrophoresis, chromatography, isoelectric focusing, substrate differentiation, selective inhibition and immunological techniques.

Electrophoresis is the method most widely used because of its simplicity and separating power. A variety of supporting media – paper, agar, polyacrylamide gel, cellulose acetate and agarose gel – have all been used [6]. Paper electrophoresis gives poor resolution and is no longer used. Agar, polyacrylamide and agarose gels provide satisfactory separations but may require laborious gel preparation and prolonged separation times. Cellulose acetate is very satisfactory because of its good resolution, ready availability, speed of separation and quantitation, convenience and easy handling. In general, electrophoretic techniques separate serum isoamylases into the two major P- and S-fractions and into two to four minor subfractions of each, all of gamma globulin mobility.

Ion-exchange chromatography using positively charged, cross-linked dextran gel (DEAE or QAE Sephadex A-50) as the ion exchanger separates isoamylases into two major bands. Demonstration of subfractions of isoamylases requires a more complex system. The serum sample is first passed through a Sephadex G-75 column to yield two peaks, each of which is then passed through DEAE Sephadex A-50. The full chromatographic technique is slow and labour-intensive. Simplified and abbreviated methods have been described which use QAE Sephadex A-50 and DEAE-cellulose mini-columns and are more applicable to the routine clinical biochemistry laboratory [7, 8].

Isoelectric focusing separates isoamylases electrophoretically in a pH gradient, each isoamylase migrating to a position determined by its "isoelectric point" (pI). pI is defined as the pH at which the positive and negative charges of a protein molecule are equal. Serum isoamylases can be separated into three major peaks: pancreatic isoamylase at pI 7.0 and salivary isoamylase at pI 5.9 and 6.5 [9]. Isoelectric focusing has been claimed to give higher resolution and better separation of isoamylases than electrophoresis [10]. However, technical complexities, time and expense have limited the application of the procedure.

Isoamylase analysis by substrate differentiation has not been widely used. Catalytic dissimilarities between pancreatic and salivary isoamylases have been shown by the different rates at which they digest various plant starches [11]. It has been observed that salivary isoamylase does not hydrolyze maltotriose, while pancreatic isoamylase splits maltotriose to glucose and maltose [12]. A further difference is that salivary isoamylase has a greater affinity for starch, while pancreatic isoamylase has greater affinity for glycogen.

Naturally-occurring inhibitors of amylase from wheat, rye, kidney beans and colocasia tuber have been known for many years. A procedure for differential isoamylase determination in serum and urine using the inhibitor from wheat has been described and is available commercially [13]. The inhibitor inhibits salivary isoamylase by 90% and pancreatic isoamylase by only 20%. A limitation of the method is the inability to analyze samples with extremely low (less than 9%) or extremely high (greater than 89%) pancreatic isoamylase; also, isoamylase subfractions cannot be demonstrated.

Attempts at immunological differentiation between human pancreatic and salivary isoamylases have usually failed because S- and P-forms cross-react. A radioimmunoassay for human salivary isoamylase with very low (1%) cross-reactivity with pancreatic isoamylase has been described [14] but most radioimmunoassays for human pancreatic isoamylase have shown substantial cross-reactivity with the salivary isoamylase [15, 16].

International reference method and standards: standardization at the international level has not been carried out and standard reference methods and materials are not available.

Enzyme effectors: amylase requires calcium ions for activity and hence can be inhibited by EDTA, oxalate and citrate. Chloride ion is required at a concentration of not less than 0.01 mol/l.

Assay

Method Design

Principle: because of the advantages of electrophoresis over other isoamylase measurement procedures, and of cellulose acetate over other separative media (see above), we have chosen to describe an electrophoretic procedure for isoamylase measurement using cellulose acetate as supporting medium [17, 18]. This procedure is particularly suitable for use in the clinical biochemistry laboratory.

The basis of electrophoretic separation is the tendency of isoamylases to migrate at different rates in an electric field at alkaline pH as a result of their differences in their net molecular charge. Isoamylases are separated on cellulose acetate membrane in a discontinuous Tris-EDTA-borate/barbitone buffer system. The isoamylases are demonstrated by incubating the membrane with substrate incorporated in an agar gel. The enzyme substrate is an insoluble starch-dye complex and amylase activity causes the release of a soluble blue dye. The isoamylases appear on the membrane as blue-staining bands which are quantitated by reflectance densitometry. Sucrose is incorporated in the gel to prevent diffusion; it is not inhibitory at the concentration used.

Optimized conditions for measurement: the method described is that found by experimental variation of reaction conditions to give the most satisfactory and convenient resolution of isoamylase fractions.

Equipment

1. Electrophoretic tank

   Any tank for use with cellulose acetate membrane is suitable but we have found the *Beckman* Microzone R101 (*Beckman-RIIC Ltd.,* High Wycombe, Bucks) most satisfactory and convenient.

2. Direct current power supply

   Any suitable apparatus capable for producing a constant voltage of 250 V.

3. Reflectance densitometer

   Any suitable densitometer which can measure accurately and reproducibly at 620 nm wavelength; we have found the *Corning* Densitometer Model 720 (*Corning Medical*) satisfactory.

4. Water-baths, 37 °C and 56 °C

5. Centrifuge

6. Spectrophotometer

Reagents and Solutions

Purity of reagents: all reagents should be of Analytical Reagent Grade.

Preparation of solutions: all solutions in re-purified water (cf. Vol. II, chapter 2.1.3.2).

1. Electrophoretic buffers:

   a) Tris/EDTA/borate buffer (Tris, 0.2 mol/l; EDTA, 8.6 mmol/l; boric acid, 0.03 mol/l; pH 9.1 at 20 °C):

      dissolve 25.2 g tris (hydroxymethyl)-aminomethane, 2.5 g ethylene diamine-tetraacetic acid (EDTA) and 1.9 g boric acid in 800 ml water, adjust pH to 9.1 at 20 °C with hydrochloric acid, 1 mol/l and dilute to 1 l with water.

   b) Barbitone buffer (sodium diethyl barbitone, 0.02 mol/l; diethyl barbituric acid, 5 mmol/l; pH 8.6 at 20 °C):

      dissolve 5.15 g sodium diethyl barbitone and 0.92 g diethyl barbituric acid in 800 ml water, adjust pH to 8.6 at 20 °C and dilute to 1 l with water.

2. Buffer mixture:

   dissolve 1 g bovine serum albumin (BSA) in 50 ml Tris/EDTA/borate buffer (1a) and 50 ml barbitone buffer (1b).

3. Substrate solution:

   several chromogenic amylase substrates are suitable. We have found Cibachron-Blue F3GA-Starch polymer in the form of Phadebas tablets (*Pharmacia,* Sweden) to be a sensitive and convenient substrate formulation. Dissolve completely 5 Phadebas tablets and 2 g sucrose with 5 ml water.

4. Agar (2% w/v):

   dissolve 2 g Agar Noble (*Difco Laboratories,* West Surrey), in 100 ml water by heating in a boiling water-bath. Distribute in 5 ml aliquots in screw-capped glass bottles.

Stability of solutions: the electrophoretic buffer solutions and buffer mixture are stable for at least one month at 4 °C as long as no microbial contamination occurs. The electrophoresis buffer in the tank can be used for three separations. Prepare the substrate solution freshly before use. The agar gel is stable at least three months at room temperature.

Procedure

Collection and treatment of sample: serum is used in preference to plasma to avoid any possible anticoagulant interference.

Stability of the enzyme in the sample: we have found total amylase activity and iso-amylase patterns in serum to be unaltered (average less than 7% variation) for at least one month at −18 °C. Others have reported stable isoamylase patterns for up to 18 months at −20 °C [19] and unaltered activity despite twenty freeze-thaw cycles [20].

Assay conditions: electrophoretic separation is carried out at a constant voltage of 250 V at 4 °C.

Measurement

1. Determine total amylase activity in the serum sample using any suitable method (cf. chapters 2.2.1 – 2.2.3).

2. Fill the anode compartment of the electrophoretic tank with Tris/EDTA/borate buffer (1a) and the cathode compartment with barbitone buffer (1b).

3. Pour the buffer mixture (2) into a rectangular plastic container. Carefully float the cellulose acetate membrane onto the buffer surface, avoiding air bubbles. Allow buffer to soak through fully to the upper surface of the membrane and immerse it completely by gentle agitation of the container. After soaking for one to two minutes, remove the membrane, lightly blot off excess buffer with lint-free absorbent paper (e.g. Postlip, *Raymond A. Lamb,* London, UK) and immediately transfer the membrane to the supporting frame in the tank, taking care that both its ends are immersed in the respective buffer compartments.

4. Apply the serum sample to the membrane 10 mm anodal to the midline with a suitable applicator. The volume of sample to be applied will vary with the apparatus and the total amylase activity. Using the *Beckman* Microzone system and applicator (0.25 μl sample per application), the number of applications required is calculated (to the nearest whole number) by dividing 2000 by the sample total amylase activity so that approximately 0.5 mU of amylase activity at 37 °C is applied. With sample amylase activity below 200 U/l 10 applications are used, and with amylase activities above 2000 U/l the sample is diluted with BSA dissolved in sodium chloride, 0.15 mol/l, at a concentration of 10 g/l. Human saliva (diluted to suitable activity as above) and acute pancreatitis serum or diluted human autopsy pancreatic homogenate or human pancreatic juice are separated in parallel with the serum samples as location markers for salivary and pancreatic isoamylases respectively.

5. Following application, close the tank lid and separate samples at a constant voltage of 250 V for 1.5 hours at 4 °C in the cold room, or, with appropriate safety precautions, in the refrigerator.

6. During separation a substrate-agar gel is prepared. Melt an aliquot of 2% agar solution (4) in a beaker of boiling water, then transfer to a 56 °C water-bath to cool. Warm the substrate solution (3) in the 37 °C water-bath. Mix the warmed substrate solution and cooled melted agar and immediately pour into a shallow plastic tray, avoiding air bubbles. Trays of dimension 100 × 70 × 2 mm are suitable for *Beckman* Microzone membrane. The gel should be poured when the agar has cooled to 56 °C; if poured while it is at a higher temperature, the substrate will settle to the bottom of the tray and will not be available for satisfactory contact with the enzyme.

7. Allow the gel to set on a level surface protected from dust. Five minutes is usually adequate.

8. When electrophoresis is completed, remove the membrane from the tank, trim off the ends soaked in the buffer, and gently layer the membrane face down onto the surface of the substrate-agar gel, avoiding air bubbles.

9. Place the membrane and gel in a moist chamber (e.g. a plastic box with moistened absorbent paper) and incubate for one hour at 37 °C.

10. Remove the membrane from the gel and completely dry under pressure between absorbent papers or in a hot air oven for 15 minutes at 56 °C.

11. When completely dry, quantitate the isoamylases by scanning at 620 nm using a reflectance densitometer capable of integrating the area under each peak and expressing the proportion of each as a percentage of the total activity. Summate for calculation of P-and S-activity subfractions.

Calculation: for calculation see also Vol. III, chapter 1.3.3, p. 88.

catalytic activity concentration of P-isoamylase in the sample:

$$b = \text{TA} \times \frac{\%\text{P}}{100} \quad \text{U/l}$$

catalytic activity concentration of S-isoamylase in the sample:

$$b = \text{TA} \times \frac{\%\text{S}}{100} \quad \text{U/l}$$

where

TA = total catalytic activity concentration of amylase in the sample
%P = percentage of P-isoamylase activity of total amylase activity, and
%S = percentage of S-isoamylase activity of total amylase activity.

Serum isoamylase patterns

$P_1P_2P_3S_1S_2$

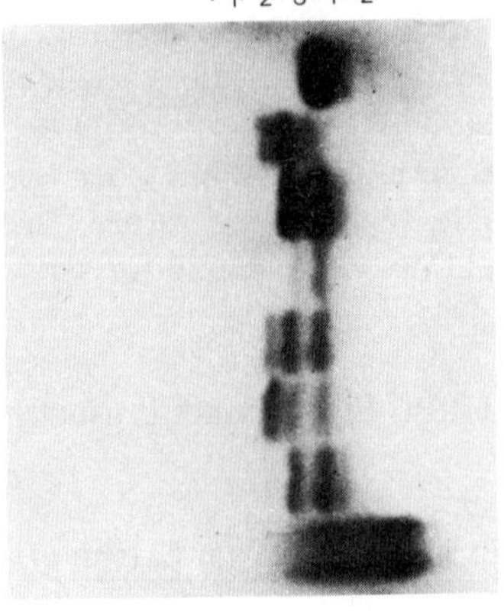

1 Saliva
2 Pancreatitis
3 Hyperamylasaemia
4 Pancreatic insufficiency
5 P_1 heterozygote
6 P_1 homozygote
7 Normal
8 Macroamylasaemia

Cathode Anode

Interpretation: isoamylases are demonstrated as sharp blue-stained bands, present only in the gamma globulin region with clear separation of pancreatic and salivary isoamylases and their subfractions. Pancreatic isoamylase shows greater mobility towards the cathode.

On cellulose acetate, up to three isoamylases in pancreatic extracts and three in saliva have been demonstrated. They are consecutively numbered from the cathode P1, P2 and P3, and S1, S2 and S3. Normal serum usually shows only P2 and S1 fractions, with the pancreatic fraction predominating, and occasionally the additional presence of S2 isoamylase. Polymorphism of P-isoamylase results in the appearance of P1 isoamylase in serum. The presence of a P1 band accompanying the usual P2 (a pattern found in less than 10% of Europeans) indicates heterozygosity for the P1 variant. In homozygotes (1% of Europeans), P1 is the principal pancreatic isoamylase and P2 is diminished. Approximately 1% of healthy Europeans may lack S1 isoamylase.

P3 isoamylase, migrating between the principal P2 and S1 fractions, does not normally appear in the serum, but appears in the serum of patients with acute pancreatitis and is highly specific for this condition [21]. It has been suggested that it originates from inflammatory intra-pancreatic proteolytic breakdown of pancreatic amylase [22].

Macroamylasaemia (present in up to 1% of healthy subjects) is generally easily recognizable on cellulose acetate by diffuse streaking in the gamma globulin region.

Validation of Method

Precision, accuracy, detection limits and sensitivity: the precision of the method will depend on the performance of the densitometer etc. Using the instrumentation described above, we have found the average relative standard deviation to be 0.05 (5%) within-batch and 0.06 (6%) between-batch, at a mean pancreatic isoamylase activity (154 U/l) representing 46% of the total activity at 37 °C (334 U/l). The recovery of both pancreatic and salivary isoamylases was found to average 100% despite a seven-fold variation in the relative activity of each and with activity of each fraction ranging between 100 to 700 U/l. A linear relationship has been found between staining intensity and isoamylase activity over the range of total activity from 266 to 1600 U/l and activity of pancreatic and salivary isoamylases from 130 to 800 U/l. Data about accuracy are not available since standard reference material is not established yet. The detection limit of the staining procedure is less than 5 U/l with a 2.5 µl sample size.

Sources of error: no special sources of error that would not also apply to the determination of total amylase activity, are known.

Specificity: pancreatic isoamylase is specifically related to and solely formed in the pancreas. Salivary isoamylase, on the other hand, lacks organ specificity, for isoamylase of salivary type has been recovered from other sources, such as lung, tumours, ovary and fallopian tubes.

Reference ranges: the reference values of serum in published data differ as a consequence of differing methodology. We studied 100 apparently healthy blood donors, 50 males and 50 females with ages ranging from 20 to 67 years. After the exclusion of three outliers (including one macroamylasaemia), serum total amylase, P- and S-isoamylase activities all showed *Gaussian* distributions. Serum total amylase and salivary isoamylase activities showed no sex difference whereas a significant sex difference was demonstrated in serum pancreatic isoamylase activity, with higher mean values observed in the females.

Table 1. Reference values for serum amylase, pancreatic and salivary isoamylase activities (30 °C)

	All subjects (97) mean ± SD	Male (48) mean ± SD	Female (49) mean ± SD
Total serum amylase (U/l)	209 ± 52	205 ± 41	213 ± 61
Pancreatic isoamylase (U/l)	115 ± 27	108 ± 27*	122 ± 25*
Salivary isoamylase (U/l)	95 ± 41	97 ± 57	93 ± 44

* male cf. female $P < 0.02$.

Similar findings have been reported by others [13] though not all [4, 19, 20]. In general, serum pancreatic isoamylase activity has been reported to account for approximately half (range 30 to 70%) of total amylase activity.

In serial studies of healthy subjects, extending over a period of one month, we have found no significant physiological variation in serum total amylase activity, isoamylase activities and isoamylase patterns. We have also found that serum isoamylase activities and composition are unaffected by fasting and are unaltered in the post-prandial state. Similar findings have been reported by others [22].

Salivary isoamylase is present at very low activity at birth and activity rises steadily with age reaching adult levels at the age of five. The pancreatic isoamylase develops later; infants of less than three months often have no detectable pancreatic isoamylase, the adult level of pancreatic isoamylase has been variously reported to be attained between the ages of two and eight [23, 24] and ten and fifteen [22].

References

[1] *A. D. Merritt, E. W. Lovrien, M. L. Rivas et al.,* Human Amylase Loci: Genetic Linkage with the Duffy Blood Group Locus and Assignment to Linkage Group 1, Am. J. Hum. Genet. *25*, 523 – 538 (1973).

[2] *R. Laxova, J. Kamaryt,* Genetic Aspects of Human Amylase Heterogeneity, FEBS Symposium *18*, 335 – 339 (1970).

[3] *R. G. Karn, J. D. Shulkin, A. D. Merritt et al.,* Evidence for Post-translational Modification of Human Salivary Amylase (AMY_1) Isozymes, Biochem. Genet. *10*, 341 – 350 (1973).

[4] *J. E. Berk, L. Fridhandler,* Hyperamylasemia: Interpretation and Newer Approaches to Evaluation, Adv. Int. Med. *26*, 235 – 264 (1981).

[5] *L. Fridhandler, J. E. Berk,* Macroamylasemia, in: *O. Bodansky, A. L. Latner* (eds.), Advances in Clinical Chemistry, Vol. *20*, Academic Press, New York 1978, pp. 267 – 286.

[6] *S. Meites, S. Rogols,* Amylase Isoenzymes, CRC Crit. Rev. Clin. Lab. Sci. *2*, 103 – 138 (1971).

[7] *L. Fridhandler, J. E. Berk,* Simplified Chromatographic Method for Isoamylase Analysis, Clin. Chim. Acta *101*, 135 – 138 (1980).

[8] *J. Štěpán, J. Škrha,* Measurement of Amylase Isoenzymes in Human Sera and Urine Using a DEAE-Cellulose Mini-column Method, Clin. Chim. Acta *91*, 263 – 271 (1979).

[9] *M. D. Levitt, C. Ellis, R. R. Engel,* Isoelectric Focusing Studies of Human Serum and Tissue Isoamylases, J. Lab. Clin. Med. *90*, 141 – 152 (1977).

[10] *T. Takeuchi, T. Matsushima, T. Sugimura,* Separation of Human α-Amylase Isozymes by Electrofocusing and Their Immunological Properties, Clin. Chim. Acta *60*, 207 – 213 (1975).

[11] *S. Meites, S. Rogols,* Serum Amylase, Isoamylases and Pancreatitis. I. Effect of Substrate Variation, Clin. Chem. *14*, 1176 – 1184 (1968).

[12] *M. J. Kaczmarek, H. Rosenmund,* The Action of Human Pancreatic and Salivary Isoamylases on Starch and Glycogen, Clin. Chim. Acta *79*, 69 – 73 (1977).

[13] *M. D. O'Donnell, O. FitzGerald, K. F. McGeeney,* Differential Serum Amylase Determination by Use of Inhibitor, and Design of a Routine Procedure, Clin. Chem. *23*, 560 – 566 (1977).

[14] *M. Boehm-Truitt, E. Harrison, R. O. Wolf et al.,* Radioimmunoassay for Human Salivary Amylase, Anal. Biochem. *85*, 476 – 487 (1978).

[15] *Y. Takatsuka, T. Kitahara, K. Matsuura et al.,* Radioimmunoassay for Human Pancreatic Amylase: Comparison of Human Serum Amylase by Measurement of Enzymatic Activity and by Radioimmunoassay, Clin. Chim. Acta *97*, 261 – 268 (1979).

[16] *S. P. Crouse, J. Hammond, J. Savory,* Radioimmunoassay of Human Pancreatic Amylase in Serum and Urine, Res. Commun. Chem. Path. Pharmacol. *29*, 513 – 525 (1980).

[17] *S. B. Rosalki,* A Direct Staining Technique for Amylase Isoenzyme Demonstration, J. Clin. Pathol. *23*, 373 – 374 (1970).

[18] *T. J. Davies,* A Fast Technique for the Separation and Detection of Amylase Isoenzymes Using a Chromogenic Substrate, J. Clin. Pathol. *25*, 266 – 267 (1972).

[19] *B. K. Gillard,* Quantitative Gel-electrophoretic Determination of Serum Amylase Isoenzyme Distributions, Clin. Chem. *25*, 1919 – 1923 (1979).

[20] *P. Leclerc, J. C. Forest,* Electrophoretic Determination of Isoamylases in Serum with Commercially Available Reagents, Clin. Chem. *28*, 37 – 40 (1982).

[21] *M. E. Legaz, M. A. Kenny,* Electrophoretic Amylase Fractionation as an Aid in Diagnosis of Pancreatic Disease, Clin. Chem. *22*, 57 – 62 (1976).

[22] *G. Skude,* On Human Amylase Isoenzymes, Scand. J. Gastroenterol. (Suppl), *12(44),* 1 – 37 (1977).

[23] *M. Otsuki, H. Yuu, S. Saeki et al.,* The Characteristics of Amylase Activity and the Isoamylase Pattern in Serum and Urine of Infants and Children, Eur. J. Pediatr. *125*, 175 – 181 (1977).

[24] *M. D. O'Donnell, N. J. Miller,* Plasma Pancreatic and Salivary-Type Amylase and Immunoreactive Trypsin Concentrations: Variations with Age and Reference Ranges for Children, Clin. Chim. Acta *104*, 265 – 273 (1980).

2.2 Cellulases

Endo-1,4-β-glucanase; 1,4-(1,3;1,4)-β-D-glucan 4-glucanohydrolase, EC 3.2.1.4 and
Exo-1,4-β-glucanase; 1,4-β-D-glucan cellobiohydrolase; EC 3.2.1.91

Klaus Buchholz, Peter Rapp, and František Zadražil

2.2.1 General

Natural cellulose-containing substrates (plant residues; lignocellulosics) in general are associated with other components, primarily hemicellulose, pectin and lignin. The action of cellulases on such substrates depends greatly on the origin of the substrate and its composition, on physical or chemical pre-treatment, and on synergistic action with other classes of enzymes (xylanases, pectinases, peroxidase, laccase, phenol-oxidases etc.) [1, 2]. Pure cellulose is most readily hydrolyzed by the combined action of endo- and exo-β-1,4-glucanases if it contains crystalline parts, as most natural substrates do.

Application of methods: evaluation of micro-organisms which produce cellulases; production of cellulases and use of cellulase systems in saccharification and conversion of substrate to animal foods or chemical feedstocks, in the food industries and in digestion aids.

Enzyme and substrate properties relevant in analysis: endo-β-1,4-glucanase(1,4-β-D-glucan 4-glucanohydrolase, EC 3.2.1.4) attacks cellulose at various sites although its action is not strictly random. Its attack generates non-reducing chain ends which are described as being susceptible to the action of exo-β-1,4-glucanases [2]. The products formed by the action of endoglucanases retain the β-glucose configuration of the substrate. Endo-β-1,4-glucanases are active towards cellodextrins, phosphoric acid-swollen cellulose, carboxymethylcellulose, hydroxyethylcellulose, but only slightly active against crystalline cellulose (such as Avicel).

Multiple forms of both fungal and bacterial endoglucanases (isoenzymes) have been described which vary according to micro-organisms and culture conditions [3, 4, 5]. Bacteria form both cell-bound and extracellular endoglucanases in contrast to fungi which produce predominantly extracellular endoglucanases.

Exo-β-1,4-glucanase(1,4-β-D-glucan cellobiohydrolase, exocellobiohydrolase, EC 3.2.1.91) catalyzes the release of cellobiose from the non-reducing end of cellulose. It acts in such a way that the configuration of the product is inverted. Exoglucanases are able to attack cellodextrins, phosphoric acid-swollen cellulose, microcrystalline cellulose (Avicel) and de-waxed cotton fibres.

The multiplicity of fungal exoglucanase is much less profound than that of endoglucanase [3, 4]. The existence of bacterial exoglucanase remains questionable, since there are no unambiguous data to support it. Besides the 1,4-β-D-glucan cellobiohydrolase, there is also a poorly-described exo-β-1,4-glucan glucohydrolase which removes glucose from the non-reducing end of cellulose chains [6].

Endo- and exoglucanase activity are both inhibited by their end-products [2]. Endoglucanases have also been found to catalyze transglycosylation reactions [7].

For the analysis of the hydrolytic potential of a system of cellulases, insoluble crystalline cellulose must be chosen as a substrate. Avicel (microcrystalline cellulose) is an appropriate material, since it is of rather high purity and is readily obtained from various suppliers. However, it must be remembered that Avicel itself is not a uniform substrate. It may contain minor amounts of xylan, amorphous parts (up to 30%) and a varying and broad spectrum of particle size (diameters ranging from 5 to 100 μm). Thus the external surfaces accessible to enzymatic hydrolysis vary over a broad range. Amorphous regions are much more susceptible to enzymatic degradation than are crystalline parts. They represent highly ordered linear bonds.

Kinetics are complex due to the inhomogeneity of substrates, and to the varying accessible external surface depending on the source and previous treatment of the substrate. The degradability of the substrate decreases drastically with conversion [8]. Thus it is not possible to determine initial reaction rates since no linear correlation of product formation and reaction time can be found, even at low conversion [9].

Furthermore, the overall reaction sequence contains different reaction paths, including adsorption of the enzyme to the solid substrate, hydrolysis, desorption, inhibition [2, 9] and inactivation of cellulases [9, 10]. The mechanism of the hydrolysis step is not well understood, and conversion of crystalline substrates to an acceptable degree requires at least two types of enzymes, endo- and exoglucanases [2]. Kinetic approaches based on first-order or *Michaelis-Menten* approaches give misleading results in general [8, 9]. Therefore, in order to obtain worthwhile analytical data in various applications, it is necessary to choose an empirical approach which allows a comparison of cellulase preparations under well-defined reaction conditions. Choosing such conditions always represents a compromise between conflicting requirements (e.g. easy and quick analysis versus high degree of conversion).

As a minimum, the following test conditions must be identical for comparison purposes: optimal pH, temperature, substrate concentration, conversion level, concentrations of the products glucose, cellobiose (time must not exceede several hours in order to avoid inactivation). If insufficient β-glucosidase activity is present within the enzyme system under test this enzyme must be added in order to convert cellobiose to glucose, since inhibition by cellobiose is generally greater by more than an order of magnitude than inhibition by glucose.

Methods of determination: a bewildering number of assays has been used to determine the activity of cellulases [11]. A very common procedure is to measure the rate of formation of reducing sugars during incubation of the enzymes with cellulose-containing substrates. However, because of the non-stoichiometric reaction of the aldehyde group of sugars with dinitrosalicylic acid, ferricyanide or the *Nelson-Somogyi*-reagent [12, 13, 14, 15] and the diversity of reducing sugars released from cellulose, including those resulting from the action of the other enzymes of the cellulase system, these assays are not suited for standardization [16, 17].

The method most often used is the filter-paper activity test [18], in which 50 mg filter paper is incubated at pH 4.8, 50 °C for 1 h with cellulases. Reducing sugars are determined by a colour-forming reaction. The method has several sources of error: first, those already mentioned; second, different degrees of conversion and, third, different extents of product inhibition depending on the varying ratios of glucose and cellobiose with their respectively low and high inhibition potentials.

The viscometric method for measurement of endo-β-1,4-glucanase activity using water-soluble carboxymethyl- or hydroxyethylcellulose as substrate is very sensitive and free from confusing interferences. No such analytical procedure exists at present for determining exo-β-1,4-glucanase activity in unfractionated samples of cultures of cellulolytic micro-organisms.

Purified preparations of exo-β-1,4-glucanases may be estimated quantitatively by a method developed by *Hsu et al.* [19].

International reference methods and standards: the Commission of Pharmaceutical Enzymes of F.I.P. has recommended an international standard method [20]. For available reference material cf. Vol. II, chapter 2.3.

Enzyme effectors: reaction products are inhibitors, glucose with low and cellobiose with high inhibition potential, in general.

2.2.2 Endo-1,4-β-glucanase

Viscosimetric Method

The viscometric methods for determining endo-β-1,4-glucanase activity are based on the almost random cleavage of substituted, water-soluble cellulose derivatives by these enzymes. The procedures are rapid, specific for endo-β-1,4-glucanases and, in the initial phase of reaction, very sensitive [21 – 23]. Exoglucanases do not interfere, other than by possible inhibition of endoglucanase activity by mono- or disaccharides produced by exoglucanases [11].

Hydroxyethylcellulose may be more suitable for the viscometric determination of endoglucanase activity [23] than sodium carboxymethylcellulose. However, the latter is most frequently used despite the dependence of viscosity on pH [24]. It is recommended to use a carboxymethylcellulose with a degree of substitution higher than 0.5 but lower than 1.0 [25]. Suitable dilutions of enzymes are required to obtain satisfactory linear relationships between activity and enzyme concentration [21].

Assay

Method Design

Principle: the decrease in viscosity of the substrate solution is proportional to the catalytic concentration of the enzyme in the sample.

The endoglucanase activity in arbitrary units of a given amount of enzyme is defined by the slope of the straight line obtained by plotting the ratios of specific viscosities, $\eta_{sp,t_o}/\eta_{sp,t_x}$ (η_{sp,t_x} = specific viscosity after different incubation times) against reaction time multiplied by 1000 [11]. The method developed for converting the viscosity changes to absolute units has not become widely used, since it requires careful analysis of the carboxymethylcellulose used [11, 25].

Optimized conditions for measurement: most of the fungal endo-β-1,4-glucanases have pH optima of about 5.0, whereas many endoglucanases of bacterial origin exhibit optimum activity around pH 7.0. For measurement of the latter phosphate buffer, 20 mmol/l, pH 7.0, is recommended [26], but in the following method a 0.5% (w/v) solution of carboxymethylcellulose in sodium acetate buffer, 50 mmol/l, pH 5.0, is used for the determination of fungal endoglucanase activity.

Temperature conversion factors: not available.

Equipment

KPG-viscometer of the *Cannon-Fenske* type, KPG-viscometer of the *Ostwald* type or KPG-Micro-*Ubbelohde*-viscometer in a thermostatted water-bath; stopwatches or automatic viscosity measuring system with a built-in quartz clock and light barriers.

Reagents and Solutions

Purity of reagents: sodium carboxymethylcellulose should be free from inhibitors such as glucose and cellobiose; suitable commercial preparations are available.

Preparation of solutions: use re-purified water (cf. Vol. II, chapter 2.1.3.2).

1. Carboxymethylcellulose solution (acetate buffer, 50 mmol/l, pH 5.0; CMC*, 5.0 g/l):

   add 32.24 ml NaOH, 1 mol/l, and 50 ml acetic acid, 1 mol/l, to water to make a total volume of 1 l. Dissolve 5 g Na-CM-cellulose and 0.3 g NaN_3 in this buffer solution by gentle stirring at room temperature. If exact measurement of the efflux time is not possible due to bubbles, omit NaN_3.

Stability of solution: store the solution at 0 – 4 °C and prepare it freshly every week. If NaN_3 is omitted prepare the solution freshly every day.

* The degree of substitution should be in the range of 0.5 and 0.9 and the molecular weight between 100000 and 140000.

Procedure

Collection and treatment of specimen: centrifuge the culture and collect the clear supernatant. If the enzyme activity in the supernatant is low it should be dialyzed and concentrated by the hollow-fibre technique. Dissolve solid enzyme preparations in an appropriate amount of buffer and remove insoluble solids by centrifugation.

Stability of the enzyme in the sample: many fungal endo-β-1,4-glucanases rapidly lose activity at temperatures higher than 50 °C.

Details for measurement in other specimens: cell suspensions were used instead of enzyme solution to measure cell-bound endo-β-1,4-glucanase activity of bacteria by the viscometric assay [27]. Cell-bound fungal endoglucanase activity can be measured by incubating mycelial pellets with Na-CM-cellulose [28].

Assay conditions: KPG-Micro-*Ubbelohde*-viscometer No. II with an inner capillary diameter of 0.57 mm; final volume 2.6 ml; 30 °C (capillary is thermostatted in a water-bath). Before starting the assay adjust temperature of solutions to 30 °C.

Measurement

Pipette into the viscometer:		concentration in assay mixture	
Na-carboxymethylcellulose solution (1)	2.50 ml	acetate Na-CMC	50 mmol/l 5 g/l
thermostat for 10 min at 30 °C			
sample (enzyme solution or water for determination of η_{sp,t_o})	0.10 ml	volume fraction	0.038
mix, record the efflux time at intervals as short as possible during the following 5 min.			

Calculation: the specific viscosity at the time $t = 0$, η_{sp,t_o}, is obtained by measuring the efflux time t_o of the mixture of 2.5 ml Na-CM-cellulose solution and 0.1 ml water. Calculate the specific viscosity after different reaction time of the enzyme according to

$$\eta_{sp,t_x} = \frac{t_x - t_o}{t_o}$$

t_x = efflux time of the reaction mixture of 2.5 ml Na-CM-cellulose solution and 0.1 ml enzyme solution after different time of reaction.

To calculate the enzyme activity, plot the ratios $\eta_{sp,t_0}/\eta_{sp,t_x}$ against reaction time. A straight line is obtained for enzyme solutions of sufficient dilution. The slope of the line multiplied by 1000 is chosen as the arbitrary unit. For conversion into international units see [11].

Validation of Method

Precision, accuracy and sensitivity: the relative standard deviation referred to the mean value of viscometric measurement of endo-β-1,4-glucanase activity in several fungal cultures ranges from 0.1 to 4.0% [11]. Data on accuracy are not available. The sensitivity is Δt_{efflux} = 0.01 s (AVS/G viscometer, *Schott*, Mainz).

Sources of error: inhibition of endo-β-1,4-glucanase activity due to higher concentrations of glucose and/or cellobiose. Comparison of the endoglucanase activities of mixtures of isoenzymes which differ in composition do not necessarily reflect the relative enzyme concentration in each mixture [16].

Specificity: the endo-β-1,4-glucanase activity is measured specifically. The concomitant presence of exo-β-1,4-glucanases is not likely to interfere, other than by possible inhibition of endoglucanase activity by glucose and cellobiose produced by the exo-acting enzymes [11].

Reference ranges: in supernatants of cultures of *Trichoderma viride* an endoglucanase activity of 3.69 arbitrary units × 10^4/ml at 40 °C was measured [11].

2.2.3 Endo- and Exo-glucanase

Measurement by Degradation of Crystalline Cellulose

Assay

Method Design

Principle: Avicel serves as a substrate for testing the potential of an enzyme system with respect to hydrolysis of partially crystalline cellulosic substrates. The substrate at a given concentration (25 g/l) is incubated with a sample of the enzyme system in

Erlenmeyer flasks in a shaking thermostatted water-bath at 50 °C (alternatively at 40 °C, where the reaction rate is lower by a factor of two, approximately).

The reaction time should allow the exact determination of 5% conversion (in the range 0.5 – 5 h), corresponding to glucose, 1.39 g/l (7.72×10^{-3} mol/l) (considering the addition of one molecule of water per molecule of product). The reaction time required for 5% conversion is inversely proportional to the enzyme activity.

The concentration of the intermediate product, cellobiose, must be added to that of glucose (doubling the figure for mol/l, or g/l corrected by the factor 360/342). Its concentration should not exceede 10^{-3} mol/l, since cellobiose is a much stronger inhibitor than glucose. β-Glucosidases (free from 1,4-β-glucanase activity) can be added in order to avoid cellobiose accumulation.

Detailed analysis of products, e.g. by low or high pressure liquid chromatography or gas chromatography, is essential. If these methods are not available, glucose must be determined exactly by an enzyme assay (cf. Vol. VI), and the sum of products by measurement of reducing sugars [18].

Optimized conditions for measurement: the optimal pH must be determined experimentally: it is in the range of pH 4.0 – 5.5 in general, and within pH 4.5 – 5.0 in most cases. All other parameters are the results of conflicting requirements. Thus, with rising temperature, both the reaction rate and the rate of enzyme inactivation increase. 50 °C is a well-accepted compromise for different enzyme systems. Substrate concentration is set to 25 g (Avicel) per litre, Na-citrate (or Na-acetate) buffer to 0.05 mol/l. The amount of enzyme added must produce 5% conversion within a maximum of 6 h.

Equipment

Shaking water-bath, *Erlenmeyer* flasks, 50 or 100 ml, high-pressure liquid chromatograph with refractometer and amino-substituted column, e.g. Lichrosorb $RPNH_2$ (*Merck*) (or other equivalent chromatographic system). pH meter; centrifuge or filter.

Reagents and Solutions

Purity of reagents: suitable commercial preparations are available.

Preparation of solutions: all solutions in re-purified water (cf. Vol. II, chapter 2.1.3.2).

1. Buffer (citrate or acetate, 0.1 mol/l; pH 4.8 or 4.5):

dissolve 21.0 g citric acid monohydrate (or 6 g glacial acetic acid) in ca. 500 ml water, add 200 ml NaOH (160 ml, respectively), 1 mol/l, add HCl, 0.1 mol/l, to obtain the pH wanted (12% for pH 4.8 or 28.1% for pH 4.5 respectively), make up with water to 1 litre. Add 1 droplet of toluene (or 0.04%) for protection against microbial contamination.

2. Substrate suspension:

   put 0.5 g Avicel in a 50 ml *Erlenmeyer* flask and add 10 ml buffer (1).

3. Eluent:

   prepare a mixture of pure 70% acetonitrile and 30% water for chromatography (or 85% and 15%, respectively, for optimal resolution).

4. Standard solutions:

   0.1 g glucose and 0.04 g cellobiose in 25 ml buffer (1), add 25 ml water; prepare standard chromatograms with dilution ratios of 1, 2, 5, 10.

Stability of solutions: buffer (1) and substrate solution (2) are stable for 3 days in the refrigerator. The eluent (3) must be kept in a closed flask in order to avoid evaporation of acetonitrile.

Procedure

Collection and treatment of sample: control pH of enzyme solution, bring to pH of assay, centrifuge if solution contains solids. Solid enzymes: dissolve in buffer (1) to 1 – 10 g/l, centrifuge in order to obtain a clear solution. The sample must be analyzed for its sugar content (correct measurements if necessary).

Collect the clear supernatant. If the enzyme activity in the supernatant is low it should be dialyzed and concentrated by the hollow fibre technique. Dissolve solid enzyme preparations in 10 to 100 ml of buffer, remove insoluble solids by centrifugation or membrane filtration (do not use cellulose derivatives). The sample must be analyzed for its sugar content. Bring to pH of assay.

Stability of the enzyme in the sample: enzymes free from proteases and microbial contamination should be stable for about 3 days in the refrigerator. Impure preparations should be used immediately after sample preparation.

Assay conditions: pH optimum of the enzyme system (generally in the range of 4.5 to 4.8); temperature 50 °C in closed shaking water-bath. Incubation volume 20 ml. A sample blank is run (1 min incubation time).

Measurement

Incubation

Equilibrate in *Erlenmeyer* flasks:			concentration in incubation mixture
substrate suspension	(2)	10 ml	Avicel 25 g/l buffer 0.025 – 0.05 mol/l
close with foil, wait for 5 min			
sample buffer	(1)	0.5 – 10 ml 0 – 9.5 ml	
close with foil, put into shaking thermostat and mix for 1 min; remove a sample quickly with solids well suspended; analyze as blank; take samples of 0.5 – 1.0 ml at appropriate intervals, e.g. every 20 min, or 1 h, up to 5% conversion; centrifuge samples quickly, analyze 10 to 100 µl by chromatography.			

Remarks: take samples quickly by means of an appropriate *Finn* pipette from the shaken flask (narrow plastic tips should be cut in order to allow unrestricted uptake of the suspension); take care that well-mixed suspension is removed in order to ensure constant substrate concentration in the suspension during the assay. In general 4 samples should be taken every 20 min (10 min with very active enzyme preparations), then at every additional h of reaction time until 5% conversion is attained (with known enzyme activities approximately 4 samples near 5% conversion are sufficient). Dilute or concentrate enzyme samples (by ultrafiltration) if reaction times are less than 1 h or longer than 6 h.

High-pressure liquid chromatography with amino-silanized carrier:

a pre-column is recommended for protection of the analytical column; analytical column 250 × 4 mm (e.g. filled with Lichrosorb $RPNH_2$, *Merck*), flow rate of eluent 1 ml/min, inject samples of 10 – 100 µl for sensitive refractometer. (If problems arise with separation of products or interference by buffer, the eluent (3) should be modified to 85% acetonitrile and 15% water).

Calculation

Graphical interpolation of data: plot experimental points of product concentration as a function of reaction time (ordinate: glucose (x mol/l) + cellobiose (2y mol/l); abscissa: reaction time).

Example: at 1.0 h reaction time 1.26 g/l = 7.0×10^{-3} mol/l of glucose plus 0.205 g/l = 0.6×10^{-3} mol/l of cellobiose are liberated (equivalent to 1.2×10^{-3} mol glucose, giving in sum 8.2×10^{-3} mol product calculated as glucose). Draw a (connecting) line through experimental points; take the time for 5% conversion (7.7×10^{-3} mol/l or 1.39 glucose g/l, allowing for the addition of water to hydrolysis products).

Reaction rate at 5% conversion, at 50 min reaction time: $7.7 \times 10^{-3}/50 = 1.54 \times 10^{-4}$ mol $\times$ l^{-1} $\times$ min^{-1}.

Catalytic activity of sample: divide by x ml enzyme sample (e.g. 1 ml), multiply by 20 ml of sample: 1.54×10^{-4}: (1/20) = 3.08×10^{-3} mol $\times$ l^{-1} $\times$ min^{-1} activity of enzyme sample.

If the enzyme sample contains sugar, all measurements must be corrected appropriately. High sugar concentrations must be removed by dialysis.

Validation of Method

Precision, accuracy: limit of error ±10%. If conversion to 5% of substrate is not attained within 6 h, the enzyme solution must be concentrated by ultrafiltration.

Sources of error: inactivation (heat, proteases, microbial contamination); sugars in enzyme preparations, quality of substrates (very different content of crystalline cellulose or particle diameters).

Specificity: depends on combined action of different enzymes, including the occurrence of isoenzymes and β-glucosidases.

The activity characterizes the potential of an enzyme system for the degradation of complex and/or crystalline substrates.

Reference ranges: a protein complex from fungal cellulases (*Penicillium funicolosum,* precipitate of extracellular protein) exhibits an activity in the range of 5×10^{-4} mol/min per g (or per litre, respectively, for a cellulase solution of 1 g/l).

References

[1] *T. K. Kirk, T. Higuchi, Hou-min-Chang,* Lignin Biodegradation: Microbiology, Chemistry, and Potential Applications, Volume *1* and *2*, CRC Press, Inc. Boca Raton, Florida 1980, pp. 241 – 255.

[2] *M. R. Ladisch, K. W. Lin, M. Voloch, G. T. Tsao,* Process Considerations in the Enzymatic Hydrolysis of Biomass, Enzyme Microb. Technol. *5*, 82 – 102 (1983).

[3] *C.-S. Gong, G. T. Tsao,* Cellulase and Biosynthesis Regulation, Annual Reports on Fermentation Processes *3*, pp. 111 – 140 (1979).

[4] *V. S. Bisaria, T. K. Ghose,* Biodegradation of Cellulosic Materials: Substrates, Microorganisms, Enzymes and Products, Enzyme Microb. Technol. *3*, 90 – 104 (1981).

[5] *T. Yoshikawa, H. Suzuki, K. Nisizawa,* Biogenesis of Multiple Cellulase Components of Pseudomonas fluorescens var. cellulosa, I. Effects of Culture Conditions on the Multiplicity of Cellulase, J. Biochem. *75*, 531–540 (1974).
[6] *Y.-H. Lee, L. T. Fan, L.-S. Fan,* Kinetics of Hydrolysis of Insoluble Cellulose by Cellulase, in: Advances in Biochemical Engineering, Vol. *17*, Springer Verlag Berlin, Heidelberg, New York 1980, pp. 131–168.
[7] *G. Okada, K. Nisizawa,* Enzymatic Studies on a Cellulase System of Trichoderma viride, III. Transglycosylation Properties of Two Cellulase Components of Random Type, J. Biochem. *78*, 297–306 (1975).
[8] *Y.-H. Lee, L. T. Fan,* Kinetic Studies of Enzymatic Hydrolysis of Insoluble Cellulose: Analysis of the Initial Rates, Biotechnol. Bioeng. *24*, 2383–2406 (1982).
[9] *K. Buchholz, J. Puls, B. Gödelmann, H. H. Dietrichs,* Hydrolysis of Cellulosic Wastes, Process Biochem. Dec./Jan. 37–43 (1980/81).
[10] *E. T. Reese, M. Mandels,* Stability of the Cellulases of Trichoderma reesei under Use Conditions, Biotechnol. Bioeng. *22*, 323–335 (1980).
[11] *G. Canevascini, C. Gattlen,* A Comparative Investigation of Various Cellulase Assay Procedures, Biotechnol. Bioeng. *23*, 1573–1590 (1981).
[12] *G. L. Miller, R. Blum, W. E. Glennon, A. L. Burton,* Measurement of Carboxymethylcellulase Activity, Anal. Biochem. *2*, 127–132 (1960).
[13] *N. Nelson,* A Photometric Adaption of the Somogyi Method for the Determination of Glucose, J. Biol. Chem. *153*, 375–380 (1944).
[14] *M. Somogyi,* Notes on Sugar Determination, J. Biol. Chem. *195*, 19–28 (1952).
[15] *T. M. Wood, S. I. McCrae,* The Purification and Properties of the C_1 Component of Trichoderma koningii Cellulase, Biochem. J. *128*, 1183–1192 (1972).
[16] *W. A. Lindner, C. Dennison, G. V. Quicke,* Pitfalls in the Assay of Carboxymethylcellulase Activity, Biotechnol. Bioeng. *25*, 377–385 (1983).
[17] *W. L. Marsden, P. P. Gray, G. J. Nippard, M. R. Quinlan,* Evaluation of the DNS Method for Analysing Lignocellulosic Hydrolysates, J. Chem. Tech. Biotechnol. *32*, 1016–1022 (1982).
[18] *M. Mandels, R. Andreotti, C. Roche,* Measurement of Saccharifying Cellulase, Biotechnol. Bioeng. Symp. No. *6*, 21–33 (1976).
[19] *T.-A. Hsu, C.-S. Gong, G. T. Tsao,* Kinetic Studies of Cellodextrins Hydrolyses by Exocellulase from Trichoderma reesei, Biotechnol. Bioeng. *22*, 2305–2320 (1980).
[20] *R. Ruyssen, A. Lauwers* (eds.), Pharmaceutical Enzymes, E. Story-Scientia, Gent 1978, pp. 155–166.
[21] *M. A. Hulme,* Viscometric Determination of Carboxymethylcellulase in Standard International Units, Arch. Biochem. Biophys. *147*, 49–54 (1971).
[22] *K. E. Almin, K.-E. Eriksson, B. Pettersson,* Extracellular Enzyme System Utilized by the Fungus Sporotrichum pulverulentum (Chrysosporium lignorum) for the Breakdown of Cellulose, Eur. J. Biochem. *51*, 207–211 (1975).
[23] *R. Guignard, P.-E. Pilet,* Viscosimetric Determination of Cellulase Activity: Critical Analyses, Plant and Cell Physiol. *17*, 899–908 (1976).
[24] *H. F. Mark, N. G. Gaylord, N. M. Bikales,* Encyclopedia of Polymer Science and Technology, Vol. *3*, Wiley-Interscience, Inc., New York, London, pp. 520–539 (1965).
[25] *K. E. Almin, K.-E. Eriksson,* Influence of Carboxymethyl Cellulose Properties on the Determination of Cellulase Activity in Absolute Terms, Arch. Biochem. Biophys. *124*, 129–134 (1968).
[26] *K. Osmundsvăg, J. Goksøyr,* Cellulases from Sporocytophaga myxococcoides, Purification and Properties, Eur. J. Biochem. *57*, 405–409 (1975).
[27] *W. Stoppok, P. Rapp, F. Wagner,* Formation, Location, and Regulation of Endo-1,4-β-Glucanases and β-Glucosidases from Cellulomonas uda, Appl. Environ. Microbiol. *44*, 44–53 (1982).
[28] *K.-E. Eriksson, S. G. Hamp,* Regulation of Endo-1,4-β-glucanase Production in Sporotrichum pulverulentum, Eur. J. Biochem. *90*, 183–190 (1978).

2.3 Lysozyme (Muramidase)

Mucopeptide *N*-acetylmuramoylhydrolase, EC 3.2.1.17

Boje Weisner

General

The bacteriolytic effect of lysozyme was first described by *Fleming* in 1922 [1] in relation to non-pathogenic bacteria, and more recently in relation to pathogenic bacteria in the presence of complement in the defence reaction. Today, lysozyme attracts considerable medical interest, since in a variety of illnesses it occurs at elevated concentrations and activity in the serum, faeces, urine and cerebrospinal fluid, as well as in other body fluids [2].

The enzyme also occurs in animals and less frequently in plant tissues [3]. Large quantities of lysozyme can be obtained from chicken egg-white. In man it derives principally from the neutrophil granulocytes, monocytes, and macrophages [4, 5].

Application of method: in biochemistry and in clinical chemistry.

Enzyme properties relevant in analysis: in man, only one biologically active form is identified by ion-exchange chromatography in healthy persons; two separate peaks can be distinguished in cases of chronic lymphatic leukaemia. Consequently, this points to an altered nitrogen metabolism [6, 7, 8]. Lysozyme is a basic polypeptide composed of 120 amino acids and with a molecular weight of 15000 [3]. The dissolution of bacterial cell wall results from the cleavage of $1 \rightarrow 4$ glycosyl bonds in glycoproteins [9].

Lysozymes of different origins can be divided into the following classes [3]:

1. primary muramidases with only slight chitinase activity (e.g. chicken egg-white lysozyme, human lysozyme),
2. pure muramidases (e.g. goose-egg lysozyme),
3. primary chitinases with slight muramidase activity (e.g. plant lysozyme).

Methods of determination: the majority of quantitative methods rely upon the ability of lysozyme to lyse freeze-dried *Micrococcus luteus* in suspension at a known concentration. This results in a reduction in turbidity which can be measured photometrically (turbidimetrically) [10]. The "lysoplate" method [11] has proved especially useful for large numbers of samples. Today, immunological methods are also available, such as radial immunodiffusion, electro-immunodiffusion (specific antibody is commercially available) and radioimmunoassay [12], in which enzyme concentrations can be determined on the basis of an antigen-antibody reaction.

International reference methods and standards: standardization at the international level is not known so far. For available reference material cf. Vol. II, chapter 2.3.

Enzyme effectors: specific activators of lysozyme are not known, although there is evidence of lysozyme-inhibiting activity in serum [13 – 15].

Assay

Method Design

Principle: lysozyme cleaves bacterial cell walls by hydrolyzing glycosidic bonds. The turbidity of a bacterial cell suspension is thereby reduced to an extent proportional to the catalytic activity concentration of lysozyme in the assay mixture. The turbidimetric method described here is adapted from [16].

Optimized conditions for measurement: pH, ionic strength and temperature are important to obtain optimum measuring conditions [3, 16]. Both chicken egg-white lysozyme and human lysozyme are commercially available, though it should be noted that human lysozyme is some 4 times more active than the chicken enzyme. In the following method human lysozyme is used as standard, with *Micrococcus luteus* as substrate. The temperature optimum for this system is in the range of 25 °C – 37 °C.

Temperature conversion factors: the following factors were found for measurements at different temperatures (related to the value measured at 25 °C):

°C	20	25	30	35	40
F	1.24	1	0.89	0.61	0.45

Equipment

Variable or fixed-wavelength spectrophotometer capable of exact measurements at 450 or 546 nm; thermostatted cuvette holder; automatic cuvette changer; stopwatch.

Reagents and Solutions

Purity of reagents: the commercially available preparations of *Micrococcus luteus* and human or chicken egg-white lysozyme can be used without further purification. All other reagents should be of analytical reagent grade.

Preparation of solutions (for about 20 determinations): all solutions in re-purified water (cf. Vol. II, chapter 2.1.3.2).

1. Phosphate/NaCl/azide solution (phosphate, 0.067 mol/l, pH 6.3; NaCl, 0.0154 mol/l; azide, 0.008 mol/l):

   dissolve 7.02 g KH_2PO_4 + 2.6 g Na_2HPO_4 + 0.9 g NaCl + 0.52 g sodium azide in 500 ml water, adjust to pH 6.3 with HCl, 0.1 mol/l, and dilute to 1000 ml with water.

2. *Micrococcus luteus* suspension (0.2 g/l):

   suspend 0.006 g *Micrococcus luteus* lyophilisate in 30 ml solution (1).

3. Lysozyme stock solution (1 mg/ml, 40100 units/ml):

   dissolve 2 mg human lysozyme in 2 ml water.

4. Lysozyme standard solutions:

   a) dilute 0.1 ml solution (3) with 2.9 ml NaCl, 0.0154 mol/l;
   b) mix 0.5 ml solution (4a) with 0.5 ml water (668 k units/l);
   c) mix 0.5 ml solution (4a) with 1.5 ml water (334 k units/l);
   d) mix 0.25 ml solution (4a) with 1.75 ml water (167 k units/l);
   e) mix 0.25 ml solution (4a) with 3.75 ml water (83.5 k units/l);
   f) mix 0.25 ml solution (4a) with 7.75 ml water (42 k units/l).

Stability of solutions: solution (1) is stable at 0 – 4 °C. Solution (2) is stable for 24 hours at +4 °C. Solution (3) can be kept for up to 6 days at +4 °C, and for several weeks at −20 °C. Solutions (4a) – (4f) are stable for 4 hours at room temperature.

Procedure

Collection and treatment of specimen: serum, plasma, urine, cerebrospinal fluid or other body fluids should be centrifuged immediately.

Stability of the enzyme in the sample: serum and urine samples should be analyzed within a few hours. At +15 °C to 25 °C a decline in activity is already noticeable after 8 hours. Samples can be kept for several weeks at −20 °C, for 6 days at +4 °C.

Details for measurement in tissue: the optimum pH and ionic strength vary for different lysozymes. It is similar for human and chicken egg-white lysozymes. The optimum pH for human tissue lysozyme is 8 – 8.2, ionic strength 0.025 – 0.0375 [3].

Assay conditions: wavelength 546 nm; light path 10 mm; final volume (semi-micro cuvette) 1.525 ml; 25 °C, thermostatted cuvette holder. Measure against air. Before starting the assay, adjust temperature of solutions to 25 °C.

Measurement

Pipette successively into the cuvette:	standard	sample	concentration in assay mixture
standard solutions (4a) – (4e)	0.025 ml	–	lysozyme 984 – 15737 U/l
sample	–	0.025 ml	volume fraction 0.016
Micrococcus suspension (2)	1.5 ml	1.5 ml	*Micrococcus* 0.2 g/l phosphate 67 mmol/l NaCl 15.4 mmol/l
mix thoroughly with a plastic spatula, start stopwatch and read absorbance after exactly 30 s (A_1), 120 s (A_2) and 210 s (A_3).			

The measurement of standard solutions (4a) – (4e) is necessary only once for each series of determinations.

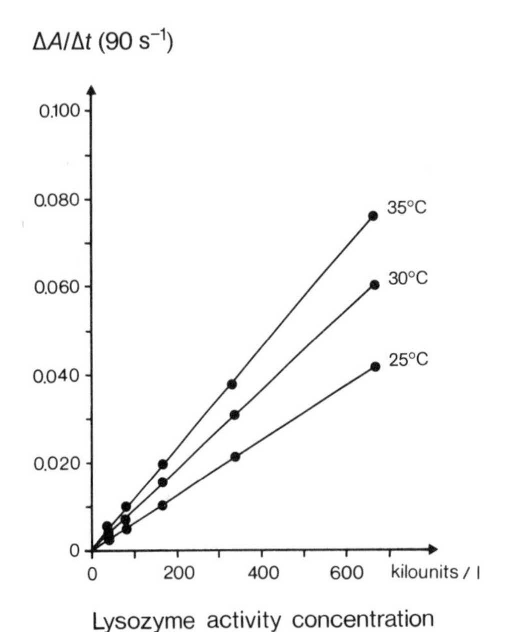

Fig. 1. Calibration curve for determination of catalytic activity concentration of lysozyme.

Calibration curve: calculate $\Delta A/\Delta t$ (90 s^{-1}) for each standard solution by computing $\Delta A_1/\Delta t = (A_2 - A_1)/\Delta t$ and $\Delta A_2/\Delta t = (A_3 - A_2)/\Delta t$ and taking the mean of both. Plot $\Delta A/\Delta t$ *vs.* catalytic activity concentration (units/l) of the corresponding standard solution. At 25 °C to 35 °C the calibration curves are virtually linear up to an activity of 400 kilo units/l (Fig. 1). Within this range the reaction proceeds linearly with time for up to 300 s. Beyond this period of time the reaction rate gradually declines (Fig. 2).

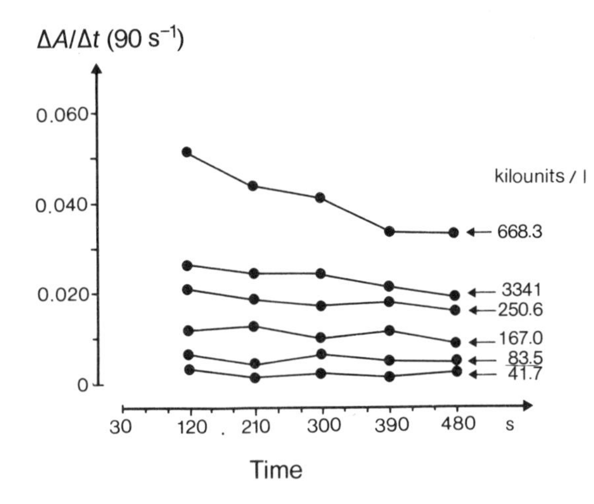

Fig. 2. Progress curves of the lysozyme reaction.

Calculation: one unit is defined as that amount of enzyme which causes a $\Delta A/\Delta t$ of 0.001/min (450 nm; light path 10 mm; potassium phosphate buffer, 0.1 mol/l, pH 7.0; 25 °C; substrate as above).

Calculate the mean $\Delta A/\Delta t$ as described above. Read the catalytic activity concentration of lysozyme in the sample from the calibration curve.

Validation of Method

Precision, accuracy, detection limits and sensitivity: for values of 246 kilo units/l there is a day-to-day imprecision of ± 14.8 kilo units/l. Data about accuracy are not available. The lowest detectable value corresponds to 20 kilo units/l with a day-to-day imprecision of ± 2.4 kilo units/l.

Sources of error: the measured lysis of *Micrococcus luteus* is known to be influenced by hormones, macromolecules and even by other peptidoglycan glycosidases [2, 9].

Specificity: lysozyme activity is measured specifically in the absence of other hydrolyses with similar activity (see above).

Reference ranges: measured at 25 °C: serum, up to 400 kilo units/l = 10 mg/l; cerebrospinal fluid, up to 25 kilo units/l = 0.6 mg/l ([14]; using human lysozyme for calibration). The enzyme normally is not detectable in faeces and urine ([17]; using chicken egg-white lysozyme for calibration).

Reasons for elevated lysozyme activities

Serum: leukaemia, bacterial infections [17, 18];
Urine: leukaemia, infections of the urinary tract [17];
Faeces: intestinal infections, tumours [17];
Cerebrospinal fluid: bacterial meningitis [14, 18],
Viral meningitis, multiple sclerosis and other neurological disorders are associated with only a slight increase in lysozyme activity in the cerebrospinal fluid [14].

References

[1] *A. Fleming,* On a Remarkable Bacteriologic Element Found in Tissues and Secretions, Proc. R. Soc., Ser. B *93*, 306 – 317 (1933).

[2] *E. F. Osserman, R. E. Canfield, S. Beychok* (eds.), Lysozyme, Academic Press, New York 1974.

[3] *P. Jollès, I. Bernier, J. Berthou, D. Charlemagne, A. Faure, J. Hermann, J. Jollés, J.-P. Rérin, J. Saint-Blancard,* From Lysozyme to Chitinases: Structural, Kinetic, and Crystallographic Studies, in: *E. F. Osserman, R. E. Canfield, S. Beychok* (eds.), Lysozyme, Academic Press, New York, London 1974, pp. 31 – 54.

[4] *N. E. Hansen, H. Karle, A. Jensen, E. Bock,* Lysozyme Activity in Cerebrospinal Fluid, Acta Neurol. Scandinav. *55*, 418 – 424 (1977).

[5] *P. Jollès, M. Sternberg, G. Mathé,* The Relationship Between Serum Lysozyme Levels and Blood Leukocytes, Isr. J. Med. Sci. *1*, 445 – 447 (1965).

[6] *P. Jollès,* Lysozymes: A Chapter of Molecular Biology, Angew. Chem. Int. Ed. Engl. *8*, 227 – 239 (1969).

[7] *A. Mouton, J. Jollès,* On the Identity of Human Lysozymes Isolated from Normal Tissues or Secretions, FEBS Lett. *4*, 377 (1969).

[8] *J.-P. Périn, P. Jollès,* Etude comparée des lysozymes de leucocytes de sujets sains et de malades attaints de leucémie myéloide chronique, Clin. Chim. Acta *42*, 77 – 84 (1972).

[9] *J. M. Ghuysen,* Use of Bacteriolytic Enzymes in Determination of Wall Structure and Their Role in Cell Metabolism, Bacteriol. Rev. *32* Suppl., 425 – 464 (1968).

[10] *D. J. Prockop, W. D. Davidson,* A Study of Urinary and Serum Lysozyme in Patients with Renal Diseases, New Engl. J. Med. *270*, 269 – 274 (1964).

[11] *E. F. Osserman, D. P. Lawlor,* Serum and Urinary Lysozyme (Muramidase) in Monocytic and Monomyelocytic Leukemia, J. Exp. Med. *124*, 921 – 952 (1966).

[12] *M. J. Thomas, A. Russo, P. Craswell, M. Ward, I. Seinhardt,* Radioimmunoassay for Serum and Urinary Lysozyme, Clin. Chem. *27*, 1223 – 1226 (1981).

[13] *J. F. Harrison, M. Swingler,* The Effect of Serum Macromolecules on the Lysis of *Micrococcus Lysodeicticus* Cells by Lysozyme, Clin. Chim. Acta *31*, 149 – 154 (1971).

[14] *U. Kauerz, B. Weisner,* Lysozymeactivity and -concentration in Cerebrospinal Fluid and Serum. Comparison of Methods, Enzyme Inhibition, Berichte der ÖGKG, III International Congr. of Clinical Enzymology, Salzburg, Sept. 6 – 9, Jahrgang 4, Heft 3 (1981).

[15] *D. Shugar,* The Measurement of Lysozyme Activity and the Ultra-Violet Inactivation of Lysozyme, Biochim. Biophys. Acta *8*, 302–309 (1952).
[16] *J. Saint-Blancard, P. Chuzel, Y. Mathieu, J. Perrot, P. Jollès,* Influence of pH and Ionic Strength on the Lysis of Micrococcus Lysodeicticus by Six Human and Avian Lysozymes, Biochim. Biophys. Acta *220*, 300–306 (1970).
[17] *W. Dick,* Lysozym. Grundlagen und diagnostische Bedeutung, Fortschr. Med. *26*, 1230–1234 (1982).
[18] *P. E. Perillie, S. C. Finck,* Lysozyme Measurements in Acute Leukemia: Diagnostically Useful? In: *E. F. Osserman, R. E. Canfield, S. Beychok* (eds.), Lysozyme, Academic Press, New York 1974, pp. 359–372.

2.4 Sialidase (Neuraminidase)

Acylneuraminyl hydrolase, EC 3.2.1.18

Roland Schauer and Ulrich Nöhle

General

Sialidases or neuraminidases (earlier "receptor destroying enzyme", RDE) are widespread in micro-organisms (viruses, mycoplasmae, bacteria, fungi and protozoa) as well as in mammalian and avian tissues [1–3], but have not been found in plants. Isolation and crystallization of a sialidase was first achieved from *Vibrio cholerae* by *Schramm & Mohr* and *Ada & French* in 1959 [4]. Sialidases studied so far show differences in molecular weight, subunits (e.g. in viruses), substrate specificity, pH, ion requirements and response to inhibitors.

The cellular location of sialidases differs according to their function. Whereas in bacteria sialidases are usually excreted into the medium as soluble enzymes, they occur in bound form in viruses and in lysosomes, synaptosomes, *Golgi* and plasma membranes of mammalian cells. Soluble activity in low concentrations has been found in cytoplasm, human blood serum and milk [3].

Sialidases release α-glycosidically-linked sialic acids from sialyl-oligosaccharides and sialylglycoconjugates ("complex carbohydrates"). As sialic acids usually occupy terminal positions in oligosaccharide chains, sialidases initiate the catabolism of these molecules. Thus, sialidases are important enzymes involved in the metabolism of complex carbohydrates. However, they have also been recognized to play important pathophysiological roles in viral (e.g. influenza) and bacterial (e.g. gas oedema) infections, in some forms of anaemia, in genetic diseases such as "sialidoses" and in the metabolism of cancer cells [3, 5].

Application of the method: in biochemistry and clinical chemistry.

Enzyme properties relevant in analysis: liberation of sialic acids from substrates proceeds linearly with time, at least up to 30 min. The substrate specificity of sialidases is a complex subject and can be divided into two main areas:

- the classification of complex carbohydrates into oligosaccharides, polysaccharides, glycoproteins, gangliosides and proteoglycans
- the nature of the glycosidic linkage between the sialic acid and the neighbouring monosaccharide, which can be 2–3, 2–4, 2–6, 2–8 or 2–9.

Moreover, the type of the neighbouring sugar itself (D-galactose, N-acetyl-D-galactosamine, N-acetyl-D-glucosamine and sialic acid), the type of sialic acid (N-acetyl- or N-glycolylneuraminic acid and their O-acylated derivatives) and the position of sialic acid in the oligosaccharide chain (in terminal, linear, or side-branch positions) have profound effects on the action of sialidases. For example, the 2–3 linkage is the most easily hydrolyzable bond (except in the case of the *Arthrobacter ureafaciens* sialidase, which cleaves 2–6 bonds more rapidly), O-acetyl groups of sialic acids markedly hinder enzyme activity, and sialic acid bound to the internal galactose as in ganglioside G_{M1} can be hydrolyzed at only a very slow rate by most sialidases. Although the activity of sialidases from different origins is similar with respect to these chemical parameters, there exist differences between the sialidases with regard to the general nature of the complex carbohydrate, e.g. oligosaccharide, glycoprotein or ganglioside [1–3, 5].

Furthermore, sialidases differ in their requirements for the nature and concentration of ions as well as in their pH optima and K_m values. Different activators and inhibitors (see below) are also known [2, 6]. Research in this field is still in rapid progress.

These facts, and differences in the purity of sialidases the activity of which has to be determined, renders it impossible to describe a "standard sialidase assay". Therefore, the experimenter has to select a suitable substrate for the enzyme under study [2, 7]. Therefore, sialidase assays with three different substrates covering a relatively large variety of sialidases are described below. The substrates are:

a) sialyllactose, Neu5Acα(2-3,6)Galβ(1-4)Glc, b) gangliosides, c) fetuin.

Methods of determination: the activity of the enzyme can be estimated

1. by measuring the sialic acids liberated from substrates a) – c) photometrically by the periodic acid/thiobarbituric acid assay according to *Warren* [8]. This method can best be used for measuring sialidase activity in pure or enriched enzyme preparations or in biological materials with high sialidase activity, e.g. bacteria and viruses. When assaying sialidase in tissue homogenates where normally only low activities occur, the detection limits of the photometric test according to *Warren* are not always satisfactory and the assay may furthermore be disturbed by interfering substances (see below). In this case, enzyme activity should be estimated;

2. by the use of substrates a) – c) in a radioactive form:
   a) [^{3}H]sialyllactitol, Neu5Acα(2-3,6)Galβ(1-4)Glc-1-[^{3}H]ol,
   b) [^{3}H]gangliosides (G_{D1a} or a ganglioside mixture from bovine brain),
   c) [^{3}H]fetuin.

The last two substrates are labelled in their sialic acid moieties.

Sialidase activity can be determined by measuring the radioactivity liberated from these compounds in a liquid scintillation counter:

a) [^{3}H]lactitol or [^{3}H]glucitol after separation from the remaining [^{3}H]sialyllactitol by anion-exchange chromatography [9]
b) c) the C_7- or C_8-analogues of [^{3}H]Neu5Ac remaining in the supernatant of the incubation mixture after precipitation of the remaining [^{3}H]gangliosides or [^{3}H]fetuin [5, 10].

High-performance (pressure) liquid chromatography (HPLC) represents another method for the detection of liberated sialic acids and enables separate measurement of different N,O-acylated sialic acids [11].

Use of synthetic substrates for sialidase investigations, e.g. Neu5Ac-4-methylumbelliferyl-α-ketoside [5, 12] and fluorimetric measurement of the liberated 4-methylumbelliferone, should be treated with care, as these substrates do not give information about substrate specificity and thus about the type of sialidase studied. However, this fluorimetric assay represents a quick and sensitive test for screening of sialidase activity in biological materials and for activity determinations during purification of the enzyme.

International reference method and standards: neither standardization nor the existence of reference standard materials is known.

Enzyme effectors: there exists a great variety of activators (e.g. Ca^{2+}, Mg^{2+}, Zn^{2+}, NaCl) and inhibitors (e.g. 2-deoxy-2,3-dehydro-N-acetylneuraminic acid, Hg^{2+}, N-(4-nitrophenyl)-oxamic acid and polyanionic compounds). Details are given in ref. [1, 2, 6].

Assay

Method Design

Principle

(a) Sialyllactose + H_2O $\xrightarrow{\text{sialidase}}$ lactose + sialic acid

(b) Fetuin + H_2O $\xrightarrow{\text{sialidase}}$ asialofetuin + sialic acid

(c) Gangliosides + H_2O $\xrightarrow{\text{sialidase}}$ glycosylceramides (or G_{M1}) + sialic acid

The amount of sialic acid liberated from the substrate is a measure of enzyme activity. The equilibrium of the reaction is in favour of the hydrolytic cleavage. Blanks are always necessary in order to subtract free sialic acid normally occurring in low concentrations in tissue homogenates or to exclude other interfering substances.

Optimized conditions for measurement: the most important characteristics of the various sialidases are their different K_m values (0.05 – 1 mmol/l at 37 °C), pH optima (pH 3.1 – 6.5), ion requirements and substrate specificities, as was mentioned above. When studying a sialidase which has not been characterized before, it is therefore recommended to vary the following conditions of the assay in order to obtain optimal activity:

- use of different types of glycoconjugates (glycoproteins or gangliosides) or oligosaccharides,
- use of isomeric substrates, e.g. (2-3)- or (2-6)sialyllactose,
- adjustment of substrate concentration to about 1 mmol/l,
- change of pH between 3.0 and 7.0 in steps of 0.5,
- change of salt concentration,
- addition of activating divalent cations,
- addition of detergents (non-ionic or ionic),
- elimination of possible inhibitors (in mammalian tissues e.g. 2-deoxy-2,3-dehydro-N-acetylneuraminic acid or proteoglycans),
- purification of the enzyme.

Temperature conversion factors: according to the literature sialidase activity is always estimated at 37 °C. However, the enzyme exhibits residual activity at 0 °C.

Equipment

For method 1: spectrophotometer or filter photometer capable of exact measurement at 549 and 532 nm.

For method 2: liquid scintillation counter, preferably with quench correction.

Reagents and Solutions

Purity of reagents: sialyllactose, gangliosides and fetuin must be free from uncombined sialic acids. The radioactive substrates should be free from radioactive contaminants. The sialic acid content of the sialidase substrates can be estimated with the orcinol/Fe^{3+}/HCl ("Bial") reagent, described in ref. [13].

In order to establish a calibration curve for the determination of sialic acid by the periodic acid/thiobarbituric acid assay, pure crystalline Neu5Ac should be used to give a molar decadic absorption coefficient of 57000.

Preparation and stability of solutions (for about 100 determinations): all solutions in re-purified water (cf. Vol. II, chapter 2.1.3.2).

For incubation

1. Incubation buffer (acetate, 0.15 mol/l, pH 4.5; NaCl, 0.45 mol/l; $CaCl_2$, 0.027 mol/l):

   dissolve 0.123 g sodium acetate, 0.15 mol/l, 0.26 g NaCl and 29.9 mg $CaCl_2$ in 10 ml water. Store at 4 °C.

2. Sialyllactose (5 mmol/l):

   dissolve 6.14 mg sialyllactose in 2 ml water. This substance, containing mainly 2 – 3 sialyllactose, is available from *Boehringer Mannheim* and *Sigma,* or it can be prepared from colostrum according to ref. [14].

3. [^{3}H]Sialyllactitol (83000 disintegrations × s^{-1} in 2 ml):

   use solution (2) for dilution of [^{3}H]sialyllactitol prepared acc. to "Appendix", p. 206.

4. Gangliosides solution (equivalent to sialic acid, 5 mmol/l):

   dissolve gangliosides representing 3 mg sialic acid in 2 ml water. Gangliosides from bovine brain can be purchased from *Sigma*. Their isolation has been described in ref. [10] and [15].

5. [^{3}H]Gangliosides solution (equivalent to 83000 disintegrations × s^{-1} in 2 ml):

   use solution (4) for dilution of radioactive preparation acc. to "Appendix", p. 206.

6. Fetuin solution (5 mmol sialic acid/l):

   dissolve fetuin representing 3 mg sialic acid in 2 ml water. Fetuin from calf serum is available from *Sigma*.

7. [^{3}H]Fetuin (representing 83000 disintegrations × s^{-1} in 2 ml):

   the preparation corresponds to that described for the gangliosides.

For photometry

8. Periodate (0.25 mol/l):

   mix 75 ml conc. H_3PO_4 + 5.7 g H_5IO_6, make up with water to 100 ml.

9. Arsenite (0.55 mol/l):

   dissolve 5 g $NaAsO_2$ and 7.1 g Na_2SO_4, make up with water to 100 ml.

10. Thiobarbiturate (55.4 mmol/l):

    dissolve 0.9 g thiobarbituric acid and 7.1 g Na_2SO_4, make up with water to 100 ml. Store at 4 °C.

11. Cyclohexanone

12. Neu5Ac standard (0.129 mmol/l):

    dissolve 4 mg Neu5Ac in water, store at −20 °C.

For radiometry

13. Ovalbumin solution (8% w/v):

    dissolve 1.6 g ovalbumin in 20 ml water.

14. Phosphotungstic acid (5% w/v):

    dissolve 5 g phosphotungstic acid in 100 ml water.

Stability of solutions: for longer storage all substrates should be kept in the lyophilized state. Since the sialic acid glycosidic linkage is labile, the aqueous solutions of the substrates should be prepared freshly, or stored at −20 °C for a maximum of 1 week in the case of sialyllactose, or 4 weeks in the case of gangliosides and fetuin.

Incubation buffer (1) is stable for 6 weeks, solution (10) for one week at 4 °C; solutions (8) and (9) are stable for 3 months at room temperature. Solutions (13) and (14) are stable for one year at −20 °C.

Procedure

Collection and treatment of specimen: enzyme activity may be determined directly in a native fluid, e.g. in blood serum, in the supernatants of tissue homogenates, in membrane fractions, in viral and bacterial cultures and in fractions obtained during the isolation procedure of the enzyme. For homogenization and dilution of the enzyme sample use the incubation buffer (1).

Stability of the enzyme in the specimen: solubilization of membrane-bound sialidases may be achieved by short sonication or by detergents [1, 2, 5, 7, 16]. However, during this procedure they often lose activity or change pH optimum. Freezing and thawing as well as dialysis of the enzyme may lead to a slight decrease of activity. The activity of tissue homogenates rapidly decreases at 0 °C and also slowly decreases in the frozen state, mainly due to the activity of proteases. When testing sialidase activity in blood, only fresh serum or plasma free from haemolysis should be used. Purified sialidases have been observed to be stable in lyophilized form or dissolved in buffer at 0 °C or frozen at −20 °C.

Assay conditions

Incubation: volume 100 µl; 37 °C; incubation time 30 min or longer.

Photometry: the periodic acid/thiobarbituric acid assay used is a micro-adaptation of the original method of *Warren* [8]. Wavelength 549 and 532 nm, light path 10 mm; final volume 700 µl; room temperature.

Readings in a glass semi-micro cuvette against a reagent blank containing water instead of incubation mixture.

An enzyme blank is run with heat-inactivated (5 min, 100 °C) incubation mixture, in order to check the presence of interfering substances in the sample. A standard is run with 100 µl (4 µg Neu5Ac) instead of sample, in order to check the quality of reagents.

Radiometry: liquid scintillation counter with quench correction, scintillation vials holding up to 6 ml test solution and a suitable amount of scintillation fluid. Measurement for 10 min.

Measurement

Incubation

Pipette successively into an *Eppendorf* tube:			concentration in the incubation mixture	
incubation buffer	(1)	30 µl	acetate	45 mmol/l
			NaCl	135 mmol/l
			$CaCl_2$	8 mmol/l
substrate solution*	(2, 4 or 6)	20 µl	bound sialic acid	1 mmol/l
enzyme preparation		50 µl	volume fraction	0.5
mix on an *Eppendorf* shaker, incubate for 30 min (or longer), cool in an ice-bath and determine the liberated sialic acid.				

* for subsequent radiometry use substrate solutions (3, 5 or 7)!

Photometric assay

Pipette successively into test tubes:			concentration in assay mixture	
incubation mixture		100 µl		
periodate solution	(8)	20 µl	H_5IO_6	41.6 mmol/l
			H_3PO_4	2.18 mol/l
mix and allow to stand for 30 min at room temperature				
arsenite solution	(9)	200 µl	$NaAsO_2$	0.343 mol/l
			Na_2SO_4	0.312 mol/l
mix immediately (!) on a vortex mixer and shake afterwards 2 min on an *Eppendorf* shaker				
thiobarbiturate solution	(10)	200 µl	thiobarbiturate	21 mmol/l
mix and incubate 15 min at 96 °C; cool in an ice-bath				
cyclohexanone	(11)	700 µl		
extract the red colour by shaking the tube for 3 min on an *Eppendorf* shaker and separate the organic phase by centrifugation at 15000 *g* for 2 min. Transfer the organic phase to a cuvette and read the absorbances at 549 and 532 nm against the reagent blank.				

For quantitative determination the amount of sialic acid is limited to 4 µg in 100 µl incubation mixture; for semi-quantitative or qualitative detection higher amounts are allowed.

Calculation: one unit of enzyme activity is defined as the release of 1 µmol sialic acid (e.g. N-acetylneuraminic acid, Neu5Ac) per minute. Subtract readings for enzyme blank from readings for sample.

Under the conditions described above the µmoles Neu5Ac released correspond to $0.0146 \times A_{549} - 0.00537 \times A_{532}$. Thus, the catalytic concentration in the sample is

$$b = (0.0146 \times A_{549} - 0.00537 \times A_{532}) \times 20000/\Delta t \quad \text{U/l}$$

Radiometric assay

a) With [^{3}H]sialyllactitol:

after incubation, centrifuge membranes (if present) at 15000 *g* for 3 min and wash the pellet twice with 1 ml water. Rinse the combined supernatants through a small column (e.g. *Pasteur* pipette) containing 1 ml of a strong basic anion-exchanger 2X8, 200 – 400 mesh, acetate form, and wash the column with 4 ml water. The combined effluents containing [^{3}H]lactitol liberated by sialidase are analyzed for radioactivity.

b) With [^{3}H]gangliosides or [^{3}H]fetuin:

after incubation, add to incubation mixture

ovalbumin solution 200 µl,

mix and keep for 3 min in the ice-bath,

phosphotungstic acid 200 µl,

mix and allow to stand for 5 min, spin down the precipitate at 15000 *g* for 3 min. Remove the supernatant and wash the pellet with another 200 µl phosphotungstic acid. The combined supernatants containing the liberated [^{3}H]labelled C_7- and C_8-analogues of sialic acid are analyzed for radioactivity.

Liquid scintillation counting:

add a suitable amount of scintillation fluid to the test solution, count for 10 min and correct the results for quenching (e.g. due to coloured substances).

Calculation: subtract readings for enzyme blank from readings for sample. The net counting rates C_n are entered in the calculation of enzyme activity concentration:

$$b = \frac{C_n \times 0.097 \times 20000}{C_{total} \times \Delta t} \quad \text{U/l}$$

$$b = \frac{C_n \times 0.097 \times 20000 \times 16.67}{C_{total} \times \Delta t} \quad \text{nkat/l}$$

C_{total} corresponds to the counting rate of 20 µl of the substrate solution used.

0.097 is µmol sialic acid after complete hydrolysis
20000 is conversion factor from 50 µl sample to 1 litre.

Validation of Methods

Photometric method

The validity of this sialidase assay depends on the periodic acid/thiobarbituric acid assay of free sialic acids, the quality of which is discussed in the following.

Precision, accuracy, detection limits and sensitivity: for values of about 4 µg/100 µl of a pure Neu5Ac solution an inaccuracy of ± 0.1 µg/100 µl has been found for manual performance. Sensitivity is 0.1 µg Neu5Ac. The detection limit of the method described (30 min incubation time) is 0.2 µg Neu5Ac/100 µl, corresponding to 0.02 mU of sialidase. However, these data apply for a solution containing pure, standard Neu5Ac; when assaying sialic acids accompanied by other, interfering substances, as is often the case in sialidase preparations, precision may become lower and the detection limit higher.

Sources of error: the most important interference is by deoxyribose, giving rise to a chromophore with an absorption maximum at 532 nm which interferes with the absorbance at 549 nm of the chromophore derived from Neu5Ac [8, 13]. The calculation formula given above takes this fact into account, but at very low Neu5Ac concentrations corresponding to a very low sialidase activity and at high deoxyribose concentration, e.g. in tissue homogenates, the detection limit for sialic acid will rise and the values of sialidase activity will be falsified. In order to obtain information as to whether deoxyribose or other substances (see below) interfere with Neu5Ac, a spectrum of the chromophore obtained with the enzyme test solution and with the enzyme blank should be compared with that of pure Neu5Ac. Measurement against the enzyme blank (the difference in the spectra between both tubes recorded by a double-beam spectrophotometer must be identical with the spectrum of reference Neu5Ac), or prolonged time of incubation in order to increase the amount of liberated sialic acid, may give better results.

Other sources of error are unsaturated fatty acids (which might be extracted with ether after enzyme incubation), fucose, acetaldehyde and 2-keto-3-deoxyaldonic acids, all resulting in chromophores interfering with that of Neu5Ac, but usually being of minor importance [5, 8, 13].

Further errors in the quantitative determination of enzymatically-released sialic acids may be due to differences of the molar decadic absorption coefficients of different N,O-acylated sialic acids in the periodic acid/thiobarbituric acid assay [5, 13]. This may lead to errors in the calculation of sialidase activity, since the assay system described is calibrated with pure Neu5Ac. However, sialic acid O-acetyl groups do not occur, or occur only in negligible quantities, in the substrates described, and the amount of N-glycolylneuraminic acid, which is 33% less sensitive in the periodic acid/thiobarbituric acid assay, is low or even zero in the substrates used.

The occurrence of N-acetylneuraminate lyase (EC 4.1.3.3), which cleaves the liberated sialic acid into acylmannosamine and pyruvate, and other enzymes such as proteases, lipases and endoglycosidases, which react with the macromolecular substrates and alter their properties, may also cause errors in the estimation of enzyme activity.

Specificity: sialidases are known to cleave only glycosidic linkages of sialic acids. The influence of N,O-acyl substitution of sialic acids and other factors on the periodic acid/thiobarbituric acid assay has been discussed above. For the investigation of substrate specificity of sialidases with regard to different derivatives of neuraminic acid, monitoring by HPLC is recommended [11].

Reference ranges: see below.

Radiometric method

Precision, accuracy, detection limits and sensitivity: an inaccuracy of 2% has been found for values of about 10 μg released sialic acid. The detection limits, based on 30 min incubation time, are 0.3 to 0.03 μg sialic acid, representing 8.3 to 0.83 disintegrations $\times$ s^{-1} and corresponding to 3.2×10^{-5} to 3.2×10^{-6} U of sialidase. Under these conditions, the sensitivity is 0.03 μg sialic acid. When using substrates with higher specific radioactivity (about 100 times higher can easily be realized), the sensitivity and detection limit of the method may improve.

Sources of error: since sialyllactitol is eluted from the anion-exchanger described at an acetate concentration of 10 mmol/l, the ion strength of the solution containing sialyllactitol passed through the column should not be stronger than 5 mmol/l, therefore, dilution of the enzyme incubation mixture with water is necessary before chromatography. If the substrate is stored correctly, radioactivity of the blank normally does not exceed 1 – 2% of the total.

The occurrence of proteases, lipases and other enzymes (e.g. endoglycosidases), especially in crude sialidase preparations, may cause reactions with gangliosides and fetuin resulting in a modification of the substrate or the production of radioactive degradation products (glycopeptides, oligosaccharides) which are not precipitable by phosphotungstic acid and thus may mimic a higher sialidase activity.

Corresponding to the non-radioactive assay, the occurrence of sialic acids other than Neu5Ac in the radioactive substrates may lead to errors in the calculation of enzyme units.

Specificity: the radioactive method enables a specific determination of sialidase activity, especially when using [^{3}H]sialyllactitol. The specificity of the test with [^{3}H]gangliosides and [^{3}H]fetuin may be restricted if degradative enzymes other than sialidases are present, as described in the preceeding paragraph.

Reference ranges: sialidase solutions commercially available contain 0.05 – 150 U/mg protein (*Clostridium perfringens*) or about 20 U/mg protein (*Vibrio cholerae*), according to the purification grade.

Sialidase activities usually found in mammalian tissue homogenates vary from $10^{-4}-10^{-6}$ U/mg protein. In serum of healthy human individuals sialidase activities of $10^{-6}-10^{-7}$ U/ml are normal, whereas in a patient with gas oedema this enzyme activity was observed to increase to 4×10^{-3} U/ml [17].

Appendix

Preparation of [^{3}H]sialyllactitol/sialyllactose

Reduction of sialyllactose with NaB^3H_4 according to *Frisch & Neufeld* [9] results in Neu5Acα(2-3,6)Galβ(1-4)Glc-1-[^{3}H]ol (sialyllactitol) with high specific radioactivity (up to 10^{10} disintegrations $\times$ s^{-1} $\times$ mol^{-1}).

Dissolve 0.5 mg sialyllactose (ca. 0.75 µmol) in 0.5 ml sodium carbonate buffer, 0.1 mol/l, pH 9.0, and mix with 0.1 ml NaOH, 0.01 mol/l, containing 10 mCi NaB[^{3}H]$_4$ (5 mCi/µmol, *Amersham*). After 45 min at ambient temperature add unlabelled $NaBH_4$ (1 mg in 0.1 ml NaOH, 0.01 mol/l) and continue the reduction for one hour. Finally, stop the reaction and decompose excess borohydride by addition of 0.015 ml acetic acid, 10 mol/l. After lyophilization, dissolve the reduced sialyllactose in pyridinium acetate buffer, 2 mmol/l, pH 5) and remove decomposition products by chromatography on Dowex AG 1 × 2 (minus 400 mesh), using the same buffer, but at concentration of 50 mmol/l according to ref. [14].

Check the purity of the radioactive preparation by radio-thin-layer chromatography on silica gel plates (0.2 mm) using ethanol/n-butanol/pyridine/water/acetic acid (100:10:10:30:3, by vol.) and visualizing by spraying with the *"Bial"* reagent [13] (3:1 diluted with water) and subsequent heating in a glass chamber at 120°C for 20 min.

The [^{3}H]sialyllactitol preparation is diluted with unlabelled sialyllactose without change of sialidase activity [7]. For enzyme assays prepare an aqueous solution of sialyllactose of 5 mmol/l (6.14 mg/2 ml) including about 83000 disintegrations $\times$ s^{-1} [^{3}H]sialyllactitol.

Preparation of [^{3}H]gangliosides

Mild periodate oxidation followed by reduction with NaB^3H_4 of the sialic acid side chain of gangliosides results in ganglioside preparations containing a mixture of C_7- and C_8-analogues of sialic acids, which are suitable substrates for sialidases [10].

In a typical labelling experiment dissolve 93 mg gangliosides containing mainly G_{D1a} in 50 ml water at 4°C and add the same volume acetate buffer, 0.2 mol/l, pH 5.5, containing NaCl, 0.3 mol/l. Keep the solution (1 mmol/l with respect to bound Neu5Ac) in an ice-bath and add 10 ml ice-cold solution of sodium metaperiodate, 10 mmol/l in water with stirring. After 10 min stop the oxidation by the addition of 2 ml glycerol, and after another 10 min dialyze the total mixture against three changes each of 3 l phosphate buffer, 50 mmol/l, pH 7.4, containing NaCl, 0.15 mol/l, for 24 h at 4°C. Mix the dialyzed solution containing the oxidized gangliosides with 10 ml (40 mCi) NaB^3H_4, 0.1 mol/l, 40 Ci/mol, in NaOH, 10 mmol/l, at room temperature.

After 30 min add 10 ml unlabelled $NaBH_4$, 0.1 mol/l, in NaOH, 10 mmol/l. After a further 30 min dialyze the total mixture, first against two changes each of 3 l acetate buffer, 0.1 mol/l, pH 5.5 and afterwards against two changes each of 3 l water. After Dowex 50 passage, freeze-dry the labelled gangliosides.

For dilution of the radioactive material with unlabelled gangliosides, oxidize the latter in the same manner, but reduce with unlabelled borohydride.

For enzyme assays prepare an aqueous solution of gangliosides representing about 3 mg of sialic acid in 2 ml water corresponding to 83000 disintegrations $\times$ s^{-1}.

Preparation of [^{3}H]fetuin

The preparation of radioactive fetuin as well as the preparation of test solutions correspond to that described for the gangliosides.

References

[1] *A. Rosenberg, C.-L. Schengrund,* Sialidases, in: *A. Rosenberg, C.-L. Schengrund* (eds.), Biological Roles of Sialic Acid, Plenum Press, New York 1976, pp. 295 – 359.

[2] *A. P. Corfield, J.-C. Michalski, R. Schauer,* The Substrate Specificity of Sialidases from Microorganisms and Mammals, in: *G. Tettamanti, P. Durand, S. di Donato* (eds.), Sialidases and Sialidoses, Perspectives in Inherited Metabolic Diseases, Vol. *4*, Edi Ermes, Milano 1981, pp. 3 – 70.

[3] *A. P. Corfield, R. Schauer,* Metabolism of Sialic Acids, in: *R. Schauer* (ed.), Sialic Acids – Chemistry, Metabolism and Function, Springer Verlag, Wien, New York 1982, pp. 195 – 261.

[4] *A. Gottschalk* (ed.), The Chemistry and Biology of Sialic Acids and Related Substances, Cambridge University Press 1960, pp. 98 – 105.

[5] *R. Schauer,* Chemistry, Metabolism and Biological Functions of Sialic Acids, in: *R. S. Tipson, D. Horton* (eds.), Advances in Carbohydrate Chemistry and Biochemistry, Vol. *40*, Academic Press, New York 1982, pp. 131 – 234.

[6] *R. Schauer, A. P. Corfield,* Sialidases and Their Inhibitors, in: *F. G. de las Heras, S. Vega* (eds.), Medicinal Chemistry Advances, Pergamon Press, Oxford and New York 1981, pp. 423 – 434.

[7] *R. W. Veh, M. Sander,* Differentiation between Ganglioside and Sialyllactose Sialidases in Human Tissues, in: *G. Tettamanti, P. Durand, S. Di Donato* (eds.), Sialidases and Sialidoses, Perspectives in Inherited Metabolic Diseases, Vol. *4*, Edi Ermes, Milano 1981, pp. 71 – 109.

[8] *L. Warren,* The Thiobarbituric Acid Assay of Sialic Acids, J. Biol. Chem. *234*, 1971 – 1975 (1959).

[9] *A. Frisch, E. F. Neufeld,* A Rapid and Sensitive Assay for Neuraminidase: Application to Cultured Fibroblasts, Anal. Biochem. *95*, 222 – 227 (1979).

[10] *R. W. Veh, A. P. Corfield, M. Sander-Wewer, R. Schauer,* Neuraminic Acid-Specific Modification and Tritium-Labelling of Gangliosides, Biochim. Biophys. Acta *486*, 145 – 160 (1977).

[11] *A. K. Shukla, R. Schauer,* Analysis of N,O-Acylated Neuraminic Acids by High-Performance Liquid Anion-Exchange Chromatography, J. Chromatogr. *244*, 81 – 89 (1982).

[12] *R. W. Myers, R. T. Lee, Y. C. Lee, G. H. Thomas, L. W. Reynolds, Y. Uchida,* The Synthesis of 4-Methylumbelliferyl-α-ketoside of N-Acetylneuraminic Acid and its Use in a Fluorometric Assay for Neuraminidase, Anal. Biochem. *101*, 166 – 174 (1980).

[13] *R. Schauer,* Characterization of Sialic Acids, in: *S. P. Colowick, N. O. Kaplan* (eds.), Methods in Enzymology, Vol. *L*, Academic Press, New York 1978, pp. 64 – 78.

[14] *R. W. Veh, J.-C. Michalski, A. P. Corfield, M. Sander-Wewer, D. Gies, R. Schauer,* New Chromatographic System for the Rapid Analysis and Preparation of Colostrum Sialyloligosaccharides, J. Chromatogr. *212*, 313 – 322 (1981).

[15] *T. Momoi, S. Ando, Y. Nagai,* High Resolution Preparative Column Chromatographic System for Gangliosides using DEAE-Sephadex and a New Porus Silica, Iatrobeads, Biochim. Biophys. Acta *441*, 488–497 (1976).
[16] *J.-C. Michalski, A. P. Corfield, R. Schauer,* Solubilization and Affinity Chromatography of a Sialidase from Human Liver, Hoppe-Seyler's Z. Physiol. Chem. *363*, 1097–1102 (1982).
[17] *R. Schauer, J. M. Jancik, M. Wember,* Occurrence of Neuraminidase Activity in the Serum of Patients Infected with Clostridia (Gas Oedema), in: *R. Schauer, P. Boer, E. Buddecke, M. F. Kramer, J. F. G. Vliegenthart, H. Wiegandt* (eds.), Glycoconjugates, Proc. Vth Int. Symp., Georg Thieme Publ. Stuttgart 1979, pp. 362–363.

2.5 α-Glucosidases (Disaccharidases)

Sucrase (Invertase)

Sucrose α-D-glucohydrolase, EC 3.2.1.48

Maltase

α-D-Glucoside glucohydrolase, EC 3.2.1.20

Isomaltase

Dextrin 6-α-D-glucanohydrolase, EC 3.2.1.10

Trehalase

α,α-Trehalose glucohydrolase, EC 3.2.1.28

Arne Dahlqvist

General

The enzymes listed above were first separated in homogenates of pig intestinal mucosa [1] and later also in human preparations [2]. They are located in the "brush border" membrane of the enterocytes [3]. With the exception of trehalase, which seems to have absolute substrate specificity [4], the enzymes have quite complicated cross-specificity for different substrates, especially for maltose, which is hydrolyzed at a considerable rate by all the enzymes. The brush border α-glucosidases hydrolyze disaccharides in food, and also perform the final steps in starch and glycogen digestion.

Application of method: the assay method is used for biopsy samples of the small-intestinal mucosa. Specific enzyme defects are sucrase-isomaltase deficiency [5] and trehalase deficiency.

Enzyme properties relevant in analysis: sucrase and isomaltase are bound together into a complex [7]. In most subjects with deficiency of this complex both activities are missing [8]; some patients can synthesize a certain amount of isomaltase, but no sucrase [9]. There exists also a general disaccharidase deficiency in malnourished subjects and in patients with different diseases of the small intestine (e.g. gluten intolerance). However, these conditions affect the lactase activity most markedly and, although all the activities are decreased, it is usually only the lactose intolerance that is of clinical significance (for review, cf. [7]).

In addition to the brush border α-glucosidases mentioned above, there is also in the intestinal mucosa a *lysosomal α-glucosidase* with an acid pH optimum. This enzyme does not participate in the digestion of dietary carbohydrates, and it does not give any significant contribution to the activities assayed with the method described here [11].

Methods of determination: the method to be described is based on incubation with the appropriate disaccharide followed by the assay of liberated glucose with a Tris/glucose oxidase reagent [12]. Tris serves the double function of interrupting the activity of the intestinal disaccharidase at the end of the incubation period and inhibiting contaminant disaccharidases in the glucose oxidase. Due to its simplicity, the method is well suited for analysis in the ultamicro scale, and it has been widely used for the assay of the disaccharidase activities in peroral biopsies of the small-intestinal mucosa for clinical diagnosis.

Other methods are based on measurement of reducing sugars, or assay of glucose with glucose-6-phosphate dehydrogenase (EC 1.1.1.49). Whatever method is used, it is recommended that the appropriate disaccharide is used as the substrate. Chromogenic substrates should not be used without very detailed knowledge of the specificity of the enzymes to be studied.

A quantitative immunoelectrophoretic method has recently been developed for the measurement of the intestinal brush border enzymes [13, 14]. With this method the enzyme protein measurement is independent of the enzymatic activity.

International reference methods and standards: not available.

Enzyme effectors: sodium ions activate several disaccharidases of the intestine. Other monovalent cations are less potent activators or even inhibit.

Assay

Method Design

Principle

(a) $$\text{Disaccharide} + H_2O \xrightarrow{\text{disaccharidase}} \text{glucose} + \text{monosaccharide}$$

(b) $$\text{Glucose} + O_2 \xrightarrow{\text{glucose oxidase*}} \text{gluconic acid} + H_2O_2$$

(c) $$H_2O_2 + DH_2 \xrightarrow{\text{peroxidase**}} 2\,H_2O + D$$

DH_2 Hydrogen donor (chromogen)
D Dye

* β-D-Glucose: oxygen 1-oxidoreductase, EC 1.1.3.4.
** Donor: hydrogen-peroxide oxidoreductase EC 1.11.1.7.

The amount of glucose formed per unit time, determined by the brown colour at 450 nm, is a measure of the disaccharidase activity.

Optimized conditions for measurements: the relatively wide pH optimum varies slightly with the different enzymes; in most cases, however, it is around pH 6 (Fig. 1). In a few cases a lower pH optimum has been found, indicating contamination by lysosomal enzymes. Only once has a pH optimum >6 been found: two maltases in intestinal mucosa of pigs have a relatively broad pH optimum between 6.5 and 7.5. The influence of the incubation temperature has not been systematically studied. We measure activity at 37 °C; a few experiments at 25 °C gave about half the activity. We use a substrate concentration of 28 mmol/l so that the disaccharidases are more than 50% saturated. Higher substrate concentrations result in "substrate inhibition", in part due to the occurrence of transglycosidation reactions instead of hydrolysis. The sodium maleate buffer used in this method contains sufficient sodium ions for activation.

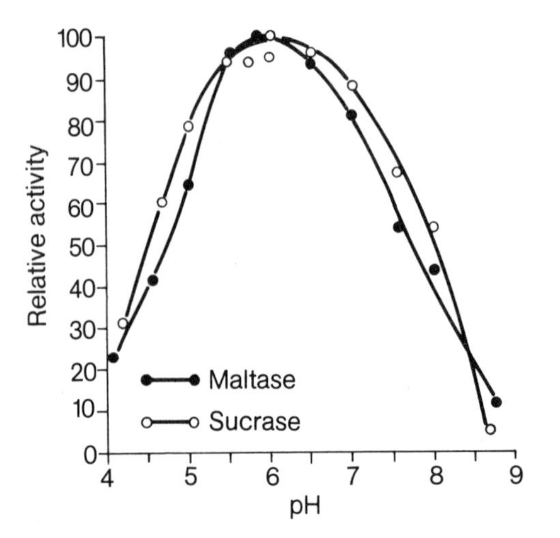

Fig. 1. pH Optima for maltase and sucrase from human duodenal mucosa.

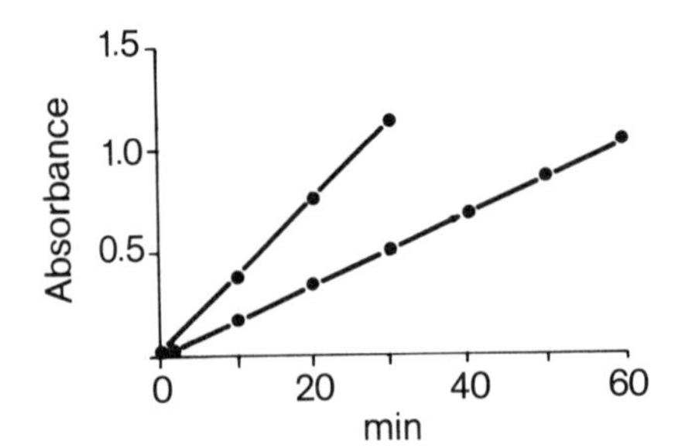

Fig. 2. Sucrase of duodenal homogenate; progress curves with two concentrations. Enzyme incubation for various times, usual colour development.

Under the conditions described here the reaction is a first order one. Until 12% of the substrate has been hydrolyzed; the amount of glucose liberated per unit time is proportional to the enzyme activity (Figs. 2, 3).

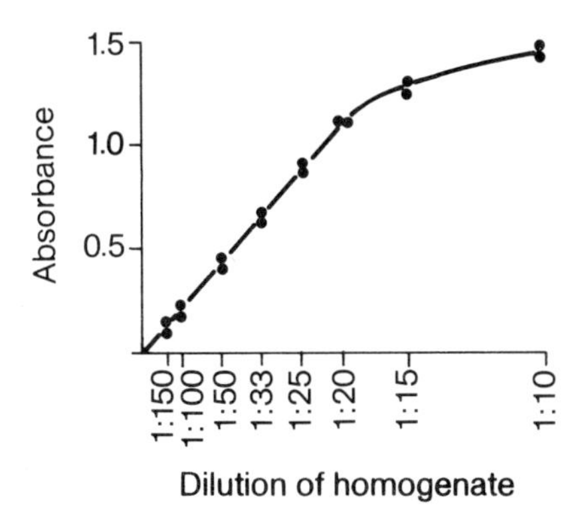

Fig. 3. Proportionality between amount of enzyme and glucose liberated (60 min). Sucrose as substrate, 1:5 homogenate of human mucosa containing 40 mg protein/ml. Final volume: 320 µl, light path: 0.6 cm; read against a reagent blank. The relationship is linear up to $A = 1.100$, equivalent to 12 µg glucose (12% hydrolysis of the substrate).

Equipment

Spectrophotometer or spectral-line photometer or filter photometer with micro cuvettes (200 µl volume) suitable for measurements at 450 nm (420 – 480) nm; *Carlsberg* constriction pipettes; small conical test tubes, 63 mm long, 8 mm internal diameter; 37 °C water-bath (constant temperature).

If no micro-equipment is available, use ten times the amounts of sample and reagents.

Reagents and Solutions

Purity of reagents: the purest commercial disaccharides can be used without further purification; if they contain free glucose they should be discarded or purified (see below). Commercial maltose usually contains several higher oligosaccharides as well as glucose. Test: mix 20 µl 1% maltose solution with 300 µl glucose reagent (solution 4); mix 20 µl water 300 µl glucose reagent as a blank; incubate for 1 h at 37 °C and then read the absorbance at 450 nm (10 mm light path). A difference in absorbance of >0.100 indicates that the substrate should be purified. We prepare isomaltose by enzymatic hydrolysis of dextran [15], followed by fractionation on carbon-Celite or Sephadex G-15 columns [16]. Isomaltose is now commercially available, but we have had no experience with these preparations.

Preparation of solutions (for about 300 determinations): all solutions in re-purified water (cf. Vol. II, chapter 2.1.3.2).

1. Maleate buffer (0.1 mol/l, pH 6.0):

   dissolve 1.16 g maleic acid in 15.3 ml NaOH, 1 mol/l, and dilute with water to 100 ml; if necessary adjust the pH to 6.0.

2. Disaccharide solution (56 mmol/l):

   dissolve 200 mg maltose monohydrate (or 210 mg trehalose dihydrate, or 190 mg sucrose or isomaltose) in maleate buffer (1) and make up to 10 ml.

3. Tris buffer (0.5 mol/l, pH 7.0):

   dissolve 61.0 g Tris in 85 ml HCl, 5 mol/l, and dilute with water to 1000 ml. If necessary adjust pH to 7.0.

4. Glucose reagent (GOD >2400 U/l, POD 1100 U/l):

   dissolve 5.6 g Glox Novum* in 100 ml Tris buffer (3).

5. Glucose standard solutions:

   prepare a series of glucose solutions in water, containing up to 10 μg glucose per 20 μl (0.5 mg/ml).

6. Sodium chloride (0.9% w/v):

   dissolve 0.9 g NaCl in water to a final volume of 100 ml.

Stability of solutions: the buffered disaccharide solutions (2) and the glucose solutions (5) shall be stored frozen (−20 °C) in small aliquots (1−2 ml). In this state they are stable for years. Thaw completely and mix well before use.

The glucose reagent (4) is stable for a few days in the refrigerator (+4 °C).

Procedure

Collection and treatment of sample: usually peroral biopsies of human small-intestinal mucosa are used. It is recommended that the tissue is wrapped in parafilm and stored in the freezer (−20 °C).

Homogenization: the tissue is weighed and transferred to a glass pestle homogenizer. We use conical homogenizers (Fig. 4).

Add 0.9% NaCl in the amount shown in Table 1. Keep the homogenizer with its content (including the glass pestle) in crushed ice for 5−10 min. Homogenize for 1−2 min at low speed (200−300 rpm) while the homogenizer is still immersed into the crushed ice. Insufficient chilling or too high motor speed may cause losses of enzymatic activity. *Never* centrifuge the homogenate. For analysis, dilutions of the homogenate are prepared according to Table 2.

---

* Glox Novum® is a ready-made reagent for glucose assay (*Kabi* Diagnostica AB, S-11287 Stockholm, Sweden). It is sold in packages of 29 g, intended for 500 ml reagent. It contains glucose oxidase, peroxidase, chromogen and buffer, but it must nevertheless be dissolved in Tris buffer for the reason given.

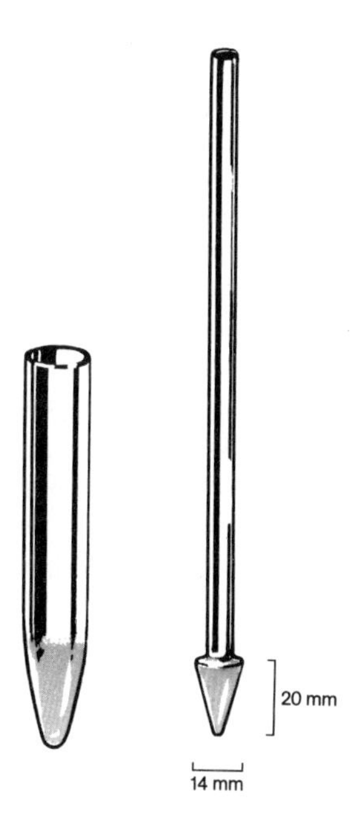

Fig. 4. Glass pestle homogenizer (conical) for mucosal biopsies.

Table 1. Approximate amount of 0.9% NaCl to be used for homogenization.

Wet weight of the biopsy	Amount of 0.9% NaCl to be added
<4 mg	0.2 ml*
4 – 8 mg	0.2 ml
12 – 20 mg	0.5 ml
20 – 40 mg	1.0 ml**
40 – 80 mg	2.0 ml**

* Different dilution (cf. Table 2).
** First add 0.5 ml, homogenize, then add the rest and homogenize again.

Table 2. Recommended dilution of the homogenates for the assay of the disaccharidase activities. If the absorbance becomes too low, or above the standard curve, a new analysis should be performed with adjusted dilution.

Activity	Biopsy weight 4 – 80 mg (Table 1)	Biopsy weight <4 mg (Table 1)
Maltase	1:50	1:20
Isomaltase	1:20	1:10
Sucrase	1:10	1:5
Trehalase	1:5	1:2
Lactase	1:5	1:2

Stability of the enzymes in the sample: the disaccharidase activities in unhomogenized tissue, stored in this way, are stable for several years. If analysis is to be performed on the same day, the tissue can be kept for several hours in the refrigerator or even at room temperature.

The analysis should preferably be performed on the same day, and the homogenate kept in ice or in the refrigerator. If the homogenate is to be stored more than 24 hours it may be frozen, but it is then recommended that it is re-homogenized after thawing. According to *Miller* [17] homogenates can be thawed, re-homogenized, assayed and re-frozen repeatedly over a period of several months, without significant change of the enzymatic activity.

Assay conditions: incubation temperature 37 °C; final volume of incubation mixture 20 µl; wavelength 450 (420 – 480) nm; light path 10 mm; final volume 320 µl. Read against reagent blank containing 20 µl water and 300 µl glucose reagent (4) after 30 min at 37 °C.

For each assay prepare a blank with sample, but in which the disaccharide solution (2) is added *after* the glucose reagent (4). The first 60 min incubation is omitted.

Glucose standards: pipette into three test tubes 20 µl of glucose standard solution (5) equivalent to 2, 6 and 10 µg glucose; add to each 300 µl glucose reagent (4) and incubate for 30 min.

Measurement

Assay

Pipette into conical test tubes (8 mm × 63 mm):			concentration in assay mixture	
mucosal homogenate disaccharide solution	 (2)	10 µl 10 µl	volume fraction disaccharide maleate	0.50 28 mmol/l 50 mmol/l
mix and incubate for 60 min,				
glucose reagent	(4)	300 µl	Tris peroxidase glucose oxidase chromogen	470 mmol/l 1095 U/l 2385 U/l 0.23 mmol/l
mix; the disaccharidase reaction is automatically stopped by the Tris buffer. Allow to stand for 30 min at 37 °C and read absorbances.				

If a sample gives a higher absorbance than that of the 10 µg glucose standard, repeat the assay with diluted homogenate.

Protein determination: protein is determined by the method of *Lowry et al.* [18] (described in detail in Vol. II, chapter 1.3), but with citrate instead of tartrate in reagent B [19]. The standard curve is made with freshly dissolved human serum albumin. For the protein assay, 50 µl of the homogenate is used with a final volume of 6.5 ml.

We have repeatedly attempted to decrease the amount of the sample used for protein determination by reducing all volumes by a factor of 20. Although excellent standard curves were obtained with the serum albumin, the mucosal samples failed to give reproducible values with the smaller volumes.

Calculation: the amount of glucose formed (sample minus blank) is measured from the standard curve.
One unit of disaccharidase activity hydrolyzes 1 µmole of disaccharide per minute at 37 °C.

Since a is µg glucose (M_r 180) formed in 60 minutes by 10 µl (0.01 ml) of a homogenate diluted by a factor F (Table 2) from a disaccharide with n glucose units ($v_i = 1$ for sucrose, $v_i = 2$ for maltose, isomaltose and trehalose), the catalytic activity concentration in the homogenate can be calculated as:

$$b = \frac{a \times \mathrm{F} \times 10^3}{60 \times 0.01 \times v_i \times 180} = \frac{a \times \mathrm{F} \times 10^3}{v_i \times 108} \quad \mathrm{U/l}.$$

For calculation of the catalytic activity content of the tissue per gram protein, use protein concentration, ρ_{Protein}, from standard curve. Then

$$z_c/m_s = \frac{b}{\rho_{\mathrm{Protein}}} \quad \mathrm{U/g}.$$

Validation of Method

Precision, accuracy, detection limits and sensitivity: standard deviation is ±2% calculated from 25 pairs of duplicate assays. Detection limit is 1 U/g protein.

Sources of error: glucose or peroxides will give colour, but this is compensated by the blank.

Specificity: the specificity of the human α-glucosidases is given in Table 3 [2]. There are five different brush-border α-glucosidases, and some of the substrates are hydrolyzed by more than one enzyme. There is also a lysosomal acid α-glucosidase, but the contribution of this enzyme to the disaccharidase activities of a mucosal homogenate is not significant [11].

Table 3. Specificity of the human small-intestinal α-glucosidases, and the contribution of each enzyme to the total activity with different substrates.

Enzyme	Substrates (% of total activity)
Isomaltase	isomaltose (>95%) maltose (~50%)
Sucrase	sucrose (100%) maltose (~25%)
Two heat-stable maltases	maltose (~25%) isomaltose (<5%)
Trehalase	trehalose (100%)

Reference ranges: values from biopsies of human small-intestinal mucosa, taken at the duodenojejunal flexure (*lig. Treitz*) are given in Table 4 [20].

Table 4. Values for the α-glucosidase activities in a group of adult Finns (n = 19).

	Units per gram protein		
	Mean	± SD	Range
Maltase	265	± 73	111 – 407
Isomaltase	85	± 26	32 – 139
Sucrase	72	± 22	35 – 131
Trehalase	28	± 8	13 – 43

References

[1] *A. Dahlqvist,* Hog Intestinal α-Glucosidases. Solubilization, Separation and Characterization. Dissertation, Univ. of Lund (1960).

[2] *A. Dahlqvist,* Specificity of the Human Small-intestinal Disaccharidases and Implications for Hereditary Disaccharide Intolerance, J. Clin. Invest. *41*, 463 – 470 (1962).

[3] *D. Miller, R. K. Crane,* The Digestive Function of the Epithelium of the Small Intestine. II. Localization of Disaccharide Hydrolysis in the Isolated Brush Border Protein of Intestinal Epithelial Cells, Biochim. Biophys. Acta *52*, 293 – 298 (1961).

[4] *A. Dahlqvist,* Characterization of Hog Intestinal Trehalase, Acta Chem. Scand. *14*, 9 – 16 (1960).

[5] *A. Prader, S. Auricchio, G. Mürset,* Durchfall infolge hereditären Mangels und intestinaler Saccharaseaktivität (Saccharose-Intoleranz). Schweiz. Med. Wochenschr. *91*, 465 – 476 (1961).

[6] *R. Bergoz,* Intestinal Trehalase Deficiency, in: *B. Borgström, A. Dahlqvist, L. Hambraeus* (eds.). Swedish Nutr. Found Symp.: Intestinal Enzyme Deficiencies and their Nutritional Implications, Almqvist & Wiksell, Uppsala (1973).

[7] *J. Kolinska, G. Semenza,* Studies on Intestinal Sucrase and on Intestinal Sugar Transport V. Isolation and Properties of Sucrase-isomaltase from Rabbit Small Intestine, Biochim. Biophys. Acta. *146*, 181 – 195 (1967).

[8] *A. Dahlqvist, S. Auricchio, G. Semenza, A. Prader,* Human Intestinal Disaccharidases and Hereditary Disaccharide Intolerance. The Hydrolysis of Sucrose, Isomaltose, Palatinose (Isomaltulose) and a 1,6-α-Oligosaccharide (Isomalto-oligosaccharide) Preparation, J. Clin. Invest. *42*, 556 – 562 (1963).

[9] *H. Skovbjerg, P. A. Krasilnikoff,* Immunoelectrophoretic Studies on Human Small-intestinal Brush Border Proteins. The Residual Isomaltase in Sucrose Intolerant Patients, Pediatr. Res. *15*, 214–218 (1981).
[10] *A. Dahlqvist,* Disturbances of the Digestion and Absorption of Carbohydrates, in: *D. J. Manners* (ed.), International Review of Biochemistry, Biochemistry of Carbohydrates II, Vol. 16, University Park Press, Baltimore (1978).
[11] *N.-G. Asp, E. Gudmand-Høyer, P. M. Christainsen, A. Dahlqvist,* Acid α-glucosidase from Human Gastrointrestinal Mucosa. – Separation and Characterization, Scand J. Clin. Lab. Invest. *33*, 239–245 (1974).
[12] *A. Dahlqvist,* Assay of Intestinal Disaccharidases, Enzymol. Biol. Clin. *11*, 52–56 (1970).
[13] *H. Skovbjerg, H. Sjöström, O. Norén, E. Gudmand-Høyer,* Immunoelectrophoretical Studies on Human Small-intestinal Brush-border Proteins. A Quantitative Study of Brush Border Enzymes from Single Small Intestinal Biopsies, Clin. Chim. Acta *92*, 315–322 (1979).
[14] *H. Skovbjerg,* Humane intestinale børstesømenzymer, Dissertation, Univ. of Copenhagen (1981).
[15] *A. Jeanes, C. A. Wilham, R. W. Jones, H. M. Tsuchiya, C. E. Rist,* Isomaltose and Isomaltotriose from Enzymic Hydrolyzates of Dextran. J. Am. Chem. Soc. *75*, 5911–5915 (1953).
[16] *A. Dahlqvist,* Hog Intestinal Isomaltase Activity, Acta Chem. Scand. *14*, 72–80 (1980).
[17] *D. L. Miller,* Disaccharidase Activities, Am. J. Clin. Nutr. *34*, 1153–1154 (1981).
[18] *O. H. Lowry, N. J. Rosebrough, A. L. Farr, R. J. Randall,* Protein Measurement with the Folin Phenol Reagent, J. Biol. Chem. *193*, 265–275 (1951).
[19] *M. Eggstein, F. H. Kreutz,* Vergleichende Untersuchungen zur quantitativen Eiweißbestimmung im Liquor und eiweißarmen Lösungen, Klin. Wochenschr. *33*, 879–884 (1955).
[20] *N. G. Asp, N. O. Berg, A. Dahlqvist, J. Jussila, H. Salmi,* The Activity of Three Different Small-intestinal β-Galactosidases in Adults with and without Lactase Deficiency, Scand. J. Gastroenterol. *6*, 755–762 (1971).

2.6 β-D-Glucosidases in Tissue

(Glucocerebrosidase, β-Glucosidase)

D-Glucosyl-*N*-acylsphingosine glucohydrolase, EC 3.2.1.45
β-D-Glucoside glucohydrolase, EC 3.2.1.21

Lydia B. Daniels and Robert H. Glew

General

Mammalian tissues contain two principal β-glucosidases which are under separate genetic control and which possess distinctive structural and kinetic properties: 1) membrane-bound, lysosomal glucocerebrosidase has a relatively acid pH optimum (4.8–5.8) and is responsible for catalyzing the hydrolysis of the glucose residue in β-linkage to ceramide in glucocerebroside, and 2) a broad-specificity β-glucosidase, with a pH optimum nearer to neutrality (5.0 – 6.5) present in abundance usually in the cytosol of most tissues (liver, spleen and kidney), whose function is unknown. In addition to β-glucosidase, glucocerebrosidase also utilizes aryl β-glucosides as substrates (e.g. 4-methylumbelliferyl-β-D-glucopyranoside, 4-nitrophenyl-β-D-glucopyranoside). Cytosolic β-glucosidase does not act on glucocerebroside, but will catalyze the hydrolysis of aryl (4-methylumbelliferone, 4-nitrophenol) derivatives of β-D-glucose, α-L-arabinose, β-D-fucose, β-D-xylose and β-D-galactose [1].

Fibroblasts contain little or no broad-specificity β-glucosidase and the β-glucosidase activity demonstrable in extracts of cultured fibroblasts is due almost entirely to lysosomal β-glucosidase. In some tissues, such as brain and leukocytes, a considerable fraction of the broad-specificity β-glucosidase activity is associated with a particulate fraction [2, 3].

Compared to the levels of total β-glucosidase activity seen in tissues, the amount of β-glucosidase activity demonstrable in body fluids is very low. Consequently, few reports concerning hydrolases or glycosidases in serum, cerebrospinal fluid or urine contain any mention of β-glucosidase activity.

Procedures for the extensive purification of placental lysosomal glucocerebrosidase [4] and cytosolic, broad-specificity β-glucosidase from human liver [1] are described in the literature.

Application of method: the most common use of β-glucosidase assays is in the diagnosis of *Gaucher's* disease, a sphingolipidosis that presents in non-neurologic and neurologic forms; tissues of all patients with this particular lipid storage disorder are profoundly deficient in glucocerebrosidase activity [5]. The results of assays of glucocerebrosidase activity carried out using either the natural substrate (glucocerebroside) or a non-physiological, commercially available one (4-methylumbelliferyl-β-D-glucopyranoside) under specified conditions [6], and extracts of fibroblasts or leukocytes as the enzyme source, can be used to identify heterozygote carriers of the disease.

Investigators engaged in studies requiring subcellular fractionation of tissues might have need for a β-glucosidase assay that would serve as a marker for lysosomes. A β-glucosidase assay specific for glucocerebrosidase would serve this purpose nicely because the enzyme is very firmly integrated into the lysosomal membrane.

Enzyme properties relevant in analysis: unless incubation conditions are rigorously controlled, the results of assays of β-glucosidase in crude homogenates of most tissues will reflect the activities of both glucocerebrosidase and the broad-specificity β-glucosidase. Incorporation of specific inhibitors and activators in the incubation medium allows the activity of each β-glucosidase isoenzyme to be estimated in the presence of the other [3, 6, 7] thereby obviating the need to separate the two forms by physical means. Lysosomal glucocerebrosidase is activated by bile salts (e.g. sodium taurocholate, sodium taurodeoxycholate), acidic phospholipids [8] and a heat-stable glycoprotein (M_r 11 000) from the spleens of *Gaucher's* disease patients [9]. Glucocerebrosidase, the lysosomal β-glucosidase, is inhibited completely and irreversibly by the active-site reagent, conduritol B epoxide [7]. In contrast, the broad-specificity β-glucosidase that does not act upon glucocerebroside is [1] inhibited by the aforementioned bile salts and phospholipids, and [2] unaffected by conduritol B epoxide. These two principal tissue β-glucosidases also can be distinguished by their lability at acid pH (pH < 4.3); glucocerebrosidase is stable at pH 4.0 whereas the broad-specificity β-glucosidase is rapidly and irreversibly inactivated by such treatment [10].

Assays for both β-glucosidases are linear for at least one hour at 37 °C provided not more than 250 units of activity are present in the incubation medium.

Methods of determination: at the present time the most rapid, sensitive, convenient, and readily available method for measuring β-glucosidase activity involves fluorimetry and the artificial substrate 4-methylumbelliferyl-β-D-glucopyranoside (MUGlc) [3, 6, 10–14]. A number of investigators in laboratories around the world have found that the basic fluorimetric procedure provides reliable estimates of glucocerebrosidase activity and total β-glucosidase activity in unfractionated homogenates of a variety of tissues.

Estimation of glucocerebrosidase activity in extracts of leukocytes or fibroblasts using the fluorimetric assay requires an incubation medium that contains sodium acetate buffer, 200 mmol/l, pH 5.5; 1.2% (w/v) sodium taurocholate and MUGlc, 5 mmol/l.

Determination of the activity of broad-specificity β-glucosidase in the presence of significant amounts of glucocerebrosidase (as is the case in crude homogenates of visceral tissues) utilizes the following incubation medium supplemented with conduritol B epoxide, 2 mmol/l: sodium acetate buffer, 200 mmol/l, pH 5.5, and MUGlc, 5 mmol/l. The conduritol B epoxide inhibits glucocerebrosidase so that activity measured under these conditions represents only broad-specificity β-glucosidase activity. The result of subtracting the amount of β-glucosidase activity observed in the presence of conduritol B epoxide from that obtained in the absence of the active-site inhibitor provides a reliable estimate of the glucocerebrosidase content of tissue homogenates [7, 13].

Glucocerebrosidase activity can also be determined directly and unambiguously using radiolabelled authentic glucocerebroside (labelled in the glucose or fatty acid moiety); however, this assay procedure [10, 15] is more costly and less convenient than the fluorimetric β-glucosidase assay. For tissues such as liver and spleen which contain large amounts of cytosolic broad-specificity β-glucosidase, glucocerebrosidase activity can be estimated accurately with the conduritol B epoxide-dependent, fluorimetric β-glucosidase assay [7] but not with the pH 5.5-taurocholate assay [16].

International reference methods and standards: at the present time, the β-glucosidase assays have not been standardized.

Enzyme effectors: glucocerebrosidase is activated by sodium taurocholate or sodium taurodeoxycholate and inhibited by conduritol B epoxide [3, 7, 10]. Broad-specificity β-glucosidase is inhibited by bile salts and certain phospholipids, including phosphatidylserine, phosphatidylcholine and phosphatidylinositol [1].

Assays

Method Design

Principle

(a) 4-Methylumbelliferyl-β-D-glucopyranoside + H_2O $\xrightarrow{\beta\text{-glucosidase}}$

D-glucose + 4-methylumbelliferone

(b) 4-Methylumbelliferone + OH^- ⟶ 4-methylumbelliferone anion + H_2O.
(weakly fluorescent) (intensely fluorescent)

The amount of substrate cleaved is equivalent to the amount of highly fluorescent 4-methylumbelliferone anion generated. Quenching the reaction with alkaline (pH 10.5) glycine : ammonium hydroxide solution accomplishes two things: 1. it prevents further enzyme-catalyzed hydrolysis of MUGlc and 2. it converts 4-methylumbelliferone to its more fluorescent anion form.

Quantitation is achieved by calibrating the fluorimeter with a standard alkaline solution of 4-methylumbelliferone. The ideal blank is one that uses heat-inactivated enzyme (boiling water-bath, 2 minutes). Since MUGlc is unstable at alkaline pH, fluorescence should be determined immediately, and certainly within 2 h after quenching the reaction.

Optimized conditions for measurement: the most critical elements of the β-glucosidase assays are the activator (sodium taurocholate) and inhibitor (conduritol B epoxide) of the lysosomal β-glucosidase. As the K_m values for MUGlc for glucocerebrosidase and broad-specificity β-glucosidase in most tissues are 1.5 mmol/l and 0.2 mmol/l, respectively, use of MUGlc, 5 mmol/l, in the final incubation medium, ensures that the reaction in each case is carried out at, or close to, a substrate concentration that saturates the enzymes. If relatively pure preparations of enzyme are being analyzed for activity, it is advisable to incorporate bovine serum albumin (10 mg per ml, final concentration) and 2-mercaptoethanol (10 mmol/l) in the assay.

Temperature conversion factors: no factors have been established.

Equipment

Turner Model 111 fluorimeter fitted with filters permitting excitation at 360 nm (*Klett* filter No. 7 – 60) and transmission of emitted light at 520 – 580 nm (*Klett* filter No. 54). A recorder is not necessary, but a stopwatch or timer is desirable. For convenience, one should use a dispenser bottle to deliver the alkaline quench buffer accurately and reproducibly.

Reagents and Solutions

Purity of reagents: the substrate, MUGlc, must be of the proper anomeric configuration and should contain a negligible amount of free methylumbelliferone, so as to minimize the fluorescence of blanks and optimize sensitivity. All other reagents should be of "analytical purity" grade.

Preparation of solutions (for about 70 determinations): all solutions in re-purified water (cf. Vol. II, chapter 2.1.3.2).

1. 4-Methylumbelliferone standard (10 mmol/l):

   dissolve 198.2 mg 4-methylumbelliferone, sodium salt, in 10 ml chloroform: methanol (2:1, v/v) in a sintered glass-stoppered volumetric flask.

2. Acetate buffer (1 mol/l, pH 5.5):

   dissolve 136.1 g $NaC_2H_3O_2 \cdot 3\ H_2O$ in 1 l water, adjust to pH 5.5 with acetic acid, 1 mol/l (57.5 ml CH_3COOH in 1 l water). Add 0.2 g azide, sodium salt, (0.02%, w/v) to prevent microbial contamination.

3. Taurocholate (12%, w/v):

   dissolve 1.2 g taurocholate, sodium salt, in 10 ml water.

4. 4-Methylumbelliferyl-β-D-glucopyranoside, MUGlc (10 mmol/l):

   dissolve 33.8 mg 4-methylumbelliferyl-β-D-glucopyranoside in 10 ml water.

5. Glycine/ammonium hydroxide quench solution (ammonium hydroxide, 0.2 mol/l, pH 10.5; glycine, 0.05 mol/l):

   dissolve 15 g glycine in 54 ml NH_4OH and 3.94 l water, adjust to pH 10.5 with NaOH, 5 mol/l.

6. Conduritol B epoxide (20 mmol/l):

   dissolve 16.2 mg conduritol B epoxide in 5 ml water.

Stability of solutions: the 4-methylumbelliferone standard solution (1) is stable at room temperature for 6 months if kept out of the light. Acetate buffer (2), when supplemented with azide, and the glycine/ammonium hydroxide quench solution (5) are stable for 6 months at room temperature. Solutions of conduritol B epoxide (6), sodium taurocholate (3), and 4-methylumbelliferyl-β-D-glucoside (4) must be stored at −20 °C; the former two are stable for 6 months, the last is stable for one month, and all three solutions withstand at least 4 cycles of freeze-thawing.

Procedures

Collection and treatment of specimen: cultured fibroblasts are prepared using standard techniques. For leukocytes, blood is collected in EGTA-containing tubes and the red cells are removed by the acid-citrate-dextran settling method (3). The leukocyte pellet is resuspended in physiological saline (0.9% w/v NaCl in water), and sonicated 4 s at setting 4 using the microtip of an ultrasonicator (*Heatsonics Inc.*). All other tissues, (e.g. fibroblasts, liver, spleen, brain) are homogenized in 7 volumes of water using 10 passes of a motor-driven *Potter-Elvejhem* homogenizer (Teflon pestle, glass vessel).

In the case of liver and spleen, the homogenate can be centrifuged 1 h at 100000 *g* to separate most of the glucocerebrosidase, which remains in the pellet, from the broad-specificity β-glucosidase which remains in the supernatant [1].

Stability of enzyme in specimen or sample: glucocerebrosidase activity in crude homogenates of leukocytes or fibroblasts is quite unstable and β-glucosidase assays on those tissues should be performed promptly; storage of crude homogenates at 4 °C or −43 °C for 18 h can result in greater than 80 percent loss of glucocerebrosidase activity. If not homogenized prior to freezing, intact leukocytes or fibroblasts can be frozen (−43 °C) without loss of glucocerebrosidase activity for at least one month.

Glucocerebrosidase and broad-specificity β-glucosidase, when extensively purified from visceral organs, lose about 15−20% of their initial activities after 3 months at 4 °C, and less than 10% after 24 h at room temperature. Bovine serum albumin, 10 mg per ml, stabilizes partially-purified broad-specificity β-glucosidase. Both enzymes can be stored for several weeks at −43 °C in a 50% glycerol solution with no loss of activity. Freezing destroys the activity of both β-glucosidases. Samples of human tissue, obtained at autopsy or surgery, can be stored at −43 °C for 6 years without apparent loss of β-glucosidase activity provided the tissue is not thawed and re-frozen.

Details for measurement in tissues: glucocerebrosidase activity can be estimated reliably using the fluorimetric, sodium taurocholate-dependent β-glucosidase assay (Assay I) only when fibroblasts and leukocytes, not crude extracts of visceral organs, serve as the enzyme source.

In cases where homogenates of visceral organs serve as the source of enzyme, two determinations of β-glucosidase activity using the fluorimetric assay supplemented with sodium taurocholate, one in the presence of conduritol B epoxide and the other in its absence, permit one to estimate the contributions of both glucocerebrosidase and broad-specificity β-glucosidase to total β-glucosidase activity (Assay II). Measurement of β-glucosidase activity in the presence of the bile salt determines both glucocerebrosidase activity and residual activity from cytosolic β-glucosidase that is not completely inhibited by sodium taurocholate. Conduritol B epoxide will inhibit essentially all of the glucocerebrosidase activity, but will not affect the activity of the cytosolic, broad-specificity β-glucosidase; therefore, performance of the taurocholate-supplemented (pH 5.5-taurocholate) β-glucosidase assay in the presence of conduritol B epoxide measures only the small amount of residual activity of the broad-specificity β-glucosidase. The difference between the two determinations provides an estimate of glucocerebroside: β-glucosidase activity.

At the present time, there is no reliable way to estimate directly the amount of β-glucosidase activity due specifically to the broad-specificity β-glucosidase when crude tissue homogenates that are also rich in glucocerebrosidase are the source of enzyme. However, once the broad-specificity β-glucosidase has been separated from lysosomal glucocerebrosidase by a variety of column chromatographic techniques [1], β-glucosidase activity can be estimated directly by fluorimetry using a pH 5.5 incubation medium containing MUGlc (Assay III).

Assay conditions: wavelength 366 nm excitation, 520 – 580 nm emission; final volume 3.0 ml; incubate at 37 °C; determine fluorescence at ambient temperature. Measure against assay containing heat-denatured enzyme blank.

Standards: dilute 4-methylumbelliferone standard solution (1) 1:1000 with water, pipette 5 to 100 µl (0.05 to 1 nmol) into assay tubes, add 2.9 ml glycine/NH_4OH solution (5), mix. Measure fluorescence as a function of nmol 4-methylumbelliferone.

Assay I. Estimation of leukocyte and fibroblast glucocerebrosidase activity using the pH 5.5-taurocholate assay

Pipette successively into a 10 × 75 mm assay tube on ice:			concentration in assay mixture	
MUGlc solution	(4)	0.050 ml	MUGlc	5.0 mmol/l
buffer	(2)	0.020 ml	acetate	0.2 mol/l
taurocholate solution	(3)	0.010 ml	taurocholate	12 g/l
water		0.010 ml		
mix thoroughly using a vortex mixer				
sample (leukocyte or fibroblast extract)		0.010 ml	volume fraction	$\varphi = 0.10$
mix thoroughly using vortex mixer, start incubation (15 – 60 min) at 37 °C				
glycine/NH_4OH	(5)	2.90 ml		
mix by inversion three times, reaction is stopped. Measure fluorescence (ΔF) immediately versus a boiled enzyme blank.				

Select the proper sensitivity setting such that the ΔF value on the fluorescence meter of the fluorimeter is within the range of 10–90. The amount of fluorescence can be altered by changing the incubation period or the amount of enzyme added to the assay medium. The ΔF of the standard solution of 4-methylumbelliferone ($\Delta F_{standard}$) is determined separately and used to calibrate the fluorimeter.

Calculation: according to eqn. (m_1) in combination with eqns. (k) or (k_1), cf. "Formulae", Appendix 3, the catalytic concentration in the sample is (t in minutes):

$$b = \frac{\Delta F_{sample}}{\Delta F_{standard}} \times c_{standard} \times \frac{1}{\varphi \times \Delta t} \quad \text{U/l}$$

Assay II. Estimation of glucocerebrosidase activity using the conduritol B epoxide-dependent fluorimetric assay

This procedure is carried out exactly as described under Assay I. In addition, a second β-glucosidase assay ("plus CBE") is performed except that the water component is replaced with 0.01 ml solution of conduritol B epoxide [6]. The standard assay (Assay I) is referred to as the "minus CBE" assay.

Calculation: the equation above is used to calculate amounts of β-glucosidase activity obtained by each of the two assay procedures. Relative glucocerebrosidase activity is obtained by subtracting the activity observed with the "plus CBE" assay (A) from that obtained with the "minus CBE" assay (B); simply, B units/ml – A units/ml = glucocerebrosidase activity (units/ml).

Assay III. Measurement of broad-specificity β-glucosidase

Pipette successively into a 10 × 75 mm assay tube on ice:			concentration in assay mixture	
MUGlc solution buffer water	(4) (2)	0.050 ml 0.020 ml 0.010 ml	MUGlc acetate	5.0 mmol/l 0.2 mol/l
mix thoroughly using a vortex mixer				
sample (tissue homogenate)		0.020 ml	volume fraction	$\varphi = 0.20$
mix thoroughly using vortex mixer, start incubation (10–60 min) at 37 °C				
glycine/NH_4OH	(5)	2.90 ml		
mix by inversion three times, reaction is stopped. Measure fluorescence (ΔF) immediately.				

Calculation: the equation under "Assay I" is used to calculate units of enzyme activity.

Validation of Method

Precision, accuracy, detection limits and sensitivity: for values of 30 units/mg protein in tissue homogenates an imprecision of ±2 units/mg protein has been found for manual performance; these assays have not been automated. Variation of levels of glucocerebrosidase or broad-specificity β-glucosidase among individuals is very large; the standard deviation calculated for the mean value of a series of measurements on leukocytes from individuals within one laboratory was 20% of that mean [13]. Other laboratories performing similar determinations report standard deviations of 12–30% [10–14]. Information about accuracy is not available since reference standards have not been established. The limit of detection is about 0.05 units/mg protein with an imprecision of ±50%. Sensitivity can be increased by varying the secondary filter to select solely the emission maximum, but this is rarely done.

Sources of error: interference due to autofluorescence of substrate is reduced to a minimum under the experimental conditions described here. Fluorescence quenching occurs when impure bile salts are used as activators.

Specificity: to measure glucocerebrosidase activity specifically and unambigously, one must use the authentic lipid substrate (glucocerebroside); however, the fluorimetric procedure provides a good approximation (±20%) of glucocerebrosidase activity when it is determined as the amount of conduritol B epoxide-sensitive tissue β-glucosidase activity in the presence of the bile salt activator, sodium taurocholate [7, 13].

Reference ranges: glucocerebrosidase, determined using the conduritol B epoxide-dependent procedure, in human leukocytes, 18.6 ± 3.7 units/mg protein; fibroblasts, 313 ± 40.5 units/mg protein; liver, 52.1 ± 27.5 units/mg protein; spleen, 33.9 ± 17.5 units/mg protein; and brain, 38.0 ± 5.6 units/mg protein. Similar values (±15%) are measured using the bile salt-dependent assay detailed here.

Broad-specificity β-glucosidase in liver, 27.5 ± 7.7 units/mg protein; spleen 2.62 ± 1.67 units/mg protein; and brain, 15.4 ± 4.7 units/mg protein. Leukocytes contain approximately 4.2 units/mg protein and fibroblasts 2.8 units/mg protein non-specific β-glucosidase activity.

Tissue β-glucosidase levels do not appear to differ significantly between adults and children.

References

[1] *L. B. Daniels, P. J. Coyle, Y. B. Chiao, R. H. Glew, R. S. Labow,* Purification and Characterization of a Cytosolic Broad-specificity β-Glucosidase from Human Liver, J. Biol. Chem. *256*, 13004 – 13013 (1981).

[2] *L. B. Daniels, P. J. Coyle, R. H. Glew, N. S. Radin, R. S. Labow,* Brain Glucocerebrosidase in *Gaucher*'s Disease, Arch. Neurol. *39*, 550 – 556 (1982).

[3] *S. P. Peters, P. Coyle, R. H. Glew,* Differentiation of β-Glucocerebrosidase from β-Glucosidase in Human Tissues Using Sodium Taurocholate, Arch. Biochem. Biophys. *175*, 569 – 582 (1976).

[4] *F. S. Furbish, H. E. Blair, J. Shiloach, P. G. Pentchev, R. O. Brady, Gaucher*'s Disease: Large Scale Purification of Glucocerebrosidase Suitable for Human Administration, Proc. Natl. Acad. Sci. USA *74*, 3560 – 3563 (1977).

[5] *R. O. Brady, J. N. Kanfer, R. M. Bradley, D. Shapiro,* Demonstration of a Deficiency of Glucocerebroside-cleaving Enzyme in *Gaucher*'s Disease, J. Clin. Invest. *45*, 1112 – 1115 (1966).

[6] *L. B. Daniels, R. H. Glew,* β-Glucosidase Assays in the Diagnosis of *Gaucher*'s Disease, Clin. Chem. *28*, 569 – 577 (1982).

[7] *L. B. Daniels, R. H. Glew, N. S. Radin, R. R. Vunnam,* A Revised Fluorometric Assay for *Gaucher*'s Disease Using Conduritol-β-epoxide with Liver as the Source of β-Glucosidase, Clin. Chim. Acta *106*, 155 – 163 (1980).

[8] *M. W. Ho, N. D. Light,* Glucocerebrosidase: Reconstitution from Macromolecular Components Depends on Acidic Phospholipids, Biochem. J. *136*, 821 – 823 (1973).

[9] *S. P. Peters, P. Coyle, C. J. Coffee, R. H. Glew, M. S. Kuhlenschmidt, L. Rosenfeld, Y. C. Lee,* Purification and Properties of a Heat-stable Glucocerebrosidase Activating Factor from Control and *Gaucher* Spleen, J. Biol. Chem. *256*, 563 – 573 (1977).

[10] *S. S. Raghavan, J. Topol, E. H. Kolodny,* Leukocyte β-Glucosidase in Homozygotes and Heterozygotes for *Gaucher*'s Disease, Am. J. Hum. Genet. *32*, 158 – 173 (1980).

[11] *E. Beutler, W. Kuhl,* The Diagnosis of the Adult Type of *Gaucher*'s Disease and its Carrier State by Demonstration of Deficiency of β-Glucosidase Activity in Peripheral Blood Leukocytes, J. Lab. Clin. Med. *76*, 747 – 755 (1970).

[12] *D. A. Wenger, C. Clark, M. Sattler, C. Wharton,* Synthetic Substrate β-Glucosidase Activity in Leukocytes: A Reproducible Method for the Identification of Patients and Carriers of *Gaucher*'s Disease, Clin. Genet. *13*, 145 – 153 (1978).

[13] *L. B. Daniels, R. H. Glew, W. F. Diven, R. E. Lee, N. S. Radin,* An Improved Fluorometric Leukocyte β-Glucosidase Assay for *Gaucher*'s Disease, Clin. Chim. Acta *115*, 369 – 375 (1981).

[14] *L. Svennerholm, G. Håkansson, S. Dreborg,* Assay of the β-Glucosidase Activity with Natural Labelled and Artificial Substrates in Leukocytes from Homozygotes and Heterozygotes with the Norbottnian Type (Type 3) of *Gaucher* Disease, Clin. Chim. Acta *106*, 183 – 193 (1980).

[15] *R. A. Synder, R. O. Brady,* The Use of White Cells as a Source of Diagnostic Material for Lipid Storage Diseases, Clin. Chim. Acta *25*, 331 – 338 (1969).

[16] *Y. Ben-Yoseph, H. L. Nadler,* Pitfalls in the Use of Artificial Substrates for the Diagnosis of *Gaucher*'s Disease, J. Clin. Pathol. *31*, 1091 – 1093 (1978).

2.7 β-Galactosidase (Lactase)

β-D-Galactoside galactohydrolase, EC 3.2.1.23

Arne Dahlqvist

General

The brush border lactase of the small-intestinal mucosa hydrolyzes dietary lactose. The enzyme activity is high at birth, but in most populations it decreases at a certain age to a low, residual value (adult lactose intolerance). There are also other β-galactosidases in the small intestine which do not seem to be involved in the hydrolysis of dietary lactose.

Application of method: the assay method is used for biopsy samples of the small intestinal mucosa. There are various conditions in which the lactase activity is low and consequently the tolerance for dietary lactose is limited, including the following (for review, cf. [1])

Congenital lactase deficiency: extremely rare. The intestinal brush border lactase activity is completely absent from birth.

Adult lactase deficiency: extremely common. Only a few populations in the world are lactase persistent, i.e. the intestinal lactase activity is high throughout life.

Secondary lactase deficiency: general disaccharidase deficiency caused by intestinal disease. The lactase activity is more depressed than that of other disaccharidases and, for clinical purposes, it is usually only the lactose intolerance that is important.

All these conditions affect only the brush border lactase, the lysosomal acid β-galactosidase and the cytoplasmic hetero-β-galactosidase remaining unchanged [2].

There is also phlorizin hydrolase (EC 3.2.1.62) in the small intestine. This activity seems to be exerted by the brush border lactase, but at a different active site in the protein molecule [3].

Enzyme properties relevant in analysis: no special properties are known.

Methods of determination: the lactase activity is measured by incubation with lactose, followed by measurement of the glucose liberated with a glucose reagent as described for the α-glucosidases (chapter 2.5).

The activity measured in this way is the sum of the activities of two enzymes, namely the brush border lactase and the lysosomal acid β-galactosidase. (The cytoplasmic hetero-β-galactosidase does not hydrolyze lactose). It is possible to measure the brush border lactase activity separately, either by using PCMB (4-chloromercuribenzoate), 0.2 mmol/l, as inhibitor of the acid β-galactosidase, or by correction of the total lactase activity according to specific measurement of the acid β-galactosidase with 2-naphthyl-β-galactoside as substrate [4]. This substrate is not hydrolyzed by the other two enzymes.

International reference methods and standards: not available.

Enzyme effectors: cf. chapter 2.5, α-glucosidases.

Assay

Method Design

Principle, Optimized conditions for measurement, Temperature conversion factors, Equipment: cf. α-glucosidases, chapter 2.5.

Reagents and Solutions

Purity of reagents: cf. α-glucosidases (chapter 2.5).

Preparation of solutions: all solutions in re-purified water (cf. Vol. II, chapter 2.1.3.2). Use solutions for determination of α-glucosidases (chapter 2.5), except solution (2). Instead prepare

2. Lactose solution (56 mmol/l):

   dissolve 200 mg lactose in maleate buffer (1) and make up to 10 ml.

Stability of solutions: as in chapter 2.5.

Procedure

Collection and treatment of sample, Stability of the enzyme in the sample, Assay conditions, Measurement: as in chapter 2.5.

Calculation: as in chapter 2.5; $\nu_i = 1$ for lactose.

Validation of Method

Precision, accuracy, detection limits and sensitivity: from 25 pairs of duplicate assays a standard deviation of ± 10% was calculated. Detection limit is 2 U/g protein.

Sources of error: cf. chapter 2.5.

Specificity: as mentioned above, the assay as described here will measure the sum of the lactase and the acid β-galactosidase. In subjects with high lactase activity, the acid β-galactosidase will account for only a few percent of the total activity measured (cf. Table 1). In subjects with lactase deficiency, the acid β-galactosidase will account for about one fifth of the total activity. (In a group of 19 subjects the mean was 21% ± 7.5% (SD) with a range of 11 – 38% of the total activity [2].) Even without specific

Table 1. Lactase activity in adult Finns with lactase persistence and adult lactase deficiency, respectively [2].

	Lactase activity units/gram protein	
	Total lactase	Specific assay of brush-border lactase
Lactase-persistent group (n = 16)		
mean	41.1	40.0
SD	± 15.7	± 15.6
range	21 – 80	20 – 79
Lactase-deficient group (n = 19)		
mean	5.1	4.1
SD	± 1.9	± 1.8
range	2.6 – 9.5	1.6 – 8.2

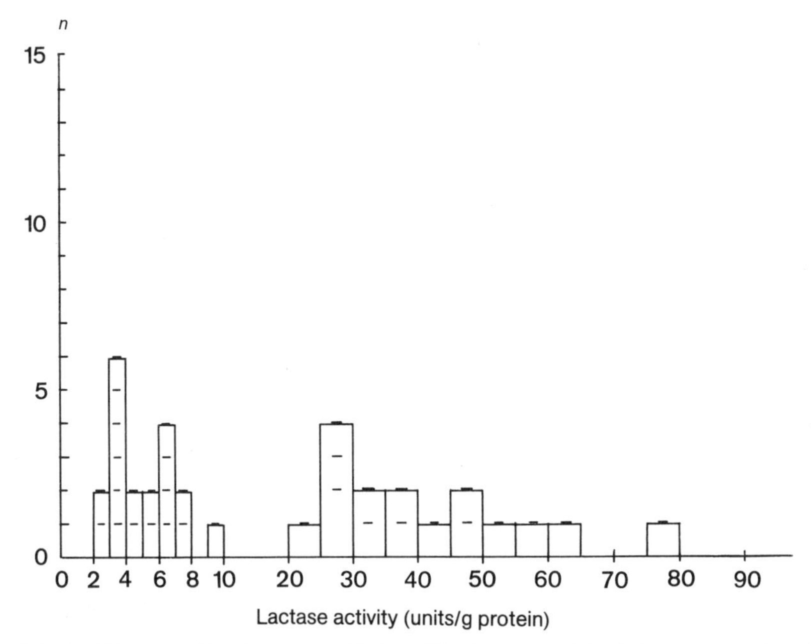

Fig. 1. Histogram of the lactase activity (n = 35), showing that the lactase-persistent and the lactase-deficient groups are two separate populations [2]. The enzyme activity was measured as total lactase, with the method described here.

assay of the brush border lactase, however, the lactase-deficient and the lactase-persistent subjects are two completely separate groups (Fig. 1) [2], and therefore there is no need for the specific assay to be used for the diagnosis of lactase deficiency.

In congenital lactase deficiency, specific assay of the different enzymes has shown that the brush-border lactase activity is completely absent [5].

Reference ranges: values from biopsies of lactase-persistent and lactase-deficient adults are given in Table 1.

Normal infants and children have the same activity per gram protein as the lactase-persistent group.

References

[1] *A. Dahlqvist,* Disturbances of the Digestion and Absorption of Carbohydrates, in: *D. J. Manners* (ed.), International Review of Biochemistry. Biochemistry of Carbohydrates II, Vol. *16,* University Park Press, Baltimore 1978.

[2] *N. G. Asp, N. O. Berg, A. Dahlqvist, J. Jussila, H. Salmi,* The Activity of Three Different Small-intestinal β-Galactosidases in Adults with and without Lactase Deficiency, Scand. J. Gastroenterol. *6,* 755 – 762 (1971).

[3] *H. Skovbjerg, H. Sjöström, O. Norén,* Purification and Characterization of Amphiphilic Lactase/Phlorizin Hydrolase from Human Small Intestine, Eur. J. Biochem. *114,* 653 – 661 (1981).

[4] *N. G. Asp, A. Dahlqvist,* Human Small Intestine β-Galactosidase, Specific Assay of Three Different Enzymes. Anal. Biochem. *47,* 527 – 538 (1972).

[5] *N. G. Asp, A. Dahlqvist, P. Kuituinen, K. Launiala, J. K. Visakorpi,* Complete Deficiency of Brush-border Lactase in Congenital Lactose Malabsorption, Lancet *II,* 329 – 330 (1973).

2.8 α-D-Mannosidase

α-D-Mannoside mannohydrolase, EC 3.2.1.24

Christopher N. Faber and Robert H. Glew

General

α-Mannosidases are exoglycosidases which hydrolyze α-D-mannosyl residues from terminal non-reducing positions of oligosaccharides. Enzymes of this family usually also cleave mannose residues linked through an α-glycosidic bond to non-physiological aglycones such as 4-methylumbelliferone [1, 2] and 4-nitrophenol [5 – 7]. α-Mannosidase activity occurs in many tissues and body fluids including liver, brain, pancreas, cerebrospinal fluid and serum [3, 4]. α-Mannosidase activity has been localized to lysosomes [5], *Golgi* apparatus [6] and cytosol [7] of rat liver.

Three major forms of α-mannosidase occur in human serum and tissues and they are identified usually on the basis of their pH optima: 1) acid (lysosomal) α-mannosidase exhibiting a pH optimum in the range of 4.0 – 4.6 [5, 8]; 2) intermediate (*Golgi*) α-mannosidase with optimum activity at pH 5.5 – 5.8 [6, 9]; and 3) neutral (cytosolic) α-mannosidase having a pH optimum in the range of 6.0 – 6.5 [7]. Acid α-mannosidase is responsible for the degradation of mannose-containing glycopolymers, including glycoproteins [5] and some species of keratan sulphate. It has been suggested that, in the *Golgi* apparatus, intermediate α-mannosidase is involved in processing core mannose residues of oligosaccharide chains during glycoprotein synthesis [6].

Application of methods: mannosidosis, a hereditary lysosomal storage disease in man, cattle and sheep, is due to a deficiency of acid (lysosomal) α-mannosidase activity [10, 11]. Patients with this disorder also have markedly diminished serum levels of acid α-mannosidase. Consequently, assessment of serum acid α-mannosidase activity provides a means for detection of both patients with mannosidosis and heterozygotes [12].

Marked changes in α-mannosidase activity are also observed in liver of persons with liver cirrhosis due to alcoholism and α_1-antitrypsin deficiency [13]. In addition, elevations in the level of activity of acid- and intermediate pH optimum α-mannosidase is seen in the serum of patients with alcoholic liver diesease (*C. N. Faber & R. H. Glew,* unpublished observations). Similar elevation in these two isoenzymes have also been seen in the serum of pyridoxine-deficient weanling rats (*C. N. Faber & R. H. Glew,* unpublished observations).

Enzyme properties relevant in analysis: as mentioned above, there are multiple forms of α-mannosidase in human serum which are classified according to their pH optima; specifically, acid-, intermediate-, and neutral pH optimum isoenzymes. Intermediate α-mannosidase is the predominant activity in human plasma [12, 14]. This enzyme in serum is rapidly inactivated by heating at 45 – 60 °C and is activated by Co^{2+} and Zn^{2+} [8, 14]. The intermediate pH optimum isoenzyme in tissue extracts is insensitive to EDTA and inactivated markedly by 4-chloromethoxyphenylsulphonic acid [5]. Sulphydryl reagents such as mercaptoethanol and dithiothreitol are additional activators of the tissue intermediate pH optimum isoenzyme, whereas Cu^{2+} and Fe^{2+} are potent inhibitors [6].

Acid α-mannosidase activity is present in human plasma to a much lesser extent than the intermediate form [9, 14]. This isoenzyme is thermostable at 60 °C and inhibited by Co^{2+} ions [8, 14]. Zinc ions activate acid α-mannosidase and shift the pH optimum from 4.2 to 3.7 [14]. Prolonged storage of the acid α-mannosidase from tissue extracts in the presence of EDTA inactivates the enzyme. However, activity can be restored by addition of Zn^{2+} ions [5]. 4-Chloromethoxyphenylsulphonic acid has no effect on the activity of this enzyme [5]. Sulphydryl reagents are mild activators of this tissue isoenzyme [5].

Neutral α-mannosidase has been reported to be heat-labile and activated by Co^{2+} ions [15]. This isoenzyme constitutes a small fraction of total serum α-mannosidase activity and its existence as a separate entity in serum is in question [9, 15].

In the 10 – 100 µmol/l range, swainsonine, an indolizidine alkaloid derivative produced by *Swainsona canescens,* causes total inhibition of tissue acid (lysosomal) α-mannosidase and 60% inhibition of neutral (cytosolic) α-mannosidase [16]. Indeed, ingestion of swainsonine by *Angus* cattle produces a clinical condition indistinguishable from bovine mannosidosis [17]. The effect of swainsonine on plasma α-mannosidases has not been systematically studied.

All three forms of α-mannosidase are inhibited by manno-1,5-lactone [18]. The activities of both intermediate and neutral α-mannosidase are unaffected by α-methylmannoside but the acid isoenzyme is inhibited completely by α-methylmannoside, 0.125 mol/l [15].

Chromatographic resolution of these multiple serum isoenzymes has been accomplished by a number of investigators [9, 12, 15] but a unifying nomenclature is lacking. *Hirani & Winchester* [9] suggested a classification system based on the chromatographic properties of the various α-mannosidases on gel filtration and concanavalin A-Sepharose columns. Lysosomal acid α-mannosidase can be fractionated by DEAE-cellulose chromatography into 2 to 3 structurally related forms: form A appears in the breakthrough fraction whereas species B_1 and B_2 are resolved by salt elution [9, 15]. Intermediate α-mannosidase activity in serum can be separated into 3 to 4 forms [9, 15]. The relationship of intermediate and neutral α-mannosidases in serum to tissue forms of these enzymes has not been rigorously established. However, since acid α-mannosidase activity in human serum is virtually absent in mannosidosis [12] and because its chromatographic properties are similar to those of tissue acid α-mannosidase [9], it is likely that the acid pH optimum α-mannosidase in serum is of lysosomal origin.

Schemes for the purification of the tissue forms of the three α-mannosidase isoenzymes from rat liver have been reported. Acid α-mannosidase has been purified 30000-fold to homogeneity [5]; *Golgi* α-mannosidase has been purified 5000-fold [6]; and cytosolic α-mannosidase has been obtained in a form that is 90% pure [7]. Their different chemical and physical properties, as well as a lack of immunological cross reactivity among these tissue isoenzymes, indicates that all three have distinct genetic origins [5]. Purification of the serum α-mannosidase isoenzymes to homogeneity has not been reported.

Methods of determination: fluorimetry provides the most sensitive and convenient procedure for estimating α-mannosidase activity; the substrate in the fluorimetric assay is 4-methylumbelliferyl-α-D-mannopyranoside. However, a colorimetric procedure employing 4-nitrophenyl-α-D-mannopyranoside can also be used to estimate α-mannosidase activity, where 4-nitrophenol release is estimated by its absorbance at 400 nm in alkaline medium [5 – 7]. Assays are linear for 4 – 60 minutes for 5 – 20 µl of serum per assay.

International reference methods and standards: at the present time the α-mannosidase assay has not been standardized, nor has an international standard been defined.

Enzyme effectors: $ZnCl_2$ (0.5 – 1.0 mmol/l) activates acid and intermediate α-mannosidase [5, 14] and bovine serum albumin is sometimes included in the assay medium to stabilize enzyme activity [5 – 7]. Cobalt chloride (0.5 – 10 mmol/l) activates the intermediate and neutral α-mannosidases [8, 14].

Assays

Method Design

Principle: non-physiological substrates are most often used to assay α-mannosidase because the hydrolysis products are conveniently measured.

1. *Fluorimetric assay:*

(a) 4-Methylumbelliferyl-α-D-mannopyranoside + H_2O $\xrightarrow{\alpha\text{-mannosidase}}$ D-mannose + 4-methylumbelliferone

(b) 4-Methylumbelliferone + OH^- ⟶ 4-methylumbelliferone anion + H_2O.
(weakly fluorescent) (intensely fluorescent)

The amount of substrate cleaved is equivalent to the amount of highly fluorescent 4-methylumbelliferone anion generated. Quenching the reaction with an alkaline (pH 10.5) medium accomplishes two things: 1) it stops further reaction from occurring, and 2) it converts 4-methylumbelliferone to its more fluorescent anion form. Quantitation is achieved by calibration of the fluorimeter using a standard alkaline solution of 4-methylumbelliferone. The ideal blank is one that includes heat-inactivated enzyme (boiling water-bath, 2 minutes).

2. *Colorimetric assay:*

(a) 4-Nitrophenyl-α-D-mannopyranoside + H_2O $\xrightarrow{\alpha\text{-mannosidase}}$ D-mannose + 4-nitrophenol

(b) 4-Nitrophenol + OH^- ⟶ 4-nitrophenol anion + H_2O.
(yellow)

The quantity of substrate cleaved is determined from the absorbance of the assay medium at 400 nm after the reaction has been terminated by the addition of an alkaline solution and from the absorption coefficient of 4-nitrophenolate anion (ε_{400} = 1390 $l \times mol^{-1} \times mm^{-1}$).

According to *Embury et al.* [19], the two methods give similar results, except for haemolyzed samples, in which case the colorimetric method yields lower values than those obtained fluorimetrically. Using the fluorimetric procedure, one can perform assays more rapidly and simply than by using the colorimetric assay.

Optimized conditions for measurement: the most important aspects of the assays used to estimate the activity of the various forms of α-mannosidase in serum are pH and the presence of effector substances (i.e., stabilizers, activators): acid α-mannosidase is determined at pH 4.0 – 4.6, sometimes in the presence of bovine serum albumin and zinc sulphate [5]. The intermediate-type isoenzyme is assayed at pH 5.5. Neutral α-mannosidase activity is determined at pH 6.0 – 6.5, often in the presence of bovine serum albumin and cobalt chloride [7].

Temperature conversion factors: no factors have been established.

Equipment

Colorimetric assays require either a spectrophotometer or colorimeter. Fluorimetric assays require a fluorimeter permitting excitation at 360 nm (*Klett* filter, No. 7 – 60) and measurement of emission at 520 – 580 nm (*Klett* filter, No. 54). A recorder is not required but a stopwatch is desirable. A dispenser that accurately and reproducibly delivers 1 – 3 ml (± 0.05 ml) volumes of the alkaline quench buffer is helpful.

Reagents and Solutions

Purity of reagents: the substrates, 4-methylumbelliferyl-α-D-mannopyranoside and 4-nitrophenyl-α-D-mannopyranoside, must be of the proper anomeric configuration and should contain a minimum amount of free 4-methylumbelliferone or 4-nitrophenol. The latter, which are disadvantageous because they raise the level of background in the assay, can be removed by recrystallization if necessary. All other reagents should be of analytical purity grade.

Preparation of solutions (for about 100 determinations): all solutions are prepared with re-purified water (cf. Vol. II, chapter 2.1.3.2).

1. 4-Methylumbelliferone standard (10 mmol/l):

   dissolve 198.2 mg 4-methylumbelliferone, sodium salt, in 100 ml chloroform: methanol (2:1, v/v) in a sintered glass-stoppered volumetric flask.

2. Acetate buffer (1.0 mol/l, pH 4.5, pH 5.5, pH 6.2):

   dissolve 136.1 g sodium acetate · 3 H_2O in 1 l water. Adjust to pH 4.5, 5.5 or 6.2, as desired, with glacial acetic acid.

3. Acetate buffer (0.5 mol/l, pH 4.5, pH 5.5):

   dissolve 68 g sodium acetate · 3 H_2O in 1 l water. Adjust to pH 4.5 or 5.5 with glacial acetic acid.

4. Cacodylate buffer (0.5 mol/l, pH 6.2):

   dissolve 107 g sodium cacodylate trihydrate in 1 l water. Adjust to pH 6.2 with HCl, 6 mol/l.

5. 4-Nitrophenyl-α-D-mannopyranoside solution (10 mmol/l):

   dissolve 30.1 mg 4-nitrophenyl-α-D-mannopyranoside in 10 ml water.

6. 4-Methylumbelliferyl-α-D-mannopyranoside (5 mmol/l):

   dissolve 17 mg 4-methylumbelliferyl-α-D-mannopyranoside in 10 ml water. Sonicate briefly or warm at 40°C for 5 minutes to dissolve.

7. Glycine/ammonium hydroxide buffer (glycine, 0.3 mol/l, pH 10.5):

   dissolve 22.5 g glycine in 1 l water. Adjust pH to 10.5 with concentrated ammonium hydroxide.

8. Glycine/sodium hydroxide buffer (glycine, 0.133 mol/l, pH 10.7):

   dissolve 10 g glycine, 8.8 g Na_2CO_3 and 3.9 g NaCl in 1 l water. Adjust pH to 10.7 with NaOH, 1 mol/l.

9. Bovine serum albumin (0.5 mg/ml):

   dissolve 5 mg bovine serum albumin in 10 ml water.

10. Bovine serum albumin (4 mg/ml):

    dissolve 40 mg bovine serum albumin in 10 ml water.

11. Zinc sulphate (1 mmol/l):

    dissolve 2.88 mg $ZnSO_4$ · 7 H_2O in 10 ml water.

12. Cobalt chloride (10 mmol/l):

    dissolve 23.79 mg $CoCl_2$ · 6 H_2O in 10 ml water.

Procedure

Preparation of serum: blood is collected by venepuncture into receptacles free from heparin or EDTA and allowed to clot for 45 min at 18 – 25 °C. Serum is obtained by centrifugation at 2000 rpm for 10 minutes in a clinical centrifuge. The serum specimen is then carefully removed so as to avoid contamination with red cells and is transferred to a clean, chilled test tube. Some investigators [1 – 2] prefer to use plasma obtained from blood collected in heparinized vessels.

Stability of the enzyme in serum: according to *Lombardo et al.* [1], lysosomal acid α-mannosidase is less stable in serum than in plasma. They also stress the need to use freshly prepared plasma or serum because they found that acid α-mannosidase activity decreased by 30 – 40 percent when serum was stored for 3 days at – 20 °C. They recommended using plasma instead of serum as a source of enzyme. We and others [9] have found that intermediate α-mannosidase activity is stable for at least three months when serum is stored at – 60 °C.

Details for measurements in tissues: after vigorous homogenization of liver in distilled water followed by centrifugation (100 000 *g*, 1 h), most of the lysosomal (acid) α-mannosidase activity and nearly all of the cytosolic (neutral) form of the enzyme are recovered in the supernatant fraction. Most of the *Golgi* (intermediate) isoenzyme remains in the particulate fraction from which it can be efficiently extracted in soluble form with the aid of 0.3% Triton X-100 [6]. In general, the assay conditions that are optimal for determining the activity of α-mannosidase isoenzymes in serum are suitable for estimating the corresponding tissue enzymes from liver, brain, fibroblasts and leukocytes.

Assay conditions

Colorimetric assay: wavelength 400 nm; light path 10 mm; reaction volume 0.5 ml; final volume 1.0 ml; incubation time 4 – 30 min; temperature 37 °C; blank, serum is added just before addition of the glycine NaOH quench buffer.

Fluorimetric assay: excitation wavelength 360 nm (*Klett* filter, No. 7 – 60) and emission wavelength 520 – 580 nm (*Klett* filter, No. 54); incubation tube 10 × 75 mm; path length 10 mm; reaction volume 0.1 ml; final volume 3.0 ml; incubation time 30 min; temperature 37 °C; blank, heat-inactivated enzyme (boiling water-bath, 2 min).

Measurement

1. *Fluorimetric assay*

Pipette successively into 10 × 75 mm test tubes on ice:	"acid" α-mannosidase	"intermediate" α-mannosidase	"neutral" α-mannosidase	concentration in assay mixture (100 μl)	
substrate solution (6)	20 μl	20 μl	20 μl	substrate	3 mmol/l
acetate buffer (2)				acetate	0.2 mol/l
pH 4.5	60 μl				
pH 5.5		60 μl			
pH 6.2			60 μl		
sample (serum)	20 μl	20 μl	20 μl	volume fraction	0.20
mix thoroughly using vortex mixer and incubate for 15 – 60 min at 37 °C;					
glycine/NH_4OH (7)	2900 μl	2900 μl	2900 μl	glycine	0.29 mol/l
mix thoroughly by inversion; measure fluorescence immediately against a blank with heat-inactivated enzyme.					

Select the appropriate sensitivity setting such that the ΔF on the fluorescence dial of the fluorimeter is within the range of 10 – 90. The amount of fluorescence can be altered by changing the incubation period or the amount of enzyme added to the assay medium.

The ΔF of the 4-methylumbelliferone standard ($\Delta F_{standard}$) is determined separately and used to calibrate the fluorimeter. The 4-methylumbelliferone standards are prepared as follows: dilute 4-methylumbelliferone standard solution (1) 1:100 with water, pipette aliquots (5, 10, 25, 50, 75, 100 μl) into clean 10 × 75 assay tubes, add 2.9 ml glycine/ammonium hydroxide buffer (7), and measure the fluorescence ($\Delta F_{standard}$).

Calculation: according to eqn. (m_1) in combination with eqns. (k) and (k_1) (cf. "Formulae", Appendix 3), the catalytic concentration in the sample is (t in minutes):

$$b = \frac{\Delta F_{sample}}{\Delta F_{standard}} \times c_{standard} \times \frac{1}{v \times \Delta t} \quad \text{U/l}$$

$$b = \frac{\Delta F_{sample}}{\Delta F_{standard}} \times c_{standard} \times \frac{16.67}{v \times \Delta t} \quad \text{nkat/l}.$$

v is sample volume in the assay.

2. *Colorimetric assay*

Pipette successively into 10 × 75 mm test tubes on ice:	"acid" α-mannosidase (modified from ref. [5])	"intermediate" α-mannosidase (modified from ref. [6])	"neutral" α-mannosidase (modified from ref. [7])	concentration in assay mixture
4-nitrophenyl-α-D-mannopyranoside substrate (5)	200 μl	100 μl	100 μl	"acid" assay 4 mmol/l; "intermdediate" and "neutral" assay 2 mmol/l
buffer				
acetate pH 4.5 (3)	100 μl			
acetate pH 5.5 (3)		100 μl		acetate 0.1 mol/l
cacodylate pH 6.2 (4)			100 μl	cacodylate 0.1 mol/l
bovine serum albumin				
0.5 mg/ml (9)	50 μl			BSA 50 μg/ml
4 mg/ml (10)			50 μl	BSA 400 μg/ml
activators				
$ZnSO_4$ (11)	50 μl			Zn^{2+} 0.1 mmol/l
$CoCl_2$ (12)			50 μl	Co^{2+} 1 mmol/l
water	0	200 μl	100 μl	
sample (serum)	100 μl	100 μl	100 μl	volume fraction 0.2
mix thoroughly using vortex mixer and incubate at 37 °C,				
incubation time	10 min	30 min	4–20 min	
glycine/NaOH (8)	500 μl	500 μl	500 μl	0.66 mol/l
mix thoroughly and read absorbance at 400 nm.				

Calculation: use eqn. (k) or (k_1), cf. "Formulae", Appendix 3. The absorption coefficient of 4-nitrophenol is $\varepsilon = 1.39\ l \times mm^{-1} \times mm^{-1}$. Correct sample readings for blank. The catalytic concentration in the sample is (t in minutes):

$$b = 719.4 \times \Delta A/\Delta t \quad \text{U/l}$$

$$b = 11\,989 \times \Delta A/\Delta t \quad \text{nkat/l}.$$

Earlier results were expressed in other units: one unit of enzyme catalyzes the release of 1 nmol of 4-nitrophenol per hour. The conversion factor to yield international units ($\mu mol \times min^{-1} \times l^{-1}$) is 0.0167. The figures given under "Validation of Method" are given in old units.

Validation of Method

Precision, accuracy, detection limits and sensitivity: for the fluorimetric assays, for which more data are available, and for activities in the range 20 – 200 units per ml serum, an imprecision of 5 – 10% has been obtained with manual performance. The relative standard deviation in a single laboratory is 4 – 6% of the mean value. Data concerning accuracy are not available. With the more sensitive fluorimetric assay, the lower limit of detection of α-mannosidase activity is about 4 units per ml of serum where the imprecision is 20%.

Sources of error: inaccurate estimates of specific α-mannosidase isoenzymes in the range pH 4.0 – 6.5 using unfractionated serum as the source of enzyme are usually the result of the presence of multiple, distinct species of enzyme with overlapping pH-activity profiles. Particularly for the neutral and intermediate α-mannosidases, in order to determine accurately the amount of activity contributed by each isoenzyme, one should subject serum samples to some type of fractionation procedure [9, 12, 15] prior to assaying for activity.

Since some investigators have reported a 30 – 40% loss of acid α-mannosidase activity when serum is stored at −20 °C for only a few days, it is advisable to determine activity as soon as possible.

Specificity: α-mannosidase activity is measured specifically. No other hydrolase has been shown to catalyze the hydrolysis of 4-methylumbelliferyl-α-D-mannopyranoside or 4-nitrophenyl-α-D-mannopyranoside.

Reference ranges: acid α-mannosidase activity in serum is not sex-dependent but is markedly age-dependent [2]; activity is high at birth (approximately 45 units/ml) drops to a minimum value at 10 – 14 years (12 units/ml) and rises again to a maximum at 20 – 24 years (25 units/ml).

We have found that the level of intermediate α-mannosidase activity in serum of adults is 120 units/ml with one standard deviation being 17% of the mean value (*C. N. Faber & R. H. Glew*; unpublished observations).

References

[1] *A. Lombardo, L. Caimi, S. Marchesini, G. C. Goi, G. Tettamanti,* Enzymes of Lysosomal Origin in Human Plasma and Serum: Assay Conditions and Parameters Influencing the Assay, Clin. Chim. Acta *108*, 337 – 346 (1980).

[2] *A. Lombardo, G. C. Goi, S. Marchesini, L. Caimi, M. Maro, G. Tettamanti,* Influence of Age and Sex on Five Human Plasma Lysosomal Enzymes Assayed by Automated Procedures, Clin. Chim. Acta *113*, 141 – 152 (1981).
[3] *M. A. Chester, A. Lundblad, P. K. Masson,* The Relationship Between Different Forms of Human α-Mannosidase, Biochim. Biophys. Acta *391*, 341 – 348 (1975).
[4] *B. Hultberg, J. E. Olsson,* Lysosomal Hydrolases in CSF of Patients with Multiple Sclerosis, Acta Neurol. Scand. *59*, 23 – 30 (1979).
[5] *D. J. Opheim, O. Touster,* Lysosomal α-Mannosidase of Rat Liver, J. Biol. Chem. *253*, 1017 – 1023 (1978).
[6] *D. R. P. Tulsiani, D. J. Opheim, O. Touster,* Purification and Characterization of α-D-Mannosidase from Rat Liver Golgi Membranes, J. Biol. Chem. *252*, 3227 – 3233 (1977).
[7] *V. A. Shoup, O. Touster,* Purification and Characterization of α-D-Mannosidase of Rat Liver Cytosol, J. Biol. Chem. *251*, 3845 – 3852 (1976).
[8] *B. C. Kress, A. L. Miller,* Altered Serum α-Mannosidase Activity in Mucolipidosis II and Mucolipidosis III, Biochem. Biophys. Res. Commun. *81*, 756 – 763 (1978).
[9] *S. Hirani, B. Winchester,* The Multiple Forms of α-D-Mannosidase in Human Plasma, Biochem. J. *179*, 583 – 592 (1979).
[10] *M. Carroll, N. Dance, P. K. Masson, B. G. Winchester,* Human Mannosidosis – The Enzymatic Defect, Biochem. Biophys. Res. Commun. *49*, 579 – 583 (1972).
[11] *B. G. Winchester, N. S. Van-de-Water, R. D. Jolly,* The Nature of Residual α-Mannosidase in Plasma in Bovine Mannosidosis, Biochem. J. *157*, 183 – 188 (1976).
[12] *P. K. Masson, A. Lundblad, S. Autio,* Mannosidosis, Detection of the Disease and of Heterozygotes Using Serum and Leukocytes, Biochem. Biophys. Res. Commun. *56*, 296 – 303 (1974).
[13] *D. B. Robinson, W. F. Diven, R. H. Glew,* Altered α-Mannosidase Isoenzymes in the Liver in Hepatic Cirrhosis, Enzyme *27*, 99 – 107 (1982).
[14] *S. Hirani, B. G. Winchester, A. D. Patrick,* Measurement of the α-Mannosidase Activities in Human Plasma by Differential Assay, Clin. Chim. Acta *81*, 135 – 144 (1977).
[15] *B. Hultberg, P. K. Masson, S. Sjoblad,* Neutral α-Mannosidase Activity in Human Serum, Biochim. Biophys. Acta *445*, 398 – 405 (1976).
[16] *P. R. Dorling, C. R. Huxtable, S. M. Colegate,* Inhibition of Lysosomal α-Mannosidase by Swainsonine, an Indolizidine Alkaloid Isolated from *Swainsona canescens,* Biochem. J. *191*, 649 – 651 (1980).
[17] *P. R. Dorling, C. A. Huxtable, P. Vogel,* Lysosomal Storage in Swainsona sp. Toxicosis: an Induced Mannosidosis, Neuropath. Appl. Neurobiol. *4*, 285 – 295 (1978).
[18] *G. A. Grabowski, J. U. Ikonne, R. J. Desnick,* Comparative Physical, Kinetic and Immunologic Properties of the Acidic and Neutral α-D-Mannosidase Isoenzymes from Human Liver, Enzyme *25*, 13 – 25 (1980).
[19] *D. H. Embury, C. E. Whiting, A. J. Sinclair,* A Comparison of Fluorimetric and Colorimetric Methods for the Determination of α-Mannosidase Activity in Bovine Plasma, Aust. Vet. J. *57*, 281 – 283 (1981).

2.9 β-D-Mannosidase

β-D-Mannoside mannohydrolase, E.C. 3.2.1.25

Guido Tettamanti and Massimo Masserini

General

β-D-mannosidase has been found to occur in molluscs [1], bacteria [2], hen oviduct [3], and mammals [4, 5]. The enzyme is widely distributed in human tissues and cell types with its highest concentrations in synovial fluid and fibroblasts [5].

Intracellular β-D-mannosidase is mainly located in lysosomes [6]. However, a non-lysosomal form of β-D-mannosidase, having a neutral pH optimum, has been found in mammalian liver [7].

β-D-mannosidase is involved in the catabolism of mannose containing glycoconjugates (especially glycoproteins) in which it splits β-glycosidically-linked mannose residues. These linkages are rare in mammalian systems apart from the ubiquitous

$$\begin{matrix} \text{Man}(\alpha, 1 \rightarrow 3) \searrow & \\ & \text{Man}(\beta, 1 \rightarrow 4)\text{GlcNAc}(\beta, 1 \rightarrow 4)\text{GlcNAc} \\ \text{Man}(\alpha, 1 \rightarrow 3) \nearrow & \end{matrix}$$

β-mannosidic linkage found in both acidic and neutral glycoproteins [8]. The trisaccharide Man(β, 1 → 4)GlcNAc(β, 1 → 4)GlcNAc has been found to accumulate in urine and tissues of the animals affected with "β-mannosidosis", an inborn lysosomal disease due to genetic deficiency of β-D-mannosidase. This disease has been described in nubian goats [4]. β-D-Mannosidase splits also D-mannose residues from various synthetic β-mannosides.

Application of method: in biochemistry and clinical chemistry.

Enzyme properties relevant in analysis: substrate conversion proceeds linearly with time (up to at least two hours with the commonly used synthetic substrates) with no lag phase.

Methods of determination: two different synthetic substrates are commonly used, the chromogenic 4-nitrophenyl-β-D-mannopyranoside, and the fluorigenic 4-methylumbelliferyl-β-D-mannopyranoside. The colorimetric assay is less expensive; the fluorimetric one is more sensitive and is useful if only small amounts of biological material are available.

Natural substrates for β-D-mannosidase have been employed in special investigations. In these cases trisaccharides or glycopeptides containing β-glycosidically linked D-mannose residues can be used, and liberated mannose is determined by the coupled lactate dehydrogenase/hexokinase assay [9] or by gas-liquid chromatography [11]. A radiochemical assay has also been applied [7].

The colorimetric method, using 4-nitrophenyl-β-D-mannopyranoside, is described here for enzyme from human body fluids, isolated cells and tissues. The assay procedure is described in detail for synovial fluid, where the optimal conditions are better known [5].

International reference methods and standards: neither standardization at the international level, nor reference standard material exists so far.

Enzyme effectors: special activators of β-D-mannosidase to be added to the assay mixture using synthetic substrates are not known. Ag^+ and Hg^{2+} and sodium taurocholate completely inactivate the enzyme [6].

Assay

Method Design

Principle

$$\text{4-Nitrophenyl-}\beta\text{-D-mannoside} + H_2O \xrightarrow{\beta\text{-D-mannosidase}} \text{4-nitrophenol} + \text{D-mannose.}$$

The amount of 4-nitrophenol liberated per time unit under standard conditions (see below) is a measure of the catalytic activity of β-D-mannosidase. If the reaction mixture is made alkaline after a definite time the solution becomes yellow and the reaction is stopped. The increase in absorbance at 400 nm due to the 4-nitrophenol formed is correlated with the amount of substrate hydrolyzed by means of a calibration curve.

Optimized conditions for measurement: the most detailed study on the optimal conditions for human β-D-mannosidase assay has been carried out on synovial fluid [5] (for maximal activity: acetic acid/acetate buffer, pH 3.5, 0.2 mol/l; substrate, 12 mmol/l; 37 °C). The optimal assay conditions in other body fluids and tissues are likely to be close to those reported for synovial fluid, but should be determined in preliminary experiments. The capacity of the acetate buffer must be sufficient to maintain the incubation mixture at the pH optimum (3.5) but on the other hand must allow it to be made alkaline at the end of the reaction.

Temperature conversion factors: these factors have not yet been determined.

Equipment

Colorimeter or spectrophotometer capable of exact measurement at 400 nm; thermostatted water-bath.

Reagents and Solutions

Purity of reagents: all chemicals must be of analytical grade; 4-nitrophenyl-β-D-mannopyranoside must be free from 4-nitrophenol.

Preparation of solutions (for about 100 determinations): all solutions in re-purified water (see Vol. II, chapter 2.1.3.2).

1. Sodium acetate solution (0.25 mol/l):

   dissolve 20.5 g sodium acetate in 1 l water.

2. Acetate buffer (0.25 mol/l; pH 3.5):

   dilute 14.5 ml glacial acetic acid to 1 l with water and adjust to pH 3.5 with solution (1) (about 78 ml).

3. Substrate solution (4-nitrophenyl-β-D-mannopyranoside, 30 mmol/l):

   dissolve 45.15 mg 4-nitrophenyl-β-D-mannopyranoside in 5 ml acetate buffer (2). Mix with a magnetic stirrer until substrate is completely dissolved.

4. Na_2CO_3/NaOH solution (Na_2CO_3, 2% w/v in NaOH, 0.1 mol/l):

   dissolve 4 g NaOH in about 500 ml water; after cooling add 20 g Na_2CO_3. Dilute to 1 litre with water and stir with a magnetic stirrer.

5. 4-Nitrophenol standard solution (1.0 mmol/l):

   dissolve 13.91 mg 4-nitrophenol in water, dilute to 100 ml with water and stir the mixture with a magnetic stirrer. Store aliquots of 1 ml in closed vials at −30 °C. Before each set of determinations thaw a single vial, equilibrate at room temperature, mix with a vortex mixer and use it for preparing the standard curve.

Stability of solutions: the acetate and acetate buffer solutions (1) and (2), and the alkaline solution (4), stored at 0−4 °C, are stable as long as no microbial contamination occurs. The substrate solution (3) can be stored at 0−4 °C for 8−10 h after preparation. The standard solution (5) can be stored at −30 °C for several months.

Procedure

Collection and treatment of specimen: freshly aspirated synovial fluid is stored, if necessary, at −30 °C. Blood is collected from the vein without stasis. Addition of sodium citrate (1 mg/ml) or EDTA (1 mg/ml), for preparing plasma is unobjectionable. Centrifuge for 10 min at about 3000 *g* in order to obtain plasma or serum, respectively. Use only fresh plasma or serum free from haemolysis.

Stability of the enzyme in the sample: according to [5], β-D-mannosidase is stable in synovial fluid stored at −20 °C, and also after several cycles of freezing and thawing, for at least two weeks. Studies on the stability of the enzyme in other human body fluids or tissues have not been carried out so far.

Details for measurement in other body fluids, cells and tissues: the optimum conditions described for the assay of β-D-mannosidase in human synovial fluid should also apply for determination of the enzyme activity in other body fluids, isolated cells, cultured fibroblasts from skin explants, and tissues. Some recommendations for the assay of β-D-mannosidase in human isolated cells and tissues are given in reference [5].

Assay conditions: wavelength 400 nm; light path 10 mm; incubation volume 0.125 ml; incubation temperature 37 °C; final volume for measurement 1.125 ml.

Read at room temperature against a blank consisting of the incubation mixture without the biological sample: the blank mixture is incubated and the sample, which has also been incubated separately, is added immediately before stopping the reaction. A standard curve is necessary.

Measurement

Pipette successively into cuvettes:		sample	blank	concentration in assay mixture	
acetate buffer	(2)	0.050 ml	0.050 ml	acetate	0.1 mol/l
substrate solution	(3)	0.050 ml	0.050 ml	4-nitrophenyl-β-D-manno-pyranoside	12 mmol/l
sample		0.025 ml	–	volume fraction	0.2
mix by gentle vortexing, stopper tubes and incubate for 1 h at 37 °C,					
incubated sample		–	0.025 ml		
Na_2CO_3/NaOH solution	(4)	1 ml	1 ml	Na_2CO_3 NaOH	1.78% (w/v) 0.089 mol/l
mix thoroughly. Reaction is stopped. Read absorbance within one hour.					

Standard curve: mix 1 – 25 μl standard solution (5), corresponding to 1 – 25 nmol 4-nitrophenol, with 0.1 ml acetate buffer (2) and 1.0 ml Na_2CO_3/NaOH solution (4). Use 25 μl of each of these solutions for absorbance measurement. Read absorbance and plot absorbance as a function of concentration of 4-nitrophenol (μmol/l).

Calculation: read the concentration of liberated 4-nitrophenol corresponding to the measured absorbance from the standard curve.

Under the above conditions the following relationship applies:

$$b = \frac{\mathrm{V}}{\mathrm{v}} \times \frac{\Delta c}{\Delta t} = 45 \times \frac{\Delta c}{\Delta t} \qquad \mathrm{U/l}$$

$$b = 750 \times \frac{\Delta c}{\Delta t} \qquad \mathrm{nkat/l},$$

where Δc is the concentration of liberated 4-nitrophenol (μmol/l) taken from the calibration curve, Δt is reaction time (min), V is the volume of the incubation mixture (1.125 ml) and v the sample volume (0.025 ml).

Validation of Method

Precision, accuracy, detection limits and sensitivity: these parameters have not yet been accurately studied for β-D-mannosidase assay.

Sources of error: therapeutics or other substances are not known to influence the measured activity.

Specificity: the β-D-mannosidase activity is measured specifically.

Reference ranges: values of 28 U/l and 1.8 U/l have been found in human synovial fluid and serum, respectively [5]. Studies on large populations of humans have not so far been made.

References

[1] *K. Sugahara, T. Okumura, I. Yamashina,* Purification of β-Mannosidase from a Snail, *Achatina Fulica,* and its Action on Glycopeptides, Biochim. Biophys. Acta *268*, 488 – 496 (1972).
[2] *S. Bouquelet, G. Spik, J. Montreuil,* Properties of a β-D-Mannosidase from *Aspergillus niger,* Biochim. Biophys. Acta *522*, 521 – 530 (1978).
[3] *T. Sukeno, A. L. Tarentino, T. H. Plummer, F. Maley,* Purification and Properties of α-D-, and β-D-Mannosidases from Hen Oviduct, Biochemistry *11*, 1493 – 1501 (1972).
[4] *M. Z. Jones, G. Dawson,* Caprine β-Mannosidosis, J. Biol.Chem. *256*, 5185 – 5188 (1981).

[5] *B. A. Bartholomew, A. L. Perry,* The Properties of Synovial Fluid β-Mannosidase Activity, Biochim. Biophys. Acta *315*, 123 – 127 (1973).
[6] *J. H. Labadie, N. N. Aronson,* Lysosomal β-D-Mannosidase of Rat Liver, Biochim. Biophys. Acta *321*, 603 – 614 (1973).
[7] *G. Dawson,* Evidence for Two Distinct Forms of Mammalian β-Mannosidase, J. Biol. Chem. *257*, 3369 – 3371 (1982).
[8] *G. Strecker, J. Montreuil,* Glycoproteines et Glycoproteinoses, Biochemie *61*, 1199 – 1246 (1979).
[9] *A. L. Tarentino, F. Maley,* The Purification and Properties of Aspartyl-N-acetylglucosamine Amidohydrolase from Hen Oviduct, Arch. Biochem. Biophys. *130*, 295 – 303 (1969).
[10] *C. C. Sweeley, R. Bentley, M. Makita, W. W. Wells,* Gas-liquid Chromatography of Trimethylsilyl Derivatives of Sugars and Related Substances, J. Am. Chem. Soc. *85*, 2497 – 2507 (1963).

2.10 β-D-Glucuronidase

β-D-Glucuronide glucuronosohydrolase, EC 3.2.1.31

Philip D. Stahl and William H. Fishman

General

The activity of β-glucuronidase has been measured in tissue extracts of mammals and other vertebrates, digestive juice of snails (*Helix pomatia*), molluscs (*Patella vulgata*), locusts, bacteria, and plants [1]. The enzyme catalyzes the hydrolysis of β-glucuronides and also the transfer of glucuronyl radicals to acceptor alcohols [2]. Like acid phosphatase, β-glucuronidase has been measured as a typical lysosomal enzyme during studies of subcellular fractionation.

β-Glucuronidase specifically catalyzes the release of terminal glucuronide units linked through the carbon-1 by a β-configuration. The enzyme is typically associated with lysosomes, but shows a unique subcellular distribution in rodent liver, with activity associated both with the lysosomal and microsomal fractions. The function of the microsomal enzyme is unclear: it may serve as a precursor molecule on its way to the lysosome [3 – 5] or it may fulfill a separate physiological function. Clinically, β-glucuronidase deficiency in man produces a mucopolysaccharide storage disease, MPS VII [6].

β-Glucuronidase assays are important in the quantitative determination of several hormone activities. Mouse kidneys, for example, react quantitatively to the amount of testosterone circulating in the blood, which is produced endogenously by gonadotrophin or by androgen injection [5]. The preputial glands of the rat likewise react to the action of several androgenic compounds. Rat preputial gland is an excellent source of highly purified rat β-glucuronidase.

Application of method: in biochemistry and clinical chemistry.

Enzyme properties relevant in analysis: the natural substrates appear to be dermatan sulphate and heparan sulphate. The classical substrates of the enzyme are β-D-gluco-pyranosiduronic acids containing a simple aglycone from one of the following groups: drugs or exogenous chemicals (menthol, β-naphthol, phenolphthalein, 4-nitrophenol, 8-hydroxquinoline), steroids (oestriol, oestradiol, testosterone, pregnanediol, corticosteroids and their tetrahydro derivatives) and endogenous non-steroid metabolites (bilirubin and thyroxine). For a complete list of conjugates hydrolyzed by β-glucuronidase see *Marsh* [7].

Newer substrates for the enzyme are androstendione-enol-β-D-glucuronide, glucuronic acid 1-phosphate, and naphthol AS-BI-D-glucuronide. Three substrates are widely used: the glucuronides of phenolphthalein, of 4-nitrophenol and of methyl-umbelliferone.

Methods of determination: the catalytic activity of β-glucuronidase can be measured spectrophotometrically or fluorimetrically. The enzyme has a molecular weight of 280000. Fluorimetric measurements can be carried out with 4-methyl-umbelliferyl-β-D-glucuronide [8] as substrate. Colorimetric measurements are carried out with phenolphthalein glucuronide. If the aglycone cannot be measured satisfactorily, the liberated glucuronic acid may be determined by the method of *Fishman & Green* [1].

The method using phenolphthalein glucuronide is described here for serum, tissue, urine, cerebrospinal fluid, bile, vaginal secretion and gastric juice. The assay conditions for the substrates 4-nitrophenol glucuronide, and 4-methyl-umbelliferyl-β-D-glucuronide are described.

International reference methods and standards: there is currently no international standard.

Enzyme effectors: β-glucuronidase is inhibited by saccharolactone and by glucuronic acid. The enzyme is activated by polylysine and other polycationic substances.

Assay

Method Design

Principle

Glucuronide + H_2O $\xrightarrow{\beta\text{-glucuronidase}}$ Glucuronic acid + ROH

(Glucuronide: COOH; H; O; O–R; H; OH; H; HO; H; H; OH — Glucuronic acid: COOH; H; O; OH; H; OH; H; HO; H; H; OH)

R = aglycone = phenolphthalein, 4-nitrophenol, 8-hydroxyquinoline, 4-methylumbelliferone.

The amount of aglycone liberated per unit time under standard conditions (substrate concentration, pH and temperature) is a measure of the β-glucuronidase activity. If the reaction mixture is made alkaline after a definite time, the solution becomes red (phenolphthalein), yellow (4-nitrophenol) or orange (8-hydroxyquinoline) [in the last case it is also necessary to add 4-aminoantipyrine and ferricyanide for chromogen formation]; or the intensity of fluorescence is increased.

The increase in absorbance due to colour formation, or the increase in fluorescence, is converted to amount of substrate hydrolyzed by means of a standard curve.

Optimized conditions for measurement: the optimum conditions for the enzyme in serum are described in detail by *Fishman et al.* [1]; they should be carefully observed because higher substrate concentrations are used than in earlier work. The optimum conditions for the tissue enzyme are described in [1]. The capacity of the acetate buffer must be such that catalysis is stopped following mixture with the alkaline reagent at the end of the reaction. The pH optimum of the enzyme from *E. coli* or other bacteria is pH 7.0.

Temperature conversion factors: the ratio of β-glucuronidase activity, measured at 25 °C and at 37 °C (i.e. catalytic activity at 37 °C divided by catalytic activity at 25 °C), using phenolphthalein glucuronide and pure rat preputial β-glucuronidase, is 2.77.

Equipment

Spectrophotometer or simple colorimeter suitable for measurements at 540 nm, or fluorimeter.

Reagents and Solutions

Purity of reagents: standard reagent grade chemicals are quite suitable for these assays.

Preparation of solutions (for about 100 determinations): all solutions in re-purified water (cf. Vol. II, chapter 2.1.3.2).

1. Phenolphthalein glucuronide (30 mmol/l):

   dissolve 309.6 mg phenolphthalein glucuronide, sodium salt, in 20 ml water.

2. Phenolphthalein glucuronide (10 mmol/l):

   dissolve 103.2 mg phenolphthalein glucuronide, sodium salt, in 20 ml water.

3. Acetate buffer (1.0 mol/l; pH 4.5):

   dissolve 57.9 g $CH_3OONa \cdot 3\ H_3O$ and 336 ml acetic acid in water and dilute to 1000 ml.

4. Glycine/SDS (glycine, 0.2 mol/l, SDS, 0.2% w/v; pH 11.7):

   dissolve 15.01 g glycine in 900 ml water, adjust to pH 11.7 with 50% NaOH, add 2 g SDS and dilute with water to 100 ml.

5. Phenolphthalein standard solution (100 μg/ml; 0.315 mmol/l):

   dissolve 10 mg phenolphthalein in 50 ml 95% ethanol and dilute to 100 ml with water.

Stability of solutions: all solutions are stable. Store substrate solution (1) in a refrigerator.

Procedure

Collection, treatment and stability of specimen

Serum

Preferably use fresh serum free from haemolysis. Anticoagulants such as heparin should be avoided. Citrate is acceptable.

Tissues

The enzyme activity can be measured either in homogeneous solution or as a 1% homogenate in distilled water or acetate buffer. Addition of 0.1% Triton X-100 is often useful. The total enzyme activity should only be determined in dilute solutions.

Standard method (e.g. mouse kidneys): homogenize 100 – 200 mg tissue in 2 ml acetate buffer, 0.1 mol/l, in a *Potter-Elvehjem* homogenizer with a Teflon pestle for 1 – 2 min. Transfer the homogenate to a measuring cylinder and rinse in with acetate buffer. Dilute to 10 ml with acetate buffer, mix thoroughly and take two 0.1 ml samples for the assay.

Other specimens

Urine: the phenolphthalein glucuronide method is similar to that for tissue. A higher substrate concentration is necessary because of the lower activity, and at the same time this reduces the competitive inhibition by low molecular-weight compounds contained in urine.

Cerebrospinal fluid: there is normally only low β-glucuronidase activity in cerebrospinal fluid, and therefore a long period of incubation is required. The incubation mixture is as above for urine, except that acetate buffer, pH 5.2, and substrate solution, 20 mmol/l, are used. The incubation takes 18 h. For the colour development, 1 ml glycine solution (0.2 mol/l, adjusted to pH 11.7 with NaOH) and 1 ml water are used. The measurements are made by the micro-method of *Reilly & Crawford* [9] against the reagent blank (without cerebrospinal fluid). Phenolphthalein solutions containing 0 – 3.0 μg/ml in acetate buffer (20 mmol/l; pH 5.2) are used as standards. The absorbance of a tube containing cerebrospinal fluid but without substrate solution must be subtracted from the absorbance of the experimental tube. Values of 10.3 U/l cerebrospinal fluid have been found. The fluorimetric assay with 4-methylumbelliferyl-β-D-glucuronide may also be used.

Bile: the phenolphthalein glucuronide method is similar to that for tissue, but with some modification. It is necessary to have a long incubation time and a slightly higher substrate concentration because of the low activity.

Vaginal fluid: for measurements with the macro-assay the sample is weighed into a *Potter* homogenizer, diluted to 3.0 ml with acetate buffer (0.1 mol/l; pH 4.5) and homogenized. The procedure is the same as that for tissue samples, except that the substrate concentration is 1 mmol/l. Incubation is for 18 h. The micro-method has the advantage that samples as small as 1 mg can easily be analyzed. The umbelliferyl β-glucuronide method may also be used.

Duodenal juice: the assay is essentially the same as for tissues, except for the following differences:

a) the phenolphthalein-β-D-glucuronic acid solution is 10 mmol/l;
b) the incubation is between 1 and 24 h depending on the activity;
c) the reaction is stopped by the addition of 5.0 ml glycine buffer (0.2 mol/l, pH 10.4).

The incubations are analogous to those for serum.

Assay conditions: wavelength 540 nm; light path 10 mm; incubation volume 1.00 ml; 38 °C (constant temperature water-bath); final volume for colour reaction 6.0 ml (serum), 7.5 ml (tissues), and 5.0 ml (standards), respectively. Read at room temperature against a blank containing water instead of buffer (2) and substrate solution (1). Prepare a standard curve.

Measurement

Standard curve

Pipette into test tubes:			concentration in assay mixture
acetate buffer	(3)	0.2 ml	acetate 3.33 mmol/l
phenolphthalein standard solution	(5)	up to 0.1 ml	phenol-phthalein 2–20 μg/tube; 6.3–63 nmol/tube
glycine solution	(4)	2.8 ml	glycine 73 mmol/l SDS 0.066%
water		1.8 ml	
mix and after 10 min read absorbance (A).			

Prepare a standard curve from the readings. Ordinate: A, abscissa: nmole phenolphthalein.

Measurement in serum

Pipette into test tubes:			concentration in assay mixture
substrate solution	(1)	0.2 ml	phenolphthalein glucuronide 6 mmol/l
acetate buffer	(3)	0.2 ml	acetate 200 mmol/l
water		0.4 ml	
serum		0.2 ml	volume fraction $\varphi = 0.2$
mix by gentle shaking, stopper tubes and incubate for 4 h;			
glycine/SDS solution	(4)	2.0 ml	glycine 67 mmol/l SDS 0.066%
water		3.0 ml	
mix thoroughly and measure absorbance* (A) after 10 min.			

* The intensity of the phenolphthalein colour depends on the pH. Therefore occasionally check that the final pH after dilution to 6 ml is 10.2–10.45.

Measurement in tissues

Pipette into test tubes:			concentration in assay mixture
substrate solution	(2)	0.1 ml	phenolphthalein glucuronide 1 mmol/l
acetate buffer	(1)	0.8 ml	acetate 80 mmol/l
homogenate		0.1 ml	volume fraction $\varphi = 0.1$
mix, stopper tube and incubate for 1 h (longer with lower activity). Stop reaction by immersing tubes in boiling water (1 min),			
water		1.5 ml	
centrifuge for 10 min at 1000 *g*. Use supernatant fluid;			
glycine solution	(4)	2.5 ml	glycine 91 mmol/l NaCl 91 mmol/l
water		0.5 ml	
supernatant fluid		2.0 ml	
mix and read absorbance* after 10 min.			

Calculation: read the phenolphthalein concentration corresponding to the ΔA from the standard curve. Under the above conditions the following relationships apply for the catalytic concentration of the sample (t in minutes):

$$b = \frac{1}{\Delta t \times \varphi} \times c_{\text{phen}} \quad \text{U/l}$$

$$b = \frac{16.67}{\Delta t \times \varphi} \times c_{\text{phen}} \quad \text{nkat/l}$$

where

c_{phen} concentration of phenolphthalein produced during incubation time, in µmol/l
Δt incubation time, in min
φ volume fraction of sample in incubation mixture

Fishman units can be used. These are defined as the enzyme activity which liberates 1 µg phenolphthalein in 1 h at pH 4.5 (acetate, 0.1 mol/l) from phenolphthalein glucu-

* The intensity of the phenolphthalein colour depends on the pH. Therefore occasionally check that the final pH after dilution to 6 ml is 10.2 – 10.45.

ronide: accordingly, this gives a conversion factor of 318. However, it should be noted that *Fishman* units are often related to other volumes (e.g. 100 ml). An alternate unit (1 μmole substrate hydrolyzed per hour at 37 °C) has been described by *Stahl & Touster* [10].

Alternative Methods for Serum

a) 4-Nitrophenyl-β-glucuronide as substrate

Solutions

1. Acetate buffer (0.1 mol/l; pH 4.5)
2. 4-Nitrophenyl-glucuronide (40 mmol/l)
3. 4-Nitrophenol standard (0.119 mmol/l)

Assay conditions: wavelength 420 nm; light path 10 mm; incubation volume 1.00 ml; 38 °C (constant temperature water-bath); final volume for colour reaction 6.00 ml. Read at room temperature against a blank containing water instead of substrate solution.

Measurement

Pipette successively into test tubes:		concentration in assay mixture
4-nitrophenyl-β-D-glucuronide solution (2) acetate buffer, pH 4.0 (1) water serum	0.2 ml 0.4 ml 0.2 ml 0.2 ml	4-nitrophenyl-β-D-glucuronide 8 mmol/l acetate 40 mmol/l volume fraction $\varphi = 0.2$
mix, stopper tubes and incubate for 2 h		
glycine/SDS solution (4) water	4.0 ml 1.0 ml	
mix and read absorbance A after 10 min.		

The final pH should be 10.2.

Analyze standards containing 0.02 – 0.20 ml standard solution (2 – 20 μg corresponding to 14.4 to 144 nmoles) and prepare a standard curve. Ordinate: ΔA; abscissa: concentration of 4-nitrophenol. Calculations as above.

b) 4-Methylumbelliferyl β-D-glucuronide as substrate [8]

Equipment

Fluorimeter with silica cuvettes; primary wavelength 360 nm; secondary wavelength 450 nm; *Corning* filter 5860 (primary); *Corning* filters 5543 and 3387 (secondary).

Solutions

1. Acetate buffer (200 mmol/l, pH 5.0)
2. 4-Methylumbelliferyl β-D-glucuronide (1.33 mg/ml; albumin, 10 mg/l; in buffer 1)
3. 4-Methyl umbelliferone standard (176 mg/l)

Standard (1 mmol/l)

Dissolve 17.6 mg in 100 ml glacial acetic acid. Dilute 1:10 with sodium citrate solution, 0.1 mmol/l, pH 4.5. Store at 4 °C in a brown bottle. Use 5 – 100 μl standard in the assay to construct a standard curve ranging from 0.5 – 10 nmoles per assay.

Assay

Mix 25 μl enzyme sample and 75 μl substrate solution. Incubate at 37 °C for 10 – 60 min. Terminate with 1.5 ml stopping reagent (i.e., glycine buffer, p. 249). Measure fluorescence as indicated above.

Alternative Methods for Tissues

a) 4-Nitrophenyl-β-D-glucuronide as substrate

This method is similar to that described under "Measurements in serum" with the following changes:

1. 0.3 ml acetate buffer (0.1 mol/l; pH 4.0) and 0.4 ml water are added.
2. The glycine/SDS contains only 0.1% SDS.
3. In the blank 0.1 ml homogenate is added after the glycine/SDS solution.

b) 4-Methylumbelliferyl-β-D-glucuronide as substrate [8]:

Equipment

As for serum samples.

Solutions

As for serum samples.

Assay

As for serum samples except that the enzyme preparation may have to be diluted with albumin/acetate (100 μg/ml in sodium acetate buffer, 0.2 mol/l, pH 5.0).

Validation of Methods

Precision, accuracy, detection limits and sensitivity: in recovery experiments in which phenolphthalein was added to the enzyme assay mixtures, a mean of 103.5% ± 6.1% was found. The coefficient of variation with 10 parallel determinations on sera with low activity was 6.5%, with normal activity 2.5%, and with high activity only 1.4%. The phenolphthalein glucuronide assay can detect as little as 0.1 μg rat preputial β-glucuronidase per ml. Estimates of precision and accuracy are not available.

Sources of error: injection of dyes can give abnormally high blanks in serum samples at alkaline pH. This may interfere with the determination of phenolphthalein. On administration of glucuronolactone, a strong inhibitor of β-glucuronidase, saccharolactone is formed. In these cases at least 4 h should elapse before collecting the serum; by this time the metabolic products of glucuronolactone are no longer present in the blood. Chemicals which form glucuronides, such as menthol, affect the β-glucuronidase level; 12 h should elapse before collection of the serum.

Specificity: β-glucuronidase is highly specific for the glucuronyl residue while having very little, if any, specificity for the aglycone.

Reference ranges: reference values (in *Fishman* units are given in Table 1).

Table 1. Normal values in serum (*Fishman* units).

Group	Men U/l (38 °C)	Women U/l (38 °C)
Normal adults	0.525 ± 0.142 (n = 20)	0.392 ± 0.123 (n = 22)
Children	0.495 ± 0.224 (n = 26)	0.571 ± 0.302 (n = 11)
Cancer patients	0.489 ± 0.168 (n = 18)	0.457 ± 0.268 (n = 43)
Diabetics	0.621 ± 0.345 (n = 21)	0.750 ± 0.420 (n = 23)
Cirrhosis patients	0.742 ± 0.160 (n = 37)	0.654 ± 0.392 (n = 32)
Pregnancy (1st trimester)		0.495 ± 0.246 (n = 39)
(2nd trimester)		0.658 ± 0.281 (n = 37)
(3rd trimester)		1.042 ± 0.461 (n = 52)

The number of subjects is given in parentheses.

References

[1] *W. H. Fishman,* β-Glucoronidase, in: *H. U. Bergmeyer* (ed.), Methods of Enzymatic Analysis, 2nd edit., Verlag Chemie, Weinheim, and Academic Press, New York 1974, pp. 929 – 943.
[2] *W. H. Fishman, S. Green,* Enzymic Catalysis of Glucuronyl Transfer, J. Biol. Chem. *225*, 435 – 452 (1957).
[3] *W. H. Fishman, S. S. Goldman, R. DeLellis,* Dual Localization of β-Glucuronidase in Endoplasmic Reticulum and in Lysosomes, Nature *213*, 457 – 460 (1967).
[4] *B. Mandell, P. Stahl,* Effects of Di-isopropyl Phosphorofluoridate on Rat Liver Microsomal and Lysosomal β-Glucuronidase, Biochem. J. *164*, 371 – 389 (1977).
[5] *R. Swank, K. Paigen,* Biochemical and Genetic Evidence for a Macromolecular β-Glucuronidase Complex in Microsomal Enzymes, J. Mol. Biol. *77*, 371 – 389 (1973).
[6] *W. S. Sly, F. Brot, J. Glaser, P. Stahl, B. Quinton, D. Rimoin, W. McAlister,* Birth Defects: Original Article Series *10*, 241 (1974).
[7] *C. A. Marsh,* in: *C. F. Dutton* (ed.), Glucuronic Acid, Academic Press, New York 1967.
[8] *D. H. Leaback, P. G. Walker,* Studies on Glucosaminidase, Biochem. J. *78*, 151 – 156 (1961).
[9] *C. N. Reilley, C. M. Crawford,* Principles of Precision Colorimetry a General Approach to Photoelectric Spectrophotometry, Anal. Chem. *27*, 716 – 725 (1955).
[10] *P. Stahl, O. Touster,* β-Glucuronidase of Rat Liver Lysosomes, J. Biol. Chem. *246*, 5398 – 5406 (1971).

2.11 Hyaluronidase

Hyaluronate 4-glycanohydrolase, EC 3.2.1.35

Alfred Linker

General

The term hyaluronidase refers here to an enzyme which acts on hyaluronic acid, irrespective of activity towards other substrates. Hyaluronidase was first isolated from micro-organisms [1], and later from mammalian testis which is now the main source. With one exception, all enzymes are endohexosaminidases. It has been shown that the bacterial enzyme is not a hydrolase, but acts as eliminase [2], while the hyaluronidase from testis, although a hydrolase, also has transglycosylase activity [3]. Enzymes with similar properties to the testis hyaluronidase have been obtained from tadpoles, snake venom, bee venom, numerous animal tissues, human serum and other sources. The tissue enzyme most probably originates from the lysosomes [4, 5].

The method described here is mainly suitable for tissue hyaluronidase and similar enzymes. Bacterial enzymes can be determined with this method, but better methods have been described [6]. The enzyme from leeches, which is an endoglucuronidase [7], cannot be determined by this method.

Application of method: in biochemistry and physiology. Under certain conditions it can be used to differentiate glycosaminoglycans.

Enzyme properties relevant in analysis: substrate conversion proceeds linearly with time. Enzymes from different tissue sources may well be not identical as indicated by different pH optima.

Methods of determination: activity can be measured in several ways as the substrate is a polysaccharide. A viscosimetric method is sensitive [8] but depends on the purity and molecular weight of substrate, and decrease in viscosity due to non-enzymatic reactions can be a problem. A method involving reduction of turbidity [9] is specific, sensitive and accurate but is not applicable to crude enzyme extracts. Measurement of the increase in reducing groups liberated by the enzyme is not very specific, unless appropriate controls are included, and crude extracts are difficult to assay. A simple assay for the determination of hyaluronidase in very crude biological samples has been described [10]. A very sensitive dye-binding method appears to work well with purified enzyme [11] but has not been sufficiently tested with crude materials.

International reference methods and standards: a reference method based on turbidity reduction is available: see U. S. Pharmacopea XX-NFXV combined edition, p. 376 (1980). Standard preparations can be obtained from National Formulary, Office of the Director of Revision, 2215, Constitution Ave., N. W., Washington, D.C. 20037. Good commercial preparations with known activity are also available.

Enzyme effectors: heparin inhibits the enzyme. Heavy metals and strong polyanions inhibit hyaluronidase, while polycations activate it. Some hyaluronidases are inhibited by acetate.

Assay

Method Design

Principle

Hyaluronidases are endohexosaminidases and catalyze the degradation of hyaluronic acid with the liberation of acetylglucosamine terminal groups which can be measured in a colorimetric assay [12].

In the assay, the anhydro-sugar is first formed from *N*-acetyl-glucosamine in alkaline solution. This is then converted in acid solution to the furan derivative, which reacts with 4-dimethylaminobenzaldehyde to form a coloured complex [13]. The amount of acetylglucosamine liberated per unit time is a measure of the hyaluronidase activity.

Optimized conditions for measurements: the pH optimum varies with the source of the enzyme. The enzyme from testis has a wide pH optimum between pH 4.0 and 6.0 (dependent on the buffer used). The optimum for the majority of the bacterial enzymes is around pH 5.0. Serum hyaluronidase has a rather sharp optimum at pH 4.0 and the lysosomal enzyme a sharp optimum at pH 3.5. As a compromise we use pH 4.0. If accurate measurements are required, particularly in the case of enzymes of unknown origin, the pH optimum must be determined separately. It is preferable to assay hyaluronidases at 37 °C.

Temperature conversion factors: none.

Equipment

Spectrophotometer of suitable accuracy: 37 °C water-bath; boiling water-bath.

Reagents and Solutions

Purity of reagents: only reagents of A. R. grade should be used. 4-Dimethylamino-benzaldehyde must be recrystallized if the commercial preparation is not of sufficient purity. Good commercial preparations of hyaluronic acid are available; they should contain at least 30% uronic acid as measured by the carbazole reaction [14]. A preparation of good quality can be prepared from umbilical cords [4].

Preparation of solutions (for about 15 determinations): all solutions in re-purified water (cf. Vol. II, chapter 2.1.3.2).

1. Acetate buffer (acetate, 50 mmol/l, pH 4.0; NaCl, 150 mmol/l):

   a) dilute 5.78 ml acetic acid to 1000 ml with water (100 mmol/l);

   b) dissolve 13.6 g $CH_3COONa \cdot 3\ H_2O$ in water and make up to 1000 ml (100 mmol/l); add 41.0 ml solution (a) to 9.0 ml solution (b) and, if necessary adjust the pH of the mixture to pH 4.0. Add 0.875 g NaCl and dilute to 100 ml with water.

2. Hyaluronate (1.25 g/l):

   dissolve 62.5 mg hyaluronic acid in 50 ml acetate buffer (1). The compound is not easily soluble; it is best to prepare the solution the day before use.

3. *N*-Acetylglucosamine standard solution:

   dissolve 10 mg *N*-acetylglucosamine in 10 ml acetate buffer (1); dilute a) 1 ml to 50 ml with acetate buffer (1) (20 μg/ml) and b) 1 ml to 100 ml (10 μg/ml).

4. Tetraborate (0.8 mol/l; pH 9.1):

   dissolve 24.44 g $K_2B_4O_7 \cdot 4\,H_2O$ in water and make up to 100 ml. Adjust to pH 9.1 with KOH, 5 mol/l.

5. Dimethylaminobenzaldehyde reagent (1% w/v):

   dissolve 10 g 4-dimethylaminobenzaldehyde in 100 ml acetic acid (containing 12.5% v/v HCl, 10 mol/l). Just before use dilute with 9 vol. acetic acid.

Stability of solutions: store all solutions at 0–4 °C. The buffer (1) is stable providing that there is no growth of bacteria or moulds. Prepare the hyaluronate and acetylglucosamine solutions (2) and (3) freshly each week. The dimethylaminobenzaldehyde stock solution (5) is stable for a month.

Procedure

Collection, treatment and stability of sample: collect and prepare the samples just before the assay; take up in acetate buffer (1). They can be stored for a limited time at 0–4 °C. With the exception of highly purified hyaluronidase preparations, the enzyme is very stable in a dry state at low temperature.

Assay conditions

Enzyme reaction: carry out all assays in duplicate. Incubation temperature 37 °C; incubation volume 1.0 ml.

Colour reaction: carry out in duplicate. Wavelength 585 nm; light path 10 mm; incubation volume 1 ml; volume for colour reaction 3.6 ml. Read against reagent blank with 0.5 ml acetate buffer (1) instead of incubation solution. Prepare a sample blank for each series of measurements: 0.2 ml tetraborate solution (4) + 0.8 ml hyaluronate solution (2) + 0.2 ml sample. Take 0.6 ml of this mixture for the colour reaction.

Measurement

Enzyme reaction

Pipette successively into test tubes (10 mm × 70 mm):			concentration in assay mixture	
hyaluronate solution	(2)	0.8 ml	hyaluronate acetate NaCl	1 g/l 50 mmol/l 150 mmol/l
pre-incubate for 15 min				
sample (buffered)		0.2 ml	volume fraction	0.2
mix and incubate for exactly 10 min, then add immediately 0.5 ml to the colour reaction.				

Colour reaction

Pipette successively into test tubes (10 mm × 100 mm):		standard 1	standard 2	sample	concentration in assay mixture	
tetraborate solution	(4)	0.1 ml	0.1 ml	0.1 ml	$K_2B_4O_7$	22.22 mmol/l
standard	(3b)	0.5 ml	–	–	acetyl-glucosamine	1.39 mg/l
standard	(3a)	–	0.5 ml	–		2.78 mg/l
incubation solution		–	–	0.5 ml	acetate	6.9 mmol/l
					NaCl	20.8 mmol/l
heat for 3 min in boiling water-bath and then cool with running tap-water,						
dimethylaminobenzaldehyde reagent	(5)	3.0 ml	3.0 ml	3.0 ml		8.3 g/l
mix, incubate for 20 min in a 37 °C water-bath and cool with running tap-water. If necessary, centrifuge the solutions to clear, pour into cuvettes and immediately measure absorbances.						

The absorbances of both standards must have the relationship 1:2, otherwise the assay is not functioning correctly.

Calculation: the enzyme activity is expressed as μmole N-acetylglucosamine liberated per min (U). Refer to the standards to calculate the amount of acetylglucosamine liberated in the incubation time (Δt in min). The reading of the 10 μg standard (10 mg/l) is taken for the calculations.

Subtract sample blank absorbance from absorbance of sample.

$A_{standard}$ absorbance of 10 μg standard
A_{sample} absorbance of sample

For calculation of the catalytic concentration of the sample use formula (k) or (k_1), respectively, in combination with eqn. (d_1) (cf. "Formulae", Appendix 3):

$$b = \frac{1.0}{0.2} \times 10 \times 1000 \times \frac{F}{221.2} \times \frac{A_{sample}}{A_{standard}} / \Delta t$$

$$= 226 \times F \times \frac{A_{sample}}{A_{standard}} / \Delta t \qquad U/l \quad (37\,°C)\,.$$

or

$$b = 3767 \times F \times \frac{A_{sample}}{A_{standard}} / \Delta t \qquad nkat/l \quad (37\,°C)$$

where

0.2 is volume of sample in enzymatic reaction
1.0 volume of incubation mixture for enzyme reaction
221.2 is molecular weight of *N*-acetylglucosamine
F is dilution factor of sample during pre-treatment.

The measured activity concentration can be compared with commercially available hyaluronidase preparations of known activity from testis (National Formulary Units, N.F. Units).

Validation of Method

Precision, accuracy, detection limits and sensitivity: with values of 20 U/g protein a standard deviation of ±1 U/g was found. The detection limit of the method is less than 2 mU (equivalent to 15 N.F. Units) per mg protein. For the assay of lower activities the reaction time can be prolonged, but this reduces the accuracy.

Sources of error: variation of the ionic strength of the buffer, the pH or the amount of sample can alter the final pH of the colour test for acetylglucosamine which functions only in alkaline solution. If modifications are made, the incubation mixture must be neutralized before the colour reaction is carried out. After addition of borate the pH must be 8.9 for optimum colour yield.

The sample blank is important as a time blank, and therefore should not be prepared too long before the heating step, because even in alkaline solution low enzyme activity can be present.

Turbidity is a source of error in the analysis of crude enzyme preparations or extracts; it should be removed by centrifugation before absorbance measurements.

Hyaluronic acid is not very soluble; it is inclined to swell, and the solution must be thoroughly mixed. Some hyaluronidases appear to be inhibited by acetate and the use of formate buffer has been suggested [15] in exceptional cases. Some sources from which hyaluronidase is obtained also may contain exo-enzymes such as β-glucuronidase and β-acetylhexosaminidases. These enzymes could contribute somewhat to the overall colour yield due to acetylglucosamine. Therefore, saccharolactone (1.5 mmol/l) can be added to the buffers used. This inhibits the glucuronidase and prevents sequential degradation of oligosaccharides.

Specificity of method: chondroitin 6-sulphate can serve as substrate instead of hyaluronic acid, whereas chondroitin 4-sulphate cannot.

References

[1] *K. Meyer, R. Dubos, E. M. Smyth,* The Hydrolysis of the Polysaccharide Acids of Vitreous Humor, of Umbilical Cord, and of Streptococcus by the Autolytic Enzyme of Pneumococcus, J. Biol. Chem. *118*, 71 (1937).

[2] *A. Linker, K. Meyer, P. Hoffmann,* The Production of Unsaturated Uronides by Bacterial Hyaluronidase, J. Biol. Chem. *219*, 13–25 (1956).

[3] *B. Weissmann, K. Meyer, P. Sampson, A. Linker,* Isolation of Oligosaccharides Enzymatically Produced from Hyaluronic Acid, J. Biol. Chem. *208*, 417–429 (1954).

[4] *N. N. Aronson jr., E. A. Davidson,* Lysosomal Hyaluronidase from Rat Liver, J. Biol. Chem. *242*, 437–440 (1967).

[5] *G. Vaes,* Hyaluronidase Activity in Lysosomes of Bone Tissue, Biochem. J. *103*, 802–804 (1967).

[6] *A. Linker, P. Hovingh,* in: *S. P. Colowick, N. O. Kaplan* (eds.), Methods in Enzymology, Vol. *XXVIII*, Academic Press, New York 1972, pp. 902–911.

[7] *A. Linker, K. Meyer, P. Hoffmann,* The Production of Hyaluronate Oligosaccharides by Leech Hyaluronidase and Alkali, J. Biol. Chem. *235*, 924–927 (1960).

[8] *H. E. Alburn, R. W. Whitley,* Factors Affecting the Assay of Hyaluronidase, J. Biol. Chem. *192*, 379–393 (1951).

[9] *M. B. Mathews,* in: *S. P. Colowick, N. O. Kaplan* (eds.), Methods in Enzymology, Vol. *VIII*, Academic Press, New York 1966, pp. 654–662.

[10] *P. G. Richman, H. Bear,* A Convenient Plate Assay for the Quantitation of Hyaluronidase in Hymenoptera Venoms, Anal. Biochem. *109*, 376–381 (1980).

[11] *L. C. Benchetrit, S. L. Pahya, E. D. Gray, R. D. Edstrom,* A Sensitive Method for the Assay of Hyaluronidase Activity, Anal. Biochem. *79*, 431–437 (1977).

[12] *J. L. Reissig, J. L. Strominger, L. F. Leloir,* A Modified Colorimetric Method for the Estimation of *N*-AcetylaminoSugars, J. Biol. Chem. *217*, 959–966 (1955).

[13] *D. H. Leaback, P. G. Walker,* On the Morgan-Elson Reaction, Biochim. Biophys. Acta *74*, 297–300 (1963).

[14] *Z. Dische,* A New Specific Color Reaction of Hexuronic Acids, J. Biol. Chem. *167*, 189–198 (1947).

[15] *R. W. Orkin, G. Jackson, B. P. Toole,* Hyaluronidase Activity in Cultured Chick Embryo Skin Fibroblasts, Biochem. Biophys. Res. Commun. *77*, 132–138 (1977).

2.12 α-L-Fucosidase

α-L-Fucoside fucohydrolase, E.C. 3.2.1.51

Guido Tettamanti, Massimo Masserini, Giancarlo Goi
and Adriana Lombardo

General

Several α-fucosidases (EC 3.2.1.38) occur in nature. Among them α-L-fucosidase (EC 3.2.1.51) is the most abundant and best known. α-L-fucosidase has been found to occur in bacteria, protozoa, marine organisms and mammals [1]. The enzyme appears to be ubiquitous in human tissues, fluids and cells, including placenta, cultured skin fibroblasts and foetal tissues [1, 2]. Human α-L-fucosidase is a glycoprotein and exists in various molecular forms (up to 9), which differ from each other in their carbohydrate composition and organization of quaternary structure [1, 2]. The isoenzyme profile appears to be similar in the different cell types or tissues, but undergoes developmental changes in each tissue. The serum profile differs from that of the tissues [1, 2].

Intracellular α-L-fucosidase is mainly located in lysosomes where the enzyme is involved in the catabolism of the fucose-containing glycoconjugates. These substances accumulate in the tissues and urine of patients suffering from "fucosidosis", an inborn lysosomal disease due to genetic deficiency of α-L-fucosidase [1, 2].

α-L-Fucosidase hydrolyzes L-fucose residues which are α-glycosidically linked to glycoproteins, glycolipids, oligosaccharides and synthetic α-fucosides. Bacterial α-L-fucosidase specifically hydrolyzes α-1→2 fucosidic bonds present in natural compounds and has no action on synthetic substrates [3, 4]. The human enzyme hydrolyzes fucose residues linked to the 2 position of galactose, or 3,4,6 positions of N-acetylglucosamine, or to synthetic aglycones [5].

Application of method: in biochemistry and clinical chemistry.

Enzyme properties relevant in analysis: the catalytic activity recorded in biological fluids, cells or tissue extracts is due to the existence of several isoenzymes which exhibit different catalytic activities [1, 2]. Substrate conversion proceeds linearly with time (up to at least 4 h with the commonly used synthetic substrates) with no lag phase.

Methods of determination: two different synthetic substrates are commonly used, the chromogenic 4-nitrophenyl-α-L-fucopyranoside [6] and the fluorigenic 4-methylumbelliferyl-α-L-fucopyranoside [7, 8]. The chromogenic and fluorigenic assays are interchangeable [1, 7, 8]. The colorimetric assay is less expensive; the fluorimetric assay is more sensitive and is useful if only small amounts of biological material (such as tissue biopsies, isolated cells, spinal or lacrimal fluids) are available, or if only residual α-L-fucosidase activity is to be determined (as in fucosidosis patients).

No natural substrates for α-L-fucosidase have been used, except in special investigations. In these cases fucose-containing oligosaccharides (isolated from milk) or glycopeptides (derived from blood group substances) can be employed and liberated fucose is determined by gas-liquid chromatography [9] or by fucose dehydrogenase NAD coupled assay [10]. Estimation of enzyme mass or active-site titration is not in use. Details of isoenzyme separation and determination are given in reference [1]. The fluorimetric method, using 4-methylumbelliferyl-α-L-fucopyranoside, which is of wider application, is described here for enzymes from human body fluids, isolated cells and tissues.

International reference methods and standards: neither standardization at the international level, nor reference standard material exists so far.

Enzyme effectors: specific activators of α-L-fucosidase to be added to the assay mixture using synthetic substrates are not known [1]. Bile salts (5), Hg^{2+} and Ag^{+} completely inactivate the enzyme (1).

Assay

Method Design

Principle

$$\text{4-Methylumbelliferyl-}\alpha\text{-L-fucoside} + H_2O \xrightarrow{\alpha\text{-fucosidase}} \text{fucose} + \text{4-methylumbelliferone.}$$

The amount of 4-methylumbelliferone liberated per unit time under standard conditions (see below) is a measure of the catalytic activity of α-L-fucosidase. If the reaction mixture is made alkaline after a definite time, the solution immediately becomes fluorescent and the course of the reaction is stopped. The fluorescence due to the 4-methylumbelliferone formed is correlated with the amount of substrate hydrolyzed by means of a calibration curve. Since 4-methylumbelliferyl-glycosides are weakly fluorescent, assay blanks are necessary. The method described below can easily be adapted to automated equipment [11].

Optimized conditions for measurement: the optimum conditions for the enzyme assay in serum and plasma are described in detail by *Lombardo et al.* [12] (pH 5.0; acetic acid/acetate buffer, 0.05 mol/l; substrate, 0.4 mmol/l; 37 °C). The optimum conditions for the enzyme assay in leukocytes, lymphocytes and platelets are described by *Caimi et al.* [13] and in cultured fibroblasts by *Zielke et al.* [14]. In all cases the optimum conditions are very close to those reported for serum. The capacity of the acetate buffer must be sufficient to maintain the serum or other sample reliably at the pH optimum (5.0), but on the other hand must allow the assay mixture to be made alkaline at the end of the reaction.

Temperature conversion factors: the factor for converting α-L-fucosidase activity (in serum or plasma) from 37 °C to 30 °C is 0.59 (unpublished result).

Equipment

Fluorimeter or spectrofluorimeter; thermostatted water-bath; sonicator.

Reagents and Solutions

Purity of reagents: all chemicals must be of analytical grade; 4-methylumbelliferyl-α-L-fucopyranoside must be free from 4-methylumbelliferone. 4-Methylumbelliferone should be recrystallized 3 times from methanol. The water used for preparing all solutions must be first de-ionized, then distilled in a glass apparatus, and stored in closed glass containers.

Preparation of solutions (for about 25 determinations): all solutions in specially re-purified water (cf. Vol. II, chapter 2.1.3.2).

1. Acetate buffer (0.5 mol/l; pH 5.0):

   dissolve 28.95 g sodium acetate trihydrate and 310 ml acetic acid in 500 – 600 ml water, adjust to pH 5.0 and dilute to 1000 ml.

2. Substrate solution (4-methylumbelliferyl-α-L-fucopyranoside, 0.57 mmol/l):

   dissolve 3.67 mg 4-methylumbelliferyl-α-L-fucopyranoside in 20 ml water, sonicate the mixture in a water-bath until completely dissolved (30 – 60 s) and mix with a vortex mixer before sample withdrawal. Prepare the solution immediately before use.

3. Glycine/NaCl/NaOH (glycine, 0.2 mol/l; NaCl, 0.125 mol/l; pH 10.75):

   dissolve 15.01 g glycine and 7.3 g NaCl in 600 – 700 ml water. Adjust to pH 10.75 with aqueous NaOH (1 mol/l, about 220 ml) and dilute to 1000 ml with water.

4. 4-Methylumbelliferone standard solution (0.1 mol/l):

   dissolve 1.762 g recrystallized 4-methylumbelliferone in 80 – 90 ml acetonitrile (or acetic acid/acetone, 80/20, v/v), dilute to 100 ml with the same solvent and stir with a magnetic stirrer. Store aliquots of 1 ml in closed vials at −30°C. Before each set of determinations thaw a single vial, equilibrate at room temperature, mix with a vortex mixer and use it for preparing the standard curve.

Stability of solutions: the acetate buffer solution (1) and the glycine solution (3), stored at 0−4°C, are stable as long as no microbial contamination occurs. The substrate solution (2) can be stored at 0−4°C for 8−10 h after preparation. It can be stored at −80°C for 8−10 days. The standard solution (4) can be stored at −30°C for several months.

Procedure

Collection and treatment of specimen: collect blood from the vein without stasis, add sodium citrate (1 mg/ml) and immediately centrifuge (15 min, 3000 *g*). If necessary, store the plasma at 0−4°C. Use only plasma free from haemolysis. Due to some liberation of α-L-fucosidase (probably from platelets) during blood clotting [12], plasma and not serum should be employed for the enzyme assay.

Stability of the enzyme in the sample: according to [12] α-L-fucosidase is stable in human plasma for at least 4 h at 37°C, 2 days at 0−4°C, and 20 days at −20°C.

Details for measurement in other body fluids, cells and tissues: the optimum conditions described for the assay of α-L-fucosidase in human plasma generally apply also for the enzyme determination in other body fluids, leukocytes, lymphocytes, platelets and cultured fibroblasts from skin explants. In the case of cerebrospinal fluid the optimum pH appears to be 4.5 and the saturating concentration of substrate 0.1 mmol/l. The conditions for preparing the various cell types which proved to be optimal for α-L-fucosidase assay [13] are the following: leukocytes, according to *Kolodny* and *Mumford* [15], treated as described by *Kihara et al.* [16] to remove contaminating erythrocytes; lymphocytes according to *Boyum* [17]; platelets according to *D'Amaro* [18]; cultured fibroblasts according to *Leroy et al.* [19]. Before assay the cell preparation must be suspended in distilled water (10^7 cells/ml) and submitted to sonication (three cycles; 30 s each; 90 s interval; 15 μm waves; sample kept in an ice-bath) and gently vortexed. The specimens of cerebrospinal fluid and of the various cells must be stored at −30°C, at which they are generally stable for at least 1 month. Details and recommendations for the assay of α-L-fucosidase in tissues are given in reference [1].

Assay conditions: excitation wavelength 365 nm; emission wavelength 448 nm; light path 10 mm; incubation volume 1.0 ml; incubation temperature 37°C (constant temperature water-bath); final volume for fluorescence measurement 4.0 ml.

Read at room temperature against a blank using incubation mixture without sample; the sample for this blank is incubated separately and added to the incubation mixture immediately before stopping the reaction.

During incubation and fluorescence measurement tubes must be protected from direct sunlight to prevent loss of fluorescence.

Plasma is diluted 1:10 with re-purified water before measurement. – A standard curve is needed.

Measurement

Pipette successively into test tubes:		sample	blank	concentration in assay mixture	
acetate buffer	(1)	0.1 ml	0.1 ml	acetate	50 mmol/l
substrate solution	(2)	0.7 ml	0.7 ml	glycoside	0.4 mmol/l
water		0.1 ml	0.1 ml		
sample		0.1 ml	–	volume fraction	0.10
mix by gentle vortexing, stopper tubes and incubate for 1 h at 37 °C,					
incubated sample		–	0.1 ml		
glycine/NaCl solution	(3)	3.0 ml	3.0 ml	glycine	0.15 mol/l
				NaCl	0.094 mol/l
mix thoroughly. Reaction is stopped. Read fluorescence within 1 h.					

Standard curve: make up 0.01 to 0.2 ml standard solution (4) (corresponding to 1 – 20 nmoles), to 1.0 ml with water, and add 3.0 ml glycine solution (3). Use 0.1 ml of each of these solutions for fluorescence measurement. Plot fluorescence in arbitrary units (ordinate) against concentration of 4-methylumbelliferone in μmol/l (abscissa).

Calculation: read the concentration of liberated 4-methylumbelliferone corresponding to the measured fluorescence arbitrary units from the standard curve.

Under the above conditions the following relationship applies for the 1:10 diluted sample (t in minutes):

$$b = \frac{V}{v} \times \Delta c/\Delta t = 40 \times \Delta c/\Delta t \qquad \text{U/l}$$

$$b = 667 \times \Delta c/\Delta t \qquad \text{nkat/l}$$

where Δc is the concentration of 4-methylumbelliferone (μmol/l) taken from the calibration curve. V is the volume of the incubation mixture after alkalinization (4 ml), v the sample volume (0.1 ml).

Validation of Method

Precision, accuracy, detection limits and sensitivity: for values of about 10 U/l an imprecision of ± 0.16 U/l has been found for manual performance (repetitive assays on the same sample) [12]. The relative standard deviation is 1.64%. A lower relative standard deviation (about 1.3%) has been obtained for automated procedures with catalytic activity concentrations in plasma ranging from 0.7 to 25 U/l [11]. Data on accuracy are not available since standard reference material is not established yet. For measurements with sufficiently sensitive fluorimeters the lowest measurable value, with a relative standard deviation not exceeding 10%, is 0.06 U/l. Sensitivity is found to range from a delta fluorescence arbitrary units of 7 to a delta of 50, depending on the window width used (*Perkin-Elmer* fluorimeter).

Sources of error: therapeutics or other substances are not known to interfere with the measured activity.

Specificity: the α-L-fucosidase activity is measured specifically.

Reference ranges: in normal adults the catalytic activity concentration of the enzyme in plasma ranges from 1.5 to 17 U/l [11]. Lower values may be found in children. Higher values are recorded in late pregnancy, in the neonatal period and through the first year of age [11].

References

[1] *J. A. Alhadeff, J. S. O'Brien,* Fucosidosis, in: *R. Glew, S. Peters* (eds.), Practical Enzymology of Sphingolipidoses, A. Liss Inc., New York 1977, pp. 247 – 281.

[2] *P. Durand, R. Gatti, C. Borrone,* Fucosidosis, in: *P. Durand, J. S. O'Brien* (eds.), Genetic Errors of Glycoprotein Metabolism, Edi-Ermes, Milano, Springer, Berlin 1982, pp. 49 – 87.

[3] *D. Aminoff, K. Furukawa,* Enzymes that Destroy Blood Group Specificity. I. Purification and Properties of α-L-Fucosidase from Clostridium Perfringens, J. Biol. Chem. *245*, 1659 – 1669 (1970).

[4] *M. Kochibe,* Purification and Properties of α-L-Fucosidase from Bacillus Fulminans, J. Biochem. *74*, 1141 – 1149 (1973).

[5] *G. Dawson, G. Tsay,* Substrate Specificity of Human α-L-Fucosidase, Arch. Biochem. Biophys. *184*, 12 – 23 (1977).

[6] *N. N. Jr. Aronson, C. de Duve,* Digestive Activity of Lysosomes, J. Biol. Chem. *243*, 4564 – 4573 (1968).

[7] *D. Robinson, R. Thorpe,* Fluorescent Assay of α-L-Fucosidase, Clin. Chim. Acta *55*, 65 – 69 (1974).

[8] *S. Wood,* A Sensitive Fluorimetric Assay for α-L-Fucosidase, Clin. Chim. Acta *58*, 251 – 256 (1975).

[9] *J. R. Clamp, G. Dawson, L. Hough,* The Simultaneous Estimation of L-Fucose, D-Mannose, D-Galactose, N-Acetylglucosamine and Sialic Acid in Glycopeptides and Glycoproteins, Biochim. Biophys. Acta *148*, 342 – 349 (1967).

[10] *H. Schachter, J. Sarney, E. J. McGuire, S. Roseman,* Isolation of a DPN-dependent L-Fucose Dehydrogenase from Pork Liver, J. Biol. Chem. *244*, 4785 – 4792 (1969).

[11] *A. Lombardo, G. Goi, S. Marchesini, L. Caimi, M. Moro, G. Tettamanti,* Influence of Age and Sex on Five Human Plasma Lysosomal Enzymes Assayed by Automated Procedures, Clin. Chim. Acta *113*, 141 – 152 (1981).

[12] *A. Lombardo, L. Caimi, S. Marchesini, G. Goi, G. Tettamanti,* Enzymes of Lysosomal Origin in Human Plasma and Serum: Assay Conditions and Parameters Influencing the Assay, Clin. Chim. Acta *108*, 337 – 346 (1980).
[13] *L. Caimi, A. Lombardo, S. Marchesini, G. Tettamanti,* Metodi per la diagnostica enzimatica delle sfingolipidosi (Methods for enzymatic diagnosis of sphingolipidoses), G. Ital. Clin. Chim. *3*, 349 – 381 (1978).
[14] *K. Zielke, M. L. Veath, J. S. O'Brien,* Fucosidosis: Deficiency of α-L-Fucosidase in Cultured Skin Fibroblasts, J. Exptl. Med. *136*, 197 – 199 (1972).
[15] *E. H. Kolodny, R. A. Mumford,* Human Leukocyte Acid Hydrolases: Characterization of Eleven Lysosomal Enzymes and Study of Reaction Conditions for their Automated Analysis, Clin. Chim. Acta *70*, 247 – 257 (1976).
[16] *H. Kihara, H. T. Porter, A. L. Fluharty, M. L. Scott, S. de laFlor, J. L. Trammel, R. H. Nakamura,* Metachromatic Leukodystrophy: Ambiguity of Heterozygote Identification, Am. J. Ment. Defic. *77*, 389 – 393 (1973).
[17] *A. Boyum,* Isolation of Mononuclear Cells and Granulocytes from Human Blood, Scand. J. Clin. Invest. *97*, 77 – 84 (1968).
[18] *J. D'Amaro,* The Microcomplement Fixation Test. A Solution of the Technical Problems of HLA Typing?, Vox Sanguinis *18*, 334 – 341 (1970).
[19] *J. G. Leroy, H. M. Ho, H. C. Mac Brinn, K. Zielke, J. Jacob, J. S. O'Brien,* I-Cell Disease: Biochemical Studies, Pediatr. Res. *6*, 752 – 759 (1972).

2.13 β-*N*-Acetylhexosaminidase

2-Acetamido-2-deoxy-β-D-hexoside acetamidodeoxyhexohydrolase, EC 3.2.1.52

John L. Stirling

General

β-*N*-Acetylglucosaminidase was first detected in almond emulsin by *Helferich & Iloff* in 1933 [1]. It is widely distributed in nature and has now been detected in many other plants, animals and micro-organisms (see ref. [2] for a survey). The enzyme is generally the most active of the lysosomal glycosidases and for that reason is sometimes chosen as a marker enzyme for these organelles. β-*N*-Acetylhexosaminidases from many sources exist as multiple molecular forms that differ in net charge. The two major forms of the human enzyme, β-*N*-acetylhexosaminidases A and B [3] differ in subunit composition, A having α and β subunits while B has only β subunits [4]. They also differ in their substrate specificities, A alone having activity towards the ganglioside GM_2 and substrates bearing an acidic group on either the sugar or the aglycone of the beta-linked N-acetylhexosaminide [5, 6].

Much of the interest in human β-*N*-acetylhexosaminidases relates to their involvement in the GM_2 gangliosidoses. In these autosomal recessive genetic disorders, β-*N*-acetylhexosaminidase A is missing in *Tay-Sachs* disease [7] and both A and B are missing in *Sandhoff's* disease [8]. There is further clinical interest in the total β-*N*-acetylhexosaminidase activities and isoenzyme patterns of lymphocytes and urine in patients with leukaemia [9] and a variety of kidney disorders [10].

β-*N*-Acetylglucosaminides are hydrolyzed by both β-*N*-acetylglucosaminidases, EC 3.2.1.30 and β-*N*-acetylhexosaminidases, EC 3.2.1.52, enzymes that have a wider specificity and are also able to hydrolyze β-*N*-acetylgalactosaminides. Both C4 epimers are thought to be hydrolyzed at the same active site. The description of an enzyme as a β-*N*-acetylglucosaminidase, EC 3.2.1.30, or a β-*N*-acetylhexosaminidase, EC 3.2.1.52, should properly be reserved for those purified enzymes in which it has been possible to establish the specificity. Often this is not possible where crude extracts are being assayed or the source of the enzyme has not previously been investigated. In these circumstances the enzyme is often named after the substrate used in the assay; hence the widespread description of the occurrence of β-*N*-acetylglucosaminidase reflects only the nature of the substrate used in the assay and not the underlying specificity. In fact, in all but a few cases, thorough investigation of β-*N*-acetylglucosaminidases has revealed an ability to hydrolyze β-*N*-acetylgalactosaminides, although at lower rates. Mammalian lysosomal activities and those detected in body fluids are β-*N*-acetylhexosaminidases but there is also a distinct enzyme with a higher pH optimum and non-lysosomal location in the cell that has the restricted activity of a β-*N*-acetylglucosaminidase. In human tissues this enzyme is β-*N*-acetylglucosaminidase C.

The names β-*N*-acetylhexosaminidase and *N*-acetyl-β-hexosaminidase both refer to the same enzymes as does the trivial form, hexosaminidase. In clinical literature the abbreviation of β-*N*-acetylglucosaminidase to NAG is sometimes used: here the reference is to the substrate used for the assay rather than the specificity of the enzymes which are invariably β-*N*-acetylhexosaminidases.

Application of method: in biochemistry and clinical biochemistry.

Enzyme properties relevant in analysis: in human tissues there are two main isoenzymes, β-*N*-acetylhexosaminidases A and B. The isoenzymes of body fluids are related to the tissue forms, in that they have the same pattern of expression in the GM_2 gangliosidoses but exhibit characteristic differences from them on electrophoresis and ion-exchange chromatography. Forms with isoelectric points between those of A and B are the 6 intermediate forms, I_1 and I_2. The residual activity in the tissues of patients with *Sandhoff's* disease is form S.

A simple method is available for the separation of the isoenzymes on DEAE-cellulose [11] and there is also an automated version of this method [12].

The release of products is linear with time and assays are usually performed by measuring the products formed during a fixed period of incubation.

Methods of determination: the most widely used methods for the assay of β-*N*-acetylhexosaminidase depend on the hydrolysis of either chromogenic or fluorigenic β-*N*-

acetylglucosaminides or galactosaminides. Active-site titration is not used. Determination of the immunological mass of β-*N*-acetylhexosaminidases A and B has been reported but this method is not in general use. Catalytic activity is used to determine β-*N*-acetylhexosaminidases because the methods are sensitive, rapid, reliable and inexpensive to perform. Full catalytic potential of β-*N*-acetylhexosaminidase is not revealed by the use of synthetic substrates in general use, however, and the reliable diagnosis of the GM_2 gangliosidoses requires assay of activity towards GM_2 ganglioside [13].

International reference method and standards: there has been no standardization of β-N-acetylhexosaminidase assays as yet.

Enzyme effectors: activity of β-*N*-acetylhexosaminidase A is obtained by measuring the total activity (A + B) and then subtracting the residual heat-stable B that remains after thermal inactivation of A [7]. This method is less reliable than those in which the isoenzymes are separated physically, because under certain conditions β-*N*-acetylhexosaminidase A is converted to a heat-stable form with properties similar to B. The specificity of β-*N*-acetylhexosaminidase A for hydrolysis of 4-nitrophenyl-β-*N*-acetylglucosaminide 6-sulphate should allow the direct assay of β-*N*-acetylhexosaminidase A in the presence of B [6]. Hg^{2+} and certain other metal ions have inhibitory effects. Effectors have been identified that are necessary for the hydrolysis of GM_2 ganglioside by β-*N*-acetylhexosaminidase A. The activating factor described by *Sandhoff* [14] is particularly well characterized, but two others with different properties have also been described. *Sandhoff's* factor is essential for the hydrolysis of GM_2 but does not appear to be necessary for hydrolysis of synthetic substrates. Inhibitors of β-*N*-acetylhexosaminidase are present in some samples of urine [15, 16].

Assay

Method Design

Principle

CH_2OH … CH_3 … NH–C=O–CH_3 (4-methylumbelliferyl-β-*N*-acetylglucosaminide) $\xrightarrow[+ H_2O]{\beta\text{-}N\text{-acetylhexosaminidase}}$ 4-methylumbelliferone + *N*-acetylglucosamine

4-Methylumbelliferone liberated in the course of the reaction is converted to its anionic form by the addition of an alkaline buffer to the reaction mixture, and its concentration is estimated fluorimetrically. Substrate blanks are included in each assay but enzyme blanks are rarely necessary.

In addition to the fluorimetric assay described here there are a variety of colorimetric assays. The principle of these assays is similar to the one described here but they use chromogenic substrates such as 4-nitrophenyl-β-*N*-acetylglucosaminide [16] or ω-nitrostyryl-β-*N*-acetylglucosaminide [17], with spectrophotometric measurement of the product. Both fluorimetric and colorimetric assays are readily adaptable to automated analyzers.

Optimized conditions for measurement: many variations of the *Leaback & Walker* method [18] have been described, most of them involving the use of smaller reaction volumes. With the exception of the rather specialized ultramicro method [19], assay volumes range from 3 µl to 2.0 ml and final substrate concentrations are in the range 0.2 mmol/l to 3.3 mmol/l. Even at the highest substrate concentration commonly used the enzyme is not saturated, 82% of maximal reaction rate *V* being given at 3.3 mmol/l and 22% at 0.2 mmol/l. In the method described here the final substrate concentration is 1.3 mmol/l which gives a rate of hydrolysis that is 66% of *V*. High blanks resulting from the use of the higher substrate concentrations are not a problem when small reaction volumes are used.

Citric acid/sodium phosphate or citric acid/sodium citrate buffers are commonly used at the pH optimum of pH 4.4 – 4.5. Acetate buffers are to be avoided because of the inhibitory effects of acetate on the reaction.

The volume of buffer used to stop the reaction depends to a large extent on the minimum volume necessary to read in the fluorimeter; volumes of 50 µl – 4.85 ml have been used.

Temperature conversion factors: using the fluorimetric assay method the following temperature conversion factors were found for the assay of human placental β-*N*-acetylhexosaminidase. They are given as a guide to the temperature dependence of the reaction and may not apply under other conditions, particularly at low concentrations of protein

°C	20	25	30	35	37	40
	1.89	1.35	1.00	0.76	0.68	0.58

Equipment

For assays using 4-methylumbelliferyl-β-*N*-acetylhexosaminides a fluorimeter providing excitation at 364 nm (λ_{max}) and measurement at 448 nm (λ_{max}) is required. We have found the double filter fluorimeter made by the *Locarte Co.* (London W 12) to be reliable and sensitive enough for assay of β-*N*-acetylhexosaminidases from tissues and body fluids, but those made by other manufacturers are widely used in other laboratories. For colorimetric assays using 4-nitrophenyl-β-*N*-acetylhexosaminides a visible-wavelength spectrophotometer is required.

Reagents and Solutions

Purity of reagents: for most purposes the purity of reagents is not critical and the commercially available products are satisfactory. The detectability of the assays is limited by the value of the substrate blank and, for assays at extreme detectability using the 4-methylumbelliferyl substrates, free 4-methylumbelliferone may be removed by extraction with acetone [2].

Preparation of solutions (for about 100 determinations): all solutions in re-purified water (cf. Vol. II, chapter 2.1.3.2).

1. Citrate/phosphate buffer (citrate, 60 mmol/l; phosphate, 95 mmol/l; pH 4.5; 0.1% BSA):

   a) dissolve 12.6 g citric acid monohydrate in 1000 ml water;
   b) dissolve 13.5 g Na_2HPO_4 in 1000 ml water;

   mix 50 ml of (1a) with 50 ml of (1b) and check the pH with a glass electrode. Dissolve 10 mg sodium azide and 50 mg bovine serum albumin in 50 ml buffer (1). It is essential to check whether the BSA is free from β-*N*-acetylhexosaminidase activity and if necessary the buffer containing BSA should be heated at 50°C for 2 h to inactivate the bovine enzyme.

2. Buffered substrate solution (2 mmol/l):

   dissolve 7.6 mg 4-methylumbelliferyl-2-acetamido-2-deoxy-β-D-glucopyranoside in 10 ml of buffer (1).

3. Stopping buffer:

   dissolve 37.5 g glycine in 1 litre water and titrate to pH 10.4 by adding 920 ml NaOH, 0.5 mol/l.

4. Fluorescence standard (4 μg/ml):

   dissolve 4 mg quinine sulphate in 100 ml H_2SO_4, 50 mmol/l. This stock solution is stable for several months if kept at 4° in the dark.
   Prepare a working standard by diluting the stock solution ten-fold with H_2SO_4, 50 mmol/l.

5. Methylumbelliferone standard (100 μmol/l):

   dissolve 14.7 mg 4-methylumbelliferone in 1 litre stopping buffer (3) to make a stock solution. Prepare working standards by diluting the stock solution appropriately with stopping buffer (3).

Stability of solutions: the buffer (1) may be stored at 4 °C for several weeks. Buffers showing signs of microbial contamination should be discarded. Buffered substrate solution (2) should be used within a few hours of preparation because the blank values increase slowly with time. When frozen at −20 °C the substrate is stable for several months. It is advisable to freeze the solution in small portions to eliminate the need for repeated freezing and thawing.

Procedure

Collection and treatment of specimen: serum samples [20] should be diluted 20-fold with buffer (1) before the assay. It is not necessary to use the buffer containing BSA for this.

Urine samples are assayed freshly after 20-fold dilution with water for the fluorimetric assay [15] or after passage through a column of Sephadex G-25 eluted with NaCl, 0.154 mol/l, in preparation for colorimetric assay [16].

Stability of enzyme in the specimen or sample: β-*N*-acetylhexosaminidase is generally stable for several hours or even days at 4 °C in solutions with a high protein concentration. According to [20] the enzyme in serum is stable at −20 °C for at least six months. In uninfected urine there was no change in the activity of β-*N*-acetylhexosaminidases after storage for one week at room temperature, 4 °C or −20 °C [15].

Details for measurement in tissues: tissue specimens should be homogenized in buffer, centrifuged to give a clear supernatant and diluted to a suitable concentration for assay. Leukocytes prepared from 10 ml heparinized blood with dextran should be sonicated in 1.5 – 2.0 ml water, using three 10 s bursts at 70 watts, centrifuged at 4 °C for 10 min at 600 *g* and a 150 μl sample diluted with 600 μl 0.6% BSA. Samples (50 μl) are then taken for fluorimetric assay. Cells from tissue cultures should be sonicated in buffer to give a protein concentration of 0.1 to 1.0 mg/ml and then diluted with buffer (1) containing BSA to give a suitable concentration for assay.

Assay conditions: incubating temperature 37 °C; incubation volume 150 μl; final volume for fluorescence determination 1.15 ml; fluorimeter set to excite at 364 nm and read at 448 nm; range of fluorimeter readings set with water (zero) and fluorescence standard solution (4). For each assay prepare a substrate blank in which the enzyme solution is replaced by buffer (1).

Analyze standards containing 0 to 3 μmol/l using 4-methylumbelliferone standard solution (5), and calculate a factor relating fluorescence readings to 4-methylumbelliferone concentration over the linear part of the range.

Measurement

Pipette successively into test tubes:			concentration in assay mixture	
sample buffered substrate	 (2)	50 µl 100 µl	 substrate citrate Na_2HPO_4	 1.33 mmol/l 20 mmol/l 32 mmol/l
mix and incubate at 37 °C for 10 min to 60 min				
stopping buffer	(3)	1 ml	glycine NaOH	226 mmol/l 204 mmol/l
mix and read fluorescence against blank				

Calculation: correct fluorimeter readings for blank, yielding ΔF. The catalytic concentration in (diluted) serum, urine or tissue extract is then

$$b = \frac{\mathrm{V}}{\mathrm{v}} \times \mathrm{f} \times \frac{\Delta F}{\Delta t} = 23 \times \mathrm{f} \times \frac{\Delta F}{\Delta t} \quad \mathrm{U/l}$$

$$b = 383 \times \mathrm{f} \times \frac{\Delta F}{\Delta t} \quad \mathrm{nkat/l}$$

where

ΔF is the corrected fluorimeter reading
t is the incubation time in minutes
f is the factor relating fluorescence reading to the concentration of 4-methylumbelliferone (µmol/l)
V is the total volume for fluorescence measurement (1.15 ml)
v is the sample volume (0.05 ml)

Or, for measurements in urine, using widely-used units:

$$z_c/m_s = \frac{60 \times 1000}{\rho_{\mathrm{creatinine}}} \times 23 \times \mathrm{f} \times \frac{\Delta F}{\Delta t} \quad \mathrm{nmol} \times \mathrm{h}^{-1} \times \mathrm{mg}^{-1}$$

where

$\rho_{\mathrm{creatinine}}$ is mass concentration of creatinine in diluted urine (mg/l).

Validation of Method

Precision, accuracy, detection limits and sensitivity: in the assay of β-*N*-acetylhexosaminidase from human placenta, using the method described here, the relative standard deviation was 3.3%. There is no information about accuracy since standard reference material has not been introduced. The lowest activity that can be measured by the fluorimetric assay is limited by the blank and not generally by the sensitivity of the fluorimeter; therefore, the smaller the volumes of substrate and enzyme solutions, the greater the sensitivity of the assay, provided the fluorescence of these small volumes can be read. Activities of 1 pmol/min (1×10^{-6} U) can be measured using a good quality spectrofluorimeter; with a microscope attachment the β-*N*-acetylhexosaminidase of a single fibroblast can be determined well above the limits of detection [19].

Sources of error: fluorimetric assays are subject to a variety of errors including quenching and self-quenching (cf. Vol. I, chapter 3.4); it is particularly important in the method described here to ensure that the concentration of 4-methylumbelliferone liberated in the assay falls on the linear part of the graph relating fluorescence intensity to 4-methylumbelliferone concentration. Fluorescent drugs administered to a patient might interfere with the assay and where this is considered a possibility enzyme blanks should be included.

Specificity: glycosidases are specific for both the sugar and glycosidic linkage, although they are much less specific for the aglycone. Both the β-*N*-acetylhexosaminidases and β-*N*-acetylglucosaminidase C will hydrolyze 4-methylumbelliferyl-β-*N*-acetylglucosaminides, but at pH 4.5 the contribution of the C form is small in most tissues and body fluids. Other glycosidases do not interfere with the assay.

Reference ranges: according to *Kaback* [20] the range of β-*N*-acetylhexosaminidase activity concentrations in human serum is 7 to 20 U/l. Mean urinary values [15] were 36 ± 10 nmol $\times$ h^{-1} $\times$ mg^{-1} creatinine for females and 27 ± 9 nmol $\times$ h^{-1} $\times$ mg^{-1} creatinine for males, but it should be noted that the substrate concentration was relatively low.

References

[1] *B. Helferich, A. Iloff,* Über Emulsin XIII, Darstellung und fermentative Spaltung von Glykosiden des N-Acetylglucosamins und der 2-Desoxyglucose, Hoppe-Seyler's Z. Physiol. Chem. *221*, 252–258 (1933).

[2] *D. H. Leaback,* An Introduction to Fluorimetric Analysis, 2nd ed., Koch-Light Ltd. Colnbrook, Bucks. U.K. (1974).

[3] *D. Robinson, J. L. Stirling,* N-Acetyl-beta-glucosaminidase in Human Spleen, Biochem. J. *107*, 321–327 (1968).

[4] *A. Hasalik, E. F. Neufeld,* Biosynthesis of Lysosomal Enzymes in Fibroblasts, J. Biol. Chem. *255*, 4937–4945 (1980).

[5] *K. Sandhoff, E. Conzelmann, H. Nehrkorn,* Specificity of Human Liver Hexosaminidase A and B Against Glycosphingolipids GM_2 and GA_2, Hoppe-Seyler's Z. Physiol. Chem. *358*, 779–788 (1977).
[6] *H. Kresse, W. Fuchs, J. Glössel, D. Holtfrerich, W. Gilberg,* Liberation of N-Acetylglucosamine-6-sulfate by Human Beta-N-Acetylhexosaminidase A, J. Biol. Chem. *256*, 12926–12932 (1981).
[7] *S. Okada, J. S. O'Brien, Tay-Sachs* Disease: Generalized Absence of a beta-N-Acetylhexosaminidase Component, Science *165*, 698–700 (1969).
[8] *K. Sandhoff, U. Andreae, H. Jatzkewitz,* Deficient Hexosaminidase Activity in an Exceptional Case of *Tay-Sachs* Disease with Additional Storage of Kidney Globoside in Visceral Organs, Life Sci. *7*, 278–285 (1968).
[9] *R. B. Ellis, N. T. Rapson, A. D. Patrick, M. F. Greaves,* Expression of Hexosaminidase Isoenzymes in Childhood Leukaemia, N. Engl. J. Med. *298*, 476–480 (1978).
[10] *R. G. Price,* Urinary N-Acetyl-beta-glucosaminidase (NAG) as an Indicator of Renal Disease, in: *U. C. Dubach, U. Schmidt* (eds.), Diganostic Significance of Enzymes and Proteins in Urine, Hans Huber, Bern 1979, pp. 150–163.
[11] *N. E. Dance, R. G. Price, D. Robinson,* Differential Assay of Human Hexosaminidases A and B, Biochim. Biophys. Acta *222*, 662–664 (1970).
[12] *R. B. Ellis, J. U. Ikonne, P. K. Masson,* DEAE-cellulose Microcolumn Chromatography Coupled with Automated Assay: Application to the Resolution of N-Acetyl-beta-hexosaminidase Components, Anal. Biochem. *63*, 5–11 (1975).
[13] *J. S. O'Brien,* Suggestions for a Nomenclature for the GM_2 Gangliosidoses Making Certain (Possibly Unwarrantable) Assumptions, Am. J. Hum. Genet. *30*, 672–675 (1978).
[14] *K. Sandhoff, E. Conzelmann,* Activation of Lysosomal Hydrolysis of Complex Glycolipids by Non-enzymic Proteins, Trends Biochem. Sci. *4*, 231–233 (1979).
[15] *J. M. Wellwood, R. G. Price, B. G. Ellis, A. E. Thompson,* A Note on the Practical Aspects of the Assay of N-Acetyl-beta-glucosaminidase in Human Urine, Clin. Chim. Acta *69*, 85–91 (1976).
[16] *D. Maruhn,* Rapid Colorimetric Assay of beta-Galactosidase and N-Acetyl-beta-glucosaminidase in Human Urine, Clin. Chim. Acta *73*, 453–461 (1976).
[17] *C.-T. Yuen, R. G. Price, L. Chattagoon, A. C. Richardson, P. F. G. Praill,* Colorimetric Assays for N-Acetyl-beta-D-glucosaminidase and beta-D-Galactosidase in Human Urine Using Newly-developed ω-Nitrostyryl Substrates, Clin. Chim. Acta *124*, 195–204.
[18] *D. H. Leaback, P. G. Walker,* Studies on Glucosaminidase 4. The Fluorimetric Assay of N-Acetylglucosaminidase, Biochem. J. *78*, 151–156 (1961).
[19] *H. Galjaard,* Genetic Metabolic Diseases; Early Diagnosis and Prenatal Analysis, Elsevier/North Holland Biomedical Press, Amsterdam (1980).
[20] *M. M. Kaback,* Thermal Fractionation of Serum Hexosaminidases: Applications to Heterozygote Detection and Diagnosis of *Tay-Sachs* Disease, in: *S. P. Colowick, N. O. Kaplan* (eds.), Methods in Enzymology, Vol. *XXVIII*,, Academic Press, New York 1972, pp. 862–867.

3 C–N and Anhydride Splitting Enzymes

3.1 β-Lactamase (Penicillinase, Cephalosporinase)

Penicillin amido-β-lactamhydrolase, EC 3.5.2.6

Karen Bush and Richard B. Sykes

General

β-Lactamase activity was first recorded in 1940; ironically, if not inappropriately, by the Oxford workers responsible for the isolation of penicillin [1]. Alongside the intensive search for new and improved β-lactam antibiotics and β-lactamase inhibitors, since those early days, has proceeded the study of enzymes that interact with these molecules. Throughout the 40-year history of β-lactam antibiotics, β-lactamases have continually asserted their presence as a major source of bacterial resistance to this group of antibiotics.

β-Lactamases are produced by gram-positive and gram-negative bacteria [2, 3], actinomycetes [4, 5], yeasts [6], and blue green algae [7]. The enzymes exhibit a considerable specificity, in that their only substrates are compounds containing a reactive β-lactam ring, e.g., penicillins and cephalosporins.

Unlike the situation with gram-positive organisms, gram-negative bacteria produce a plethora of β-lactamases which have been the subject of many classification schemes [8]. On the basis of substrate specificity, three groups of enzymes can be readily differentiated. "Penicillinases" preferentially hydrolyze penicillins, "cephalosporinases" preferentially hydrolyze cephalosporins and "broad-spectrum β-lactamases" hydrolyze both penicillins and cephalosporins. β-Lactamase production is both constitutive and inducible and the genes for β-lactamase production may be chromosomally or extra-chromosomally located [9].

Application of method: in biochemistry, clinical chemistry, microbiology.

Enzyme properties relevant in analysis: the catalytic activity observed in cell sonicates may be due to several different enzymes, depending on the occurrence of plasmids. Substrate conversion is linear with respect to time.

Methods of determination: catalytic activity can be measured acidimetrically [10], iodometrically [11, 12] or spectrophotometrically [13, 14]. Spectrophotometric methods tend to be more precise and sensitive. Estimations of enzyme mass or active-site titration are not used; total protein concentration in the sample may be determined using the method of *Lowry et al.* [15]. Isoelectric focusing followed by staining with nitrocefin will detect the occurrence of multiple β-lactamase activities [16].

International reference methods and standards: neither standardization at the international level nor the existence of reference standard material is known at this time.

Enzyme effectors: specific activators are not known. Organic solvents, including dimethyl sulphoxide (DMSO), will inhibit at high concentrations. Mixtures of iodine and potassium iodide are strongly inhibitory towards a variety of β-lactamases [3].

Assay

Method Design

Principle

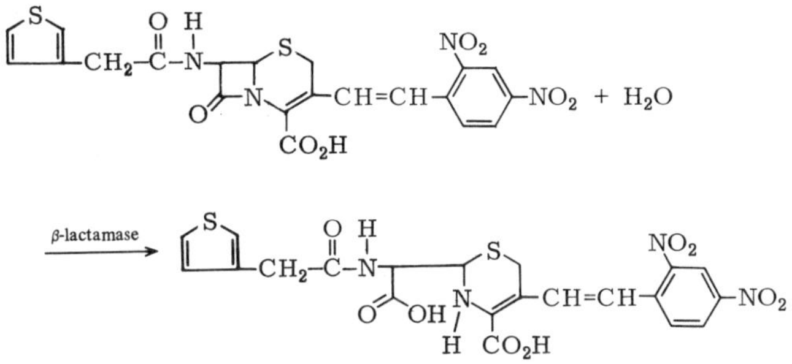

The amount of nitrocefin hydrolyzed per unit time, measured by an increase in absorbance at 495 nm due to hydrolysis of a highly conjugated cephalosporin [17], can be directly correlated with β-lactamase activity. The equilibrium is strongly shifted towards the hydrolyzed product. Although other penicillins or cephalosporins may be used as substrates for specific enzymes [9], nitrocefin has the widest spectrum of susceptibility and sensitivity of commercially available β-lactams. (For other substrates wavelengths in the UV range are generally used and would depend upon the nature of the substrate).

Blanks may be omitted for routine assays. Small blank hydrolysis rates for substrate may be observed at 37 °C. The method described below can easily be modified for use with automated analyzers.

Optimized conditions for measurement: β-lactamases are characteristically insensitive to small changes in pH or buffer composition. Although a pH optimum of 7 has been reported for many of the enzymes, assays may be performed between pH 6 and pH 8 in phosphate buffer, 0.01 to 0.2 mol/l. Substrate inhibition is commonly observed with β-lactamases: nitrocefin should not be used at concentrations greater than 0.5 mmol/l for *Michaelis-Menten* kinetics to be valid.

Temperature conversion factors: the following conversion factors have been determined for Kl (from *Klebsiella pneumoniae*) and RTEM (from *Escherichia coli*) β-lactamases to relate relative activity at various temperatures. Conversion factors for other β-lactamases may differ from these values.

°C	20	22	24	25	30	35	37
Kl β-lactamase	2.38	1.98	1.68	1.52	1.00	0.69	0.60
RTEM β-lactamase	1.93	1.83	1.52	1.44	1.00	0.73	0.65

Equipment

Spectrophotometer capable of exact measurements at 495 nm, preferably with recorder (or stopwatch); thermostatted cuvette holder.

Reagents and Solutions

Purity of reagents: reagent-grade chemicals should be used.

Preparation of solutions: (for about 50 determinations): all solutions in pyrogen-free re-purified water (cf. Vol. II, chapter 2.1.3.2).

1. Phosphate buffer (0.1 mol/l, pH 7.0):

   dissolve 13.4 g $Na_2HPO_4 \cdot 7\,H_2O$ in 500 ml water and 6.85 g KH_2PO_4 in 500 ml water. Mix 80 ml disodium hydrogen phosphate with 45 ml dihydrogen potassium phosphate solution. Check the pH and adjust to pH 7.0 if necessary with the appropriate phosphate solution.

2. Nitrocefin (0.19 mmol/l):

   weigh 5.0 ± 0.2 mg nitrocefin into a 50 ml volumetric flask, dissolve in 0.5 ml dimethyl sulphoxide, DMSO, and add phosphate buffer, 0.1 mol/l, pH 7.0, to give a final volume of 50 ml.

Stability of solutions: store all solutions at 0–4°C. Phosphate buffer should be prepared freshly at least once a week. Nitrocefin, when protected from light at 4°C, maintains greater than 90% stability after storage for two weeks.

Procedure

Collection and treatment of specimen

Gram-positive bacteria: centrifuge 25 ml fermentation broth for 30 min at about 15000 *g*. Test the supernatant for β-lactamase activity.

Gram-negative bacteria: centrifuge 50 ml fermentation broth for 30 min at about 15000 *g*. Decant the supernatant. Wash the packed cells with 5 ml phosphate buffer, 0.1 mol/l, pH 7.0. Re-suspend the cells in 5.0 ml buffer, sonicate 2 × 1.0 min at 0°C and centrifuge. The resulting supernatant contains β-lactamase activity, if enzyme is present.

Stability of the enzyme in the sample: stability is dependent upon the bacterial source and nature of the enzyme. Most β-lactamases are stable for at least one month if frozen (−20°C) in the presence of 0.02% sodium azide. Some β-lactamases from *Pseudomonas* are stable for only a few hours. Activity may be lost if sonicates are frozen in plastic tubes.

Details for measurement in other samples: Nitrocefin solution may be applied as a thin liquid layer on agar in *Petri* dishes [17] or polyacrylamide gel surfaces [16] for the qualitative determination of β-lactamase activity.

Assay conditions: wavelength 495 nm; light path 10 mm; final volume 1.1 ml; 30°C (thermostatted cuvette holder). Measure against air. Before beginning the assay, nitrocefin solution (2) should be at 30°C.

Enzyme dilutions should be prepared immediately before use because some β-lactamases are quite unstable in dilute solutions.

Measurement

Pipette successively into the cuvette:			concentration in assay mixture	
nitrocefin	(2)	1.0 ml	phosphate	100 mmol/l
			nitrocefin	0.18 mmol/l
sample		0.10 ml	volume fraction	0.091
mix well using a plastic mixing spoon; read absorbance and start stopwatch. Repeat readings exactly every 15 s or monitor on a recorder.				

The total ΔA should not exceed 1.5. If reaction proceeds too rapidly, sample may be diluted with phosphate buffer (1) or a smaller sample volume may be used. If reaction proceeds slowly, absorbance readings may be made every 1 or 2 min for at least 10 min.

Calculation: use eqn. (k) or (k_1) from "Formulae", Appendix 3. $\varepsilon_{495} = 1.406$ l × mmol^{-1} × mm^{-1}. The catalytic concentration in the sample is (Δt in min):

$$b = 782 \times \Delta A/\Delta t \quad \text{U/l}$$

$$b = 13039 \times \Delta A/\Delta t \quad \text{nkat/l.}$$

Validation of Method

Precision, accuracy, detection limits and sensitivity: for an activity of 13000 nkat/l a standard deviation of 510 nkat/l was observed; the relative standard deviation was 4.0% for ten replicates. Data concerning accuracy are not available because standard reference material has not been established. With a sensitive spectrophotometer the lowest activity which can be accurately measured is approximately 8 nkat/l with an imprecision of ±10%. Sensitivity is $\Delta A/\Delta t = 0.003/10$ min (*Gilford* spectrophotometer).

Sources of error: serum and animal tissue samples will produce a false positive assay for β-lactamase activity, due in part to the reactivity of simple thiols, cysteine and glutathione found in the albumin fraction of serum [17]. β-Lactamase from *Staphylococcus aureus* will adhere to glass surfaces [18]; therefore, disposable cuvettes should be used for assays of enzyme from this organism, or glass cuvettes should be rinsed well with ethanol between samples to prevent carryover of active enzyme.

Specificity: nitrocefin is readily hydrolyzed by a variety of microbial β-lactamases but is not known to react significantly with other microbial enzymes.

Reference ranges: strains of *Enterobacter cloacae* may exhibit β-lactamase activities as high as 8×10^6 nkat/l for the equivalent of a 100 µl sample of sonicate prepared as described above. Sonicates of *E. coli* which carry RTEM β-lactamase may exhibit activites of 1.5×10^6 nkat/l, whereas poor β-lactamase producers have activities of <20 nkat/l.

References

[1] *E. P. Abraham, E. Chain,* An Enzyme from Bacteria Able to Destroy Penicillin, Nature (London) *146*, 837 (1940).

[2] *N. Citri, M. R. Pollock,* The Biochemistry and Function of β-Lactamase (Penicillinase), Adv. Enzymol. *28*, 237–323 (1966).

[3] *M. H. Richmond, R. B. Sykes,* The β-Lactamases of Gram-Negative Bacteria and Their Possible Physiological Role, in: *A. H. Rose, D. W. Tempest* (eds.), Advances in Microbial Physiology, Vol. *9*, Academic Press, London 1973, pp. 31–88.

[4] *H. Ogawara, S. Horikawa, S. Shimada-Miyoshi, K. Yasuzawa,* Production and Property of Beta-Lactamases in *Streptomyces:* Comparison of the Strains Isolated Newly and Thirty Years Ago, Antimicrob. Agents Chemother. *13*, 865–870 (1978).

[5] *J. L. Schwartz, S. P. Schwartz,* Production of β-Lactamase by Non-Streptomyces – Actiomycetales, Antimicrob. Agents Chemother. *15*, 123–125 (1979).

[6] *R. J. Mehta, C. H. Nash,* β-Lactamase Activity in Yeast, J. Antibiotics *31*, 239–240 (1978).

[7] *D. J. Kushner, C. Brevil,* Penicillinase (β-Lactamase) Formation by Blue-Green Algae, Arch. Microbiol. *112*, 219–223 (1977).

[8] *R. B. Sykes,* The Classification and Terminology of Enzymes that Hydrolyze β-Lactam Antibiotics, J. Infect. Dis. *145*, 762–765 (1982).

[9] *R. B. Sykes, M. Matthew,* The β-Lactamases of Gram-negative Bacteria and Their Role in Resistance to β-Lactam Antibiotics, J. Antimicrob. Chemother. *2*, 115 – 157 (1976).
[10] *J. P. Hou, J. W. Poole,* Measurement of β-Lactamase Activity and Rate of Inactivation of Penicillins by a pH-Stat Alkalimetric Titration Method, J. Pharm. Sci. *61*, 1594 – 1598 (1972).
[11] *C. J. Perret,* Iodometric Assay of Penicillinase, Nature (London) *174*, 1012 – 1013 (1954).
[12] *R. B. Sykes, K. Nordström,* Microiodometric Determination of β-Lactamase Activity, Antimicrob. Agents Chemother. *1*, 94 – 99 (1972).
[13] *C. H. O'Callaghan, P. W. Muggleton, G. W. Ross,* Effects of β-Lactamase from Gram-negative Organisms on Cephalosporins and Penicillins, Antimicrob. Agents Chemother. *1968*, 57 – 63.
[14] *A. Samuni,* A Direct Spectrophotometric Assay and Determination of Michaelis Constants for the β-Lactamase Reaction, Anal. Biochem. *63*, 17 – 26 (1975).
[15] *O. H. Lowry, N. J. Rosebrough, A. L. Farr, R. J. Randall,* Protein Measurement with the *Folin* Phenol Reagent, J. Biol. Chem. *193*, 265 – 275 (1951).
[16] *M. Matthew, A. M. Harris,* Identification of β-Lactamases by Analytical Isoelectric Focusing; Correlation with Bacterial Taxonomy, J. Gen. Microbiol. *94*, 55 – 67 (1976).
[17] *C. H. O'Callaghan, A. Morris, S. M. Kirby, A. H. Shingles,* Novel Method for Detection of β-Lactamases by Using a Chromogenic Cephalosporin Substrate, Antimicrob. Agents Chemother. *1*, 283 – 288 (1972).
[18] *M. H. Richmond,* Purification and Properties of the Exopenicillinase from *Staphylococcus aureus,* Biochem. J. *88*, 452 – 459 (1963).

3.2 Arginase

L-Arginine amidinohydrolase, EC 3.5.3.1

Jean-Pierre Colombo and Liliana Konarska

General

The main source of arginase is the mammalian liver. The enzyme was first described in 1904 [1]. Its metabolic role in mammalian liver as the terminal enzyme of the urea cycle was established by *Krebs & Henseleit* in 1932 [2]. In man the enzyme occurs mainly in liver. Lower activities have been demonstrated in erythrocytes, leukocytes, platelets, kidney, skeletal and heart muscle, brain, intestine, pancreas, lung, epidermis, placenta, salivary glands, testes, plasma and fibroblasts. The metabolic function of extrahepatic arginase is still unclear.

The enzyme has also been described in other species which possess a ureotelic metabolism, but, with few exceptions, not in the liver of animals that have a uricotelic metabolism. Arginase is also found in invertebrates, as in the hepatopancreas of terrestrial and fresh water gastropods, as well as in plants and micro-organisms.

Purified liver arginase has a molecular weight around 105 000 and presents an oligomeric structure of dimeric and trimeric form. In the human liver cell the enzyme is found in the cytosol and adherent to subcellular structures. The kinetic properties determined for purified liver arginase show an optimum pH at 9.3 [3, 4] and an optimal $MnCl_2$ concentration of 2 mmol/l. Reported values for the K_m for arginine are 10.5 [4], 5.3 [5] and 7.4 [3] mmol/l.

L-Lysine exerts a competitive type of inhibition with a K_I of 4.4 mmol/l [4] to 7.1 mmol/l [3]. L-homoarginine is not a substrate for human arginase [4]. The K_m for red cell arginase is 22 mmol/l [4]. This enzyme cross-reacts with antibody against the liver enzyme [6].

An increase of liver arginase is brought about by an increased intake of protein from exogeneous sources, or provided endogenously from extrahepatic tissues as a result of starvation, disease or changes in hormonal state [7]. These elevations of arginase activity appear to represent increases in the amount of the specific enzyme protein [8].

Absence of arginase activity in human subjects leads to hyperargininaemia, a rare hereditary metabolic disease. A deficiency of arginase in this disorder has so far been demonstrated in liver, red cells and leukocytes [9–11]. Determination of arginase activity in erythrocytes is now recommended as a main diagnostic parameter in this inherited disorder [12].

Some other clinical applications of arginase determinations have been described in certain types of anaemia [13].

Application of methods: in biochemistry and clinical chemistry.

Enzyme properties relevant in analysis: arginase in human organs exists in multiple forms which can be discriminated by heat inactivation, Mn^{2+}-ion activation and DEAE-cellulose chromatography, as well as by immunological methods [14, 15, 16]. Two probably identical forms have been demonstrated in human liver and red cells [6, 17]. The catalytic activity of both forms is measured in the assay described below. Substrate conversion is linear with time and protein concentration [18].

Methods of determination: the catalytic activity of arginase can be determined by measuring a decrease in arginine concentration or an increase in the concentrations of urea and ornithine [19–22].

Assay techniques in use are based on the direct spectrophotometric assay of urea [23] or on the determination of carbon dioxide and ammonia, the products of the degradation of urea by urease [21, 24], or on measurement of ornithine by the *Chinard* reaction [25–27]. Because of its sensitivity and practicability the latter technique is described below. The results of arginine hydrolysis determined by the colorimetric methods are in good agreement with the activity measured in terms of ^{14}C-urea produced from ^{14}C-guanido arginine [28]. An immunological method for quantitative determination of the enzyme protein in liver has also been described [29].

International reference method and standards: neither standardization at the international level nor the existence of reference standard materials is known so far.

Enzyme effectors: arginase is activated and stabilized at high temperatures by bivalent metal ions such as Mn, Mg, Co, Ni and Cd [20]. Amino acids, particularly ornithine, the reaction product, and lysine, as well as adenosine and inosine, inhibit the enzyme competitively [30, 31].

Assay

Method Design

Principle

$$\text{L-Arginine} + H_2O \xrightarrow{\text{arginase}} \text{L-ornithine} + \text{urea}.$$

Arginine is hydrolyzed to ornithine and urea. Ornithine is measured after stopping the reaction with glacial acetic acid. In the presence of acetic acid ornithine gives a red colour with ninhydrin with a maximal absorbance at 515 nm. Urea does not interfere in the concentration present in the sample. Glycine buffer has often been used for arginase determination but interferes with the determination of ornithine and decreases the sensitivity of the method [25, 26]. A zero and a reagent blank are carried through the procedure to compensate for the pre-existing substrate and product concentrations.

Optimized conditions for measurement: the same measuring conditions can be applied for both liver and red cell arginases. Both require Mn^{2+}-ions for maximal activation. Binding with Mn^{2+} depends on time, temperature and Mn^{2+} concentration [5, 20]. This is achieved with manganese chloride at a final concentration of 5 mmol/l at 55 °C for 20 min, immediately prior to assay. Activation with Mn^{2+} increases the activity 2 – 6 times in erythrocytes, 4 – 5 times in liver. Arginase assay is performed at pH 9.5 in the presence of carbonate buffer. The optimal concentration of substrate used is 20 mmol/l.

Temperature conversion factors: temperature conversion is not recommended when working with crude homogenates and cell lysates.

Equipment

Spectrophotometer capable of exact measurement at 515 nm. Two water-baths set at 55 °C and 37 °C, respectively. Stopwatch.

Reagents and Solutions

Purity of reagents: analytical grade chemicals (purum or purissimum) must be used. Arginine must be free from ornithine. The absorbance of the reagent blank, which is highly dependent on the purity of both arginine and ninhydrin, should not exceed 0.03 when read at 515 nm against water in 10 mm light-path cuvette.

Preparation of solutions (for about 20 determinations): all solutions in re-purified water (cf. Vol. II, chapter 2.1.3.2).

1. Tris buffer (Tris, 5 mmol/l, pH 7.5; $MnCl_2$, 1 mmol/l):

   dissolve 605.7 mg tris(hydroxymethyl)aminomethane in 800 ml water, adjust pH to 7.5 with HCl (1 mol/l), add 198 mg $MnCl_2 \cdot 4\ H_2O$ and dilute to 1 litre with water. This buffer is used for the preparation of red cell haemolysate and liver homogenate.

2. Activator solution ($MnCl_2$, 10 mmol/l):

   dissolve 198 mg $MnCl_2 \cdot 4\ H_2O$ in 100 ml water.

3. Ornithine standard (1 mmol/l):

   dissolve 16.86 mg L-ornithine monohydrochloride and 20 mg sodium azide in 100 ml water.

4. Carbonate buffer (0.1 mol/l, pH 9.5; 37 °C):

   mix 40 ml sodium bicarbonate (0.1 mol/l, 8.4 g/l) with about 60 ml sodium carbonate, anhydrous (0.1 mol/l, 10.6 g/l) and adjust to pH 9.5.

5. Arginine solution (0.1 mol/l):

   dissolve 210.68 mg L-arginine monohydrochloride in approx. 8 ml water, adjust pH to 9.5 with NaOH, 1 mol/l, and dilute with water to 10 ml.

6. Acetic acid, glacial.

7. Ninhydrin solution (140 mmol/l):

   dissolve 2.5 g ninhydrin in a mixture of 60 ml glacial acetic acid and 40 ml phosphoric acid, 6 mol/l (404 ml concentrated 85% H_3PO_4). Heat to about 70 °C to secure dissolution of ninhydrin and cool. Store in the dark.

8. *Drabkin*'s solution:

   dissolve 200 mg $K_3[Fe(CN)_6]$, 50 mg KCN, 140 mg KH_2PO_4 and 0.5 ml Sterox in 1000 ml water [32].

Stability of solutions: store all solutions at 0–4 °C. Prepare arginine solution freshly every day. Ornithine standard (3) and ninhydrin solution (7) are stable for at least one month when stored in the dark. Tris buffer (1), carbonate buffer (4) and manganese solution (2) are stable as long as no microbial contamination occurs.

Procedure

Collection and treatment of specimen

Preparation of haemolysate: centrifuge venous heparinized (ca. 0.75 mg sodium heparinate/ml blood) or ACD (acid-citrate-dextrose) blood in a refrigerated centrifuge at 4 °C (ca. 2000 *g*). Wash the precipitated red cells three times with five volumes of cold physiological saline. Then add 0.02 ml of washed red cells to 5 ml Tris buffer (1). Use this suspension for the assay.

Haemoglobin determination: add 1 ml of haemolysate to 2 ml *Drabkin* solution (8). Mix well and let stand for 2 hours. It is advantageous to centrifuge the solution for 10 min at ca. 2000 *g*. Measure the absorbance of the supernatant at 546 nm. Calculate the concentration ρ of haemoglobin (gram per litre haemolysate) with a haemoglobin cyanide-standard or using the molar absorption coefficient $\varepsilon = 44 \text{ l} \times \text{mmol}^{-1} \times \text{mm}^{-1}$. Molecular weight of Hb is 64458 [32].

Preparation of liver homogenate: after weighing and blotting, homogenize 10–50 mg of liver tissue with 19 volumes of Tris buffer (1) with a *Potter-Elvehjem* homogenizer in an ice-bath at 1000 rpm during ca. 60 s.

Uncentrifuged homogenate is used for the assay after further appropriate final dilution, mostly 1:4000.

Stability of the enzyme in specimen or sample

Erythrocytes: the enzyme in both haemolysates and packed erythrocytes stored at 0–4 °C does not change its activity for at least two weeks, when measured after pre-activation with Mn^{2+} at 55 °C. The enzyme is stable for at least 3 months at −20 °C.

Liver: biopsy- and necropsy material should be processed immediately, whenever possible. Biopsy- and necropsy material frozen immediately at −20 °C can be stored at least for two to three weeks, or at −70 °C for several months, in our experience. Other authors claim that the storage time is shorter [23]. In frozen material the enzyme activity can no longer be referred to tissue wet weight.

Assay conditions: wavelength 515 nm; light path 10 mm; incubation volume 0.500 ml; incubation temperature 37 °C; volume fraction 0.1; volume for colour reaction 2.5 ml; before starting, adjust temperature of solutions (4) and (5) to 37 °C.

Measurement

Pipette into test tubes:	sample	sample blank	reagent blank	standard	concentration in assay mixture
activator solution (2)	0.05 ml	0.05 ml	0.05 ml	0.05 ml	Mn^{2+} 1 mmol/l
haemolysate, homogenate	0.05 ml	0.05 ml	–		vol. fraction 0.1
mix well, cover with glass marbles and incubate for 20 min at 55 °C, then cool below 37 °C;					
ornithine standard (3)	–	–	–	0.05 ml	0.1 mmol/l
water	–	–	0.05 ml	–	
carbonate buffer (4)	0.30 ml	0.30 ml	0.30 ml	0.30 ml	60 mmol/l
arginine solutions (5)	0.10 ml	–	0.10 ml	0.10 ml	20 mmol/l
mix well, incubate for exactly 10 min at 37 °C;					
acetic acid (6)	1.50 ml	1.50 ml	1.50 ml	1.50 ml	
arginine solution (5)	–	0.10 ml	–	–	
ninhydrin reagent (7)	0.50 ml	0.50 ml	0.50 ml	0.50 ml	
mix well, preferably with a vortex mixer. Place in a boiling water-bath for 60 min. Cool to room temperature and read the absorbance at 515 nm. The colour is stable for at least 1 – 2 h.					

In case of very low enzyme activity (i.e. arginase deficiency), prolong the time of incubation at 37 °C (linearity is preserved up to at least 1 – 2 h) and use more concentrated haemolysates (up to 1 : 5).

Calculations: use in combination with eqn. (d_1) formula (l) or (l_1), respectively for calculation of catalytic activity content, and formula (k) or (k_1) for calculation of catalytic activity concentration (cf. "Formulae", Appendix 3).

Subtract absorbance readings for sample blank and reagent blank from readings for sample and standard, respectively, yielding A_{sample} and $A_{standard}$.

Erythrocytes

The catalytic activity is related to g haemoglobin per litre, ρ_{Hb}, or to ml packed cells. The catalytic content is (Δt in minutes, volume fraction of sample in the assay considered):

$$z_c/m_s = \frac{1000}{\rho_{Hb}} \times c_{standard} \times \frac{A_{sample}}{A_{standard}} / \Delta t \qquad \text{U/g Hb}.$$

The catalytic concentration per ml packed cells is accordingly:

$$b = 2.5 \times 10^5 \times c_{standard} \times \frac{A_{sample}}{A_{standard}} / \Delta t \qquad \text{U/l}.$$

$$b = 4.17 \times 10^6 \times c_{standard} \times \frac{A_{sample}}{A_{standard}} / \Delta t \qquad \text{nkat/l}$$

Liver

The catalytic content per gram fresh weight is

$$z_c/m_s = \frac{1000}{\rho_{liver}} \times c_{standard} \times \frac{A_{sample}}{A_{standard}} / \Delta t \qquad \text{U/g}$$

(ρ_{liver} is concentration of liver tissue in diluted final homogenate, g/l.)

Validation of Method

Precision, accuracy, detection limits and sensitivity: with 20 haemolysates prepared from the same blood, an imprecision of ± 1.2 U/ml packed erythrocytes with an average activity of 21.8 U/ml was observed. The relative standard deviation (RSD) was 5.5%. The within-run imprecision for a series of determinations on the same haemolysate was 3.6%. Accuracy cannot be specified since standard reference material is not available yet. For measurement with commonly used spectrophotometers the detection limit is about 3 nmol ornithine per sample. This corresponds to A $\geqslant 0.03$ when measured against the reagent blank in a 10 mm light-path cuvette. At an appropriately high concentration of haemolysate and with a prolonged incubation time at 37 °C, the lowest activity which can be measured is about 0.003 U/ml packed cells.

The between-run imprecision of arginase determinations (10 homogenates from the same liver) had a RSD of 26% referred to wet weight and 19% referred to protein.

The within-run imprecision (10 determinations from the same homogenate) had a RSD of 8% referred to wet weight.

Sources of error: the substrate solution should be added accurately to all tubes, since arginine has a slight influence on colour formation. Influences on the measured activity by therapeutics, metabolites and other agents have not been tested.

Specificity: the enzyme system is specific. Theoretically, transamidinase (EC 2.1.4.1) which produces ornithine from L-arginine and glycine could interfere with arginine determination, but is practically without any influence since its pH optimum is 7 to 7.5 and the enzyme is not active at pH 9.5 used for arginase assay. The specificity of the colour formation of ornithine with ninhydrin is enhanced by the strongly acid medium.

Only proline, to a small extent, could react in the colour reaction. Citrulline, cysteine and lysine give significant increases in A_{515} when present in molar concentrations 10 times greater than those of ornithine used in the assay [25].

Reference ranges: with the method described are

Erythrocytes (n = 280):	35 – 206 U/g Hb, 37 °C
	9.1 – 53 U/ml packed cells, 37 °C
Comparison with literature:	30 – 96 U/g Hb, 37 °C, [9]
	11 – 22 U/g Hb, 37 °C, [33]
	27 – 92 U/g Hb, 37 °C, [34]
	13 – 22 U/g Hb, 37 °C, [35]
	37 – 103 U/g Hb, 37 °C, [36]
U/ml packed cells	6.0 – 19 [37]

Liver: with the method described

(n = 5)	306 – 778 U/g wet weight
	1.32 – 2.76 U/mg protein
Comparison with literature:	666 ± 371 U/g (autopsy material) [23]
	4.97 ± 2.37 U/mg protein [23]
	1433 ± 155 U/g (biopsy material) [23]
	9.65 ± 1.77 U/mg protein (biopsy material) [23]
	89 – 243 U/g wet weight [36]
	0.59 – 1.65 U/mg protein (autopsy material) [36]

References

[1] *A. Kossel, H. D. Dakin,* Über die Arginase, Hoppe-Seyler's Z. physiol. Chem. *41*, 321 – 331 (1904).
[2] *H. A. Krebs, K. Henseleit,* Urea Formation in Animal Organism, Hoppe-Seyler's Z. physiol. Chem. *210*, 33 – 66 (1932).

[3] *L. Bascur, J. Cabello, M. Véliz, A. Gonzalez,* Molecular Forms of Human-Liver Arginase, Biochim. Biophys. Acta *128*, 149–154 (1966).
[4] *J. Berüter, J. P. Colombo, C. Bachmann,* Purification and Properties of Arginase from Human Liver and Erythrocytes, Biochem. J. *175*, 449–454 (1978).
[5] *J. Cobello, C. Basilio, V. Prajoux,* Kinetic Properties of Erythrocyte and Liver Arginase, Biochim. Biophys. Acta *48*, 148–152 (1961).
[6] *J. P. Colombo, L. Konarska,* Heterogeneity of Urea Cycle Enzymes, in: *D. Scheuch, R. J. Haschen, E. Hofmann* (eds.), Multiple Forms of Enzymes, Karger, Basel 1982, pp. 92–98.
[7] *H. Aebi,* Coordinated changes in Enzymes of the Ornithine Cycle and Response to Dietary Conditions, in: *S. Grisolia, R. Baguena, F. Mayor* (eds.), The Urea Cycle, John Wiley & Sons, New York 1976, pp. 275–296.
[8] *R. T. Schimke,* The Importance of Both Synthesis and Degradation in the Control of Arginase Levels in Rat Liver, J. Biol. Chem. *239*, 3808–3817 (1964).
[9] *V. V. Michels, A. L. Bedauet,* Arginase Deficiency in Multiple Tissues in Argininemia, Clin. Genet. *13*, 61–67 (1978).
[10] *H. G. Terheggen, A. Schwenk, A. Lowenthal, M. Van Sande, J. P. Colombo,* Argininaemia with Arginase Deficiency, Lancet *ii*, 748 (1969).
[11] *B. Marescau, J. Pintens, A. Lowenthal, H. G. Terheggen, K. Adriaenssens,* Arginase and Free Amino Acids in Hyperargininemia: Leukocyte Arginine as a Diagnostic Parameter for Heterozygotes, J. Clin. Chem. Clin. Biochem. *17*, 211–217 (1979).
[12] *H. Galjaard,* Genetic Metabolic Diseases. Early Diagnosis and Prenatal Analysis, Elsevier, Amsterdam 1980, p. 315.
[13] *H. Nishibe, K. Yamagata, H. Gotoh,* A Case of Sideroblastic Anemia Associated with Marked Elevation of Erythrocytic Arginase Activity, Scand. J. Haematol. *15*, 17–21 (1975).
[14] *G. Soberon, R. Palacios,* Arginase, in: *S. Grisolia, R. Baguena, F. Mayor* (eds.), The Urea Cycle, John Wiley & Sons, New York 1976, pp. 221–235.
[15] *Z. Porembska,* Different Species of Arginase in Animal Tissues, Enzyme *15*, 198–209 (1973).
[16] *E. B. Spector, S. C. H. Rice, S. Moedjono, B. Bernard, S. D. Cederbaum,* Biochemical Properties of Arginase in Human Adult and Fetal Tissues, Biochem. Med. *28*, 165–175 (1982).
[17] *A. F. Van Elsen, J. G. Leroy,* Arginase Isoenzymes in Human Diploid Fibroblasts, Biochem. Biophys. Res. Commun. *62*, 191–198 (1975).
[18] *G. W. Brown, Jr., P. P. Cohen,* Comparative Biochemistry of Urea Synthesis. I. Methods for the Quantitative Assay of Urea Cycle Enzymes in Liver, J. Biol. Chem. *234*, 1769–1774 (1959).
[19] *R. L. Ward, P. A. Srere,* A New Spectrophotometric Arginase Assay, Anal. Biochem. *18*, 102–106 (1967).
[20] *D. M. Greenberg,* Arginase, in: *P. D. Boyer, H. Lardy, K. Myrbäck* (eds.), The Enzymes, 2nd ed., Vol. *4*, Academic Press, New York 1960, pp. 257–267.
[21] *U. T. Rüegg, A. S. Russell,* A Rapid and Sensitive Assay for Arginase, Anal. Biochem. *102*, 206–212 (1980).
[22] *M. K. Schwartz,* Clinical Aspects of Arginase, in: *S. P. Colowick, N. O. Kaplan* (eds.), Methods in Enzymology, Vol. *XVIIB*, Academic Press, New York 1971, pp. 857–861.
[23] *C. T. Nuzum, P. J. Snodgrass,* Multiple Assays of the Five Urea-Cycle Enzymes in Human Liver Homogenates, in: *S. Grisolia, R. Baguena, F. Mayor* (eds.), The Urea Cycle, John Wiley & Sons, New York 1976, pp. 325–349.
[24] *J. P. Colombo, W. Bürgi, R. Richterich, E. Rossi,* Congenital Lysine Intolerance with Periodic Ammonia Intoxication: A Defect in L-Lysine Degradation, Metabolism *16*, 910–925 (1967).
[25] *F. P. Chinard,* Photometric Estimation or Proline and Ornithine, J. Biol. Chem. *199*, 91–95 (1952).
[26] *Z. Porembska, M. Kedra,* Early Diagnosis of Myocardial Infarction by Arginase Activity Determination, Clin. Chim. Acta *60*, 355–361 (1975).
[27] *L. Konarska, L. Tomaszewski,* Studies on L-Arginase of the Small Intestins. I. Topographical Distribution and some Properties of the Small Intestine L-Arginase in the Rat, Biochem. Med. *14*, 250–262 (1975).
[28] *L. Konarska, U. Wiesmann, R. v. Fellenberg, J. P. Colombo,* Isoenzyme Pattern and Immunological Properties of Arginase in Normal and Hyperargininemia Fibroblasts, Enzyme, *29*, 44–53 (1983).

[29] *A. Sumitani,* Immunological Studies of Liver Arginase in Man and Various Kind of Vertebrates, Hiroshima J. Med. Sci. *26,* 59 − 80 (1977).
[30] *A. Hunter, C. E. Downs,* The Inhibition of Arginase by Amino Acids, J. Biol. Chem. *157,* 427 − 446 (1945).
[31] *J. L. Rosenfeld, S. P. Dutta, G. B. Chheda, G. L. Tritsch,* Purine and Pyrimidine Inhibitors of Arginase, Biochim. Biophys. Acta *410,* 164 − 166 (1975).
[32] *R. Richterich, J. P. Colombo,* Clinical Chemistry, J. Wiley & Sons, New York 1981, p. 544.
[33] *S. D. Cederbaum, K. N. F. Shaw, E. B. Spector, M. A. Verity, P. J. Snodgrass, G. I. Sugarman,* Hyperargininemia with Arginase Deficiency, Pediatr. Res. *13,* 827 − 833 (1979).
[34] *S. E. Snyderman, C. Sansaricq, W. J. Chen, P. M. Norton, S. V. Phansalkar,* Argininemia, J. Pediatr. *90,* 563 − 568 (1977).
[35] *H. G. Terheggen, A. Schwenk, A. Lowenthal, M. Van Sande, J. P. Colombo,* Hyperargininämie mit Arginasedefekt. Eine neue familiäre Stoffwechselstörung. II. Biochemische Untersuchungen, Z. Kinderheilkd. *107,* 313 − 323 (1970).
[36] *C. Bachmann, J. P. Colombo,* Diagnostic Value of Orotic Acid Excretion in Heritable Disorders of the Urea Cycle and in Hyperammonemia due to Organic Acidurias, Eur. J. Pediatr. *134,* 109 − 113 (1980).
[37] *K. Adriaenssens, D. Karcher, A. Lowenthal, H. G. Terheggen,* Use of Enzyme-loaded Erythrocytes in In-Vitro Correction of Arginase-Deficient Erythrocytes in Familial Hyperargininemia, Clin. Chem. *22,* 323 − 326 (1976).

3.3 Guanase

Guanine aminohydrolase, EC 3.5.4.3

3.3.1 UV-method, 290 nm

Giuseppe Giusti and Bruno Galanti

General

Guanase occurs in many animal tissues. In the rat the activity decreases in the order: reticulocytes, erythrocytes, lung, lactating mammary gland, liver, spleen, brain, virgin mammary gland, heart, kidney, skeletal muscle. No guanase activity was found in foetal tissues [1]. Fractionation of rat liver homogenate by centrifugation demonstrated the following distribution of total guanase activity: nuclei 15.2%, mitochondria 12.3%, cytoplasm 72.5%. In human tissues, guanase occurs mainly in liver, kidney, brain and small intestine; low or no activity has been found in heart, lung, spleen, pancreas, skeletal muscle, erythrocytes and leukocytes. Very little guanase activity is detectable in normal human serum; slight activity occurs in blood or serum of mouse, rabbit and fowl and relatively high activity in normal rat serum (for references, cf. chapter 3.4.2).

The enzyme can be purified by the method of *Kalckar* [2]; commercial preparations are available (cf. Vol. II, Chapter 2.2.1). Guanase activity in serum from patients with various liver diseases was first measured by *Hue & Free* [4], *Whitehouse, Knight et al.* [5] and *Giusti & Galanti* [6]. Enzyme concentration is high in liver and negligible in heart, pancreas and skeletal muscle; thus an increase of guanase in serum is specific for liver disease. Increases are particularly high in acute viral hepatitis and toxic liver damage; normal values or slight increases occur in obstructive jaundice, liver cirrhosis, bacterial cholangitis, myocardial infarction, acute pancreatitis and malignant tumours. Guanase in serum decreases in human and experimental kidney diseases with uraemia (cf. [3 – 10] for references).

Application of method: in biochemistry and in clinical chemistry.

Enzyme properties relevant in analysis: relative substrate specificity, determined under optimized conditions for both 8-azaguanine and guanine and expressed as the deaminase ratio (8-azaguanine/guanine) is approx. 2.50 in serum, liver, kidney and brain of man. Similar values were found in mouse and rat [11].

The optimum pH of human brain, kidney, liver and serum guanase with guanine as substrate is 7.5 – 8.0. The reaction occurs at almost maximum velocity within a wide pH range between 6.5 and 9.0 [8, 9, 11]. With 8-azaguanine as substrate the enzyme shows a sharp pH optimum between 6.0 and 6.3 [10, 12]. Guanase in serum from patients with acute viral hepatitis differs from the human tissue enzyme in apparent activation energy, but not in relative substrate specificity or pH optimum [11, 12]. With 8-azaguanine as substrate values of K_m for guanase from human serum and human liver were similar, being 0.174 and 0.144 mmol/l, respectively [10].

Guanase from rat serum and rat tissue shows the same optimum pH as the human enzyme, both for guanine and 8-azaguanine. Enzymes from guinea pig serum and tissue have a pH optimum of 8.0 for 8-azaguanine and 9.0 for guanine [11, 12].

Relative substrate specificity and/or apparent activation energy vary according to the animal source of the enzyme [11, 12]. Isoenzyme fractions from animal tissues have different dependence on substrate saturation and different activation with Mg^{2+} and GTP [13].

Substrate deamination proceeds linearly with time in all methods for measuring guanase activity.

Methods of determination: guanase activity can be assayed by monitoring the decrease in absorbance at 245 nm as result of guanine deamination [4]. If xanthine oxidase (XOD) is added to the reaction mixture, guanine deamination can be monitored by measuring the increased absorbance at 290 nm due to uric acid formation [2, 3, 6]. Because H_2O_2 is formed during the latter reaction, hydrogen peroxide can be measured in the presence of ethanol and of catalase by a NAD(P)H-coupled reaction through aldehyde dehydrogenase [7] (cf. chapter 3.3.2). The activity can also be assayed by coupling the enzyme with a xanthine oxidase-formazan reaction system [14].

Some methods of guanase determination are based on the assay by *Berthelot's* reaction of ammonia formed by the enzyme in the presence of guanine or 8-aza-guanine as substrate [8, 10, 12, 15]. The release of ammonia can also be monitored in a coupled system containing glutamate dehydrogenase and 2-oxoglutarate by measuring the rate of NADH oxidation at 339 nm [9, 16].

Radiometric assays are also available. The reaction mixture contains radioactive guanine; at the end of the reaction radioactive xanthine is separated from guanine by high voltage electrophoresis [17]. When purified XOD is added to the reaction mixture radioactive uric acid is formed and this compound can be easily separated on a small Dowex-50 cation exchange column [18]. Isoenzymes from rat liver have been separated after polyacrylamide gel electrophoresis [19]. Here we describe the spectrophotometric method of *Kalckar* [12] as modified by *Giusti et al.* [3, 6, 20].

International reference method and standards: neither standardization at the international level nor the existence of reference standard materials is known.

Enzyme effectors: special activators of guanase to be added to the assay mixture are not known. Allantoin and Mg^{2+} inhibit some guanase isoenzymes from rodent tissue [13].

Assay

Method Design

Principle

(a) $$\text{Guanine} + H_2O \xrightarrow{\text{guanase}} \text{xanthine} + NH_3$$

(b) $$\text{Xanthine} + O_2 + H_2O \xrightarrow{\text{XOD*}} \text{urate} + H_2O_2$$

Xanthine oxidase (XOD) is the indicator enzyme. The reaction is followed by spectrophotometric measurements at 290 nm of the formation of uric acid.

Optimized conditions for measurement: with guanine as substrate the enzyme shows a wide pH optimum between pH 6 and 10, with a flat peak between pH 7.5 and 8.0. In studies on human sera *Caraway* [8] and *Quast et al.* [21] observed that the reaction

* XOD, xanthine oxidase, xanthine: oxygen oxidoreductase EC 1.2.3.2.

rate does not increase appreciably when the concentration of guanine is raised from 0.07 to 0.26 mmol/l. Guanine concentrations of 0.53 mmol/l or higher inhibit enzyme activity. About 30 – 80 mU purified XOD must be added to the assay mixture. The buffered guanine solution must be saturated with oxygen. Because of the low guanase activity in normal human sera the reaction should be carried out at 37 °C.

Temperature conversion factors: the temperature conversion factors for human serum and human liver guanase are reported in Tables 1 and 2 as calculated by *Galanti et al.* [11].

Table 1. Temperature conversion factors for serum guanase of patients suffering from acute viral hepatitis.

Reaction temperature, °C	Reference temperature 25 °C	30 °C	37 °C
25	1.000	1.175	1.584
30	0.851	1.000	1.347
37	0.631	0.741	1.000
45	0.478	0.561	0.757
50	0.407	0.478	0.644
55	0.338	0.397	0.535

Table 2. Temperature conversion factors for normal human liver guanase.

Reaction temperature, °C	Reference temperature 25 °C	30 °C	37 °C
25	1.000	1.230	1.697
30	0.813	1.000	1.379
37	0.589	0.724	1.000
45	0.417	0.512	0.707
50	0.339	0.416	0.575
55	0.270	0.332	0.458

Equipment

Spectrophotometer suitable for accurate measurements at 290 nm with constant temperature cuvette holder; water-bath, stopwatch.

Reagents and Solutions

Purity of reagents: all reagents are of analytical grade; for commercial preparations of guanine see Vol. II, chapter 2.2.2; for commercial xanthine oxidase preparations see

Vol. II, chapter 2.2.1. Relative to its activity, XOD should contain less than 0.01% guanase and uricase activity*.

Preparation of solutions (for 150 determinations): prepare all solutions with re-purified water (cf. Vol. II, chapter 2.1.3.2).

1. Guanine stock solution (guanine 1.03 mmol/l):

   dissolve 16 mg guanine in a few ml NaOH, 1 mol/l, and dilute to 100 ml with water.

2. Tris buffer (0.1 mol/l; pH 8.0):

   dissolve 6.055 g Tris in about 200 ml water, add 268 ml HCl, 0.1 mol/l, adjust the pH and dilute to 500 ml with water.

3. Buffered guanine solution (Tris, 90 mmol/l; guanine, 0.103 mmol/l):

   dilute 1 volume guanine stock solution (1) with 9 volume Tris buffer (2). Prepare daily.

4. Xanthine oxidase (1.0 – 1.5 U/ml):

   dilute the commercial preparation with Tris buffer (2) to obtain an activity between 1.0 and 1.5 U/ml (25 °C).

Stability of solutions: store all solutions and suspensions stoppered, at 0 – 4 °C. Prepare fresh guanine stock solution (1) each week. Prepare fresh buffered guanine solution (3) daily. The Tris buffer (2) is stable until bacterial decomposition occurs. Both the commercial XOD stock suspension and XOD dilute suspension (4) are stable for several months.

Procedure

Collection and treatment of specimen: collect blood from a vein. Addition of oxalate (1 – 2 mg/ml), citrate (1 – 1.5 mg/ml), fluoride (2 mg/ml), EDTA (1 – 2 mg/ml) or heparin (0.2 mg/ml) does not interfere with the assay. Centrifuge at approx. 3000 *g* to obtain plasma or serum. If necessary, store samples at 0 – 4 °C. Moderate haemolysis does not affect the results in human serum, as human erythrocytes do not contain guanase. In contrast, haemolysis must be avoided with rat plasma because rat erythrocytes have high guanase activity.

---

* We have always used purified XOD from *Boehringer Mannheim*, Germany, specific activity approx. 0.4 U/mg. This preparation contains less than 0.01% of adenosine deaminase, guanase, and uricase in relation to the specific activity of XOD.

Stability of the enzyme in the sample: radiometric methods, which can detect small decreases of enzyme activity in normal human serum, indicate that serum stored for 6 h at room temperature (21 °C) undergoes a 12% decrease of guanase activity. Enzyme samples can be stored at −20 °C for 2 days without loss of activity and 24−48 h at 4 °C with a loss of activity of approx. 5% [22]. We use only fresh plasma or serum which has been stored for less than 24 h at 4 °C.

Details for measurement in other specimen: for enzyme assays in homogenates it is best to measure the enzyme activity in the supernatant fluid after high-speed centrifugation (15000−20000 *g*; 15−30 min; 0 °C). The samples should be checked for uricase (see below). Changes in absorbance during the reactions due to turbidity can be detected at 320 nm, because the guanase and XOD reactions cause no absorbance change at this wavelength.

Assay conditions: wavelength 290 nm; light path 10 mm; final volume 3.10 ml; temperature 37 °C (thermostatted cuvette holder). Read against a reference cuvette (3.0 ml Tris buffer (2), 0.05 ml sample)*: Before the assay equilibrate the guanine solution (3) to 37 °C and saturate with O_2.

Measurement

Pipette successively into a cuvette:			concentration in assay mixture	
buffered guanine solution	(3)	3.00 ml	guanine	0.1 mmol/l
			Tris	87 mmol/l
dilute XOD suspension	(4)	0.05 ml	XOD	ca. 20 U/l
mix and wait for 5 min				
sample		0.05 ml	volume fraction	0.016
mix, read absorbance and start stopwatch; after exactly 10, 20 and 30 min read absorbance again.				

The values at 290 nm should not exceed $\Delta A/\Delta t$ = 0.300/30 min (ΔA = 0.360 indicates that half the substrate has been consumed). Dilute active serum 2-or 3-fold with isotonic saline and/or read absorbance at shorter intervals. If the activity is low, so that $\Delta A/\Delta t$ is less than 0.001/min, read at 40, 50 and 60 min.

* Using a *Beckman*-DU spectrophotometer or similar instrument with thermostatted holder for 4 cuvettes, 3 determinations may be performed simultaneously against a reference cuvette.

The assay can also be performed as a semi-micro method using a semi-micro cuvette (e.g. *Hellma* 104 QS). The assay system then consists of 1 ml buffered guanine solution (3), 5 µl XOD stock suspension (containing ca. 20 mU of enzyme) and 20 µl serum.

Calculations: according to *Kalckar* [2] the specific absorbance increase for the overall reaction guanine → xanthine → uric acid is $\varepsilon_{290nm} = 0.725\ l \times mmol^{-1} \times mm^{-1}$. Therefore, the catalytic concentration of guanase in the sample is

Macro assay: $b = 8552 \times \Delta A/\Delta t$ U/l

Micro assay: $b = 7070 \times \Delta A/\Delta t$ U/l

Validation of Method

Precision, accuracy, detection limits and sensitivity: by repeating determinations nine times on each of 5 serum samples, a mean coefficient of variation of ±6.22% was found. These assays were made on sera of patients with acute viral hepatitis. Repeated determinations on the same enzyme preparation with 50 U/l* gave values ranging from 45 to 56 U/l. Data on accuracy are not available since standard reference material has not yet been established. The lowest value which can be measured with a sensitive spectrophotometer at 290 nm is approximately 5 U/l with an imprecision ±20%. Sensitivity is $\Delta A/\Delta t = 0.015/30$ min with a *Beckman* DU_2 spectrophotometer with secondary electronic multiplier and constant temperature cuvette holder.

Sources of error: therapeutic doses of allopurinol may interfere with the assay [7]. Effects of other drugs and therapeutic measure are unknown. The assay is not affected by the presence of endogenous XOD. The presence of uricase in the sample prevents quantitative formation of uric acid in the assay mixture. To test for uricase in the sample add 10 µg uric acid in 3 ml Tris buffer (2) and check whether a decrease in absorbance occurs at 290 nm. Uricase activity has been reported in serum of mice with experimental liver necrosis [23]. Guanase activity can be measured in samples containing uricase activity by the method of *Quast et al.* [21], in which the xanthine oxidase is inhibited by borate and the decrease in absorbance at 245 nm is monitored; alternatively, the colorimetric method of *Ellis & Goldberg* [10] can be used.

Specificity: the method is specific. The assay system can be checked by adding purified uricase (urate: oxygen oxidoreductase, EC 1.7.3.3) after the main reaction: the oxidation of the uric acid formed in the main reaction results in a decrease in absorbance at 290 nm.

---

* The assay was performed with serum albumin solutions to which a known amount of purified guanase had been added.

Reference ranges: with the method described we found a 95% reference interval for normal human sera of 0.5 – 4.5 U/l at 37 °C. These figures were calculated from the results for 120 normal subjects. Utilizing the NADH-linked method, *Wolthers et al.* [16] found a mean value of 0.5 ± 0.2 U/l at 37 °C and a similar value was reported by *Heinz et al.* [7] (cf. chapter 3.3.2). *Ellis & Goldberg* gave an upper limit of 2.5 U/l at 37 °C with the spectrophotometric method [9] and a mean value of 1.30 ± 0.35 U/l with the colorimetric method [10]. These results all indicate that normal human serum has a low guanase activity. Radiometric methods, which usually use a low concentration of substrate, gave lower normal values: 0.29 ± 0.22 U/l according to *Kuzmits et al.* [22] and 0.36 ± 0.07 U/l according to *Al-Khalidi* et al. [18].

Giusti & Galanti found a mean value of 20.1 ± 14.5 U/l at 37 °C with the method described here in about 150 patients with acute viral hepatitis. In 13 cases of severe viral hepatitis with coma a mean value of 97.2 ± 55.4 U/l at 37 °C was reported [20].

Normal values or slightly increased values occur in obstructive jaundice, liver cirrhosis, bacterial cholangitis, myocardial infarction, acute pancreatitis and malignant tumours (cf. [3 – 10] for references).

Alternative procedure: a colorimetric method for assaying guanase activity in presence of 8-azaguanine as substrate, is reported at the end of chapter 3.4.2 "Adenosine Deaminase", p. 315.

References

[1] *H. B. Hindman, W. E. Knox,* Guanase in Reticulocytes and other Rat Tissues, Enzyme *23*, 395 – 403 (1978).

[2] *H. H. Kalckar,* Differential Spectrophotometry of Purine Compounds by Means of Specific Enzymes. III. Studies of the Enzymes of Purine Metabolism, J. Biol. Chem. *167*, 461 – 475 (1947).

[3] *G. Giusti,* Guanase, in: *H. U. Bergmeyer* (ed.), Methods in Enzymatic Analysis, 2nd ed., Verlag Chemie, Weinheim, and Academic Press, New York 1974, pp. 1086 – 1091.

[4] *A. C. Hue, A. H. Free,* An Improved Method for the Determination of Guanase in Serum and Plasma, Clin. Chem. *11*, 708 – 715 (1965).

[5] *J. L. Whitehouse, E. M. Knights, C. L. Santos, A. C. Hue,* Clinical Evaluation of Serum Guanase as a Liver Function Test, Abstract. Clin. Chem. *10*, 632 (1964).

[6] *G. Giusti, B. Galanti,* Attività Guanasica del Siero Umano nella Epatite Virale ed in Altre Epatopatie, Boll. Soc. Ital. Biol. Sper. *41*, 1567 – 1571 (1965).

[7] *F. Heinz, S. Reckel, J. R. Kalden,* A New Spectrophotometric Assay for Enzymes of Purine Metabolism. II Determination of Guanase Activity, Enzyme *24*, 247 – 254 (1979).

[8] *W. T. Caraway,* Colorimetric Determination of Serum Guanase Activity, Clin. Chem. *12*, 187 – 193 (1966).

[9] *G. Ellis, D. M. Goldberg,* Kinetic Coupled Optical Assays for Guanase Activity at 340 Nanometers Utilizing Guanine and 8-Azaguanine as Substrates, Biochem. Med. *6*, 380 – 391 (1972).

[10] *G. Ellis, D. M. Goldberg,* Assay of Human Serum and Liver Guanase Activity with 8-Azaguanine as Substrate, Clin. Chim. Acta *37*, 47 – 52 (1972).

[11] *B. Galanti, M. Russo, S. Nardiello, G. Giusti,* Activation Energy, Relative Substrate Specificity and Optimum pH of Guanase from Human, Rat, Mouse and Guinea Pig Sera and Tissues, Enzyme *20*, 90 – 97 (1975).

[12] *B. Galanti, M. Russo, S. Nardiello, G. Giusti,* Further Observations on The Properties of Serum and Tissue Guanase from Man and Some Animal Species, Enzyme *21*, 342 – 348 (1976).

[13] *K. S. Kumar, A. Sitaramayya, P. S. Krishnan,* Modulation of Guanine Deaminase, Biochem. J. *131*, 683–687 (1973).
[14] *J. B. Park, S. W. Kimm,* A New Colorimetric Method for the Determination of Guanine Aminohydrolase Activity, Soul Uitae Chapchi *17*, 136–142 (1976).
[15] *C. Gakis, G. Giusti,* Metodo Colorimetrico per la Determinazation dell'Attività Guanasi del Siero, con Utilizzazione di 8-Azaguanina, come Substrato, Studi Sassaresi *47*, 447–451 (1969).
[16] *B. G. Wolthers, B. J. Nusse, J. Bootsma, A. Groen,* Determination of Serum Guanase by means of a NADH-linked Reaction, Clin. Chim. Acta *41*, 223–230 (1972).
[17] *C. A. Van Bennekom, J. P. Laarhoven, C. H. M. M. De Bruyn, T. L. Oei,* A Simple and Sensitive Radiochemical Assay for Plasma Guanase, J. Clin. Chem. Clin. Biochem. *16*, 245–248 (1978).
[18] *U. A. S. Al-Khalidi, S. Aftimos, S. Musharrafich, N. N. Khuri,* A Method for the Determination of Plasma Guanase on Finger-tip Blood, Clin. Chim. Acta *29*, 381–384 (1970).
[19] *S. W. Kim, B. Y. Lee, H. K. Jung,* A Study on Staining, Determination and Regulatory Mechanism of Guanine Deaminase, Soul Uitae Chapchi *20*, 153–162 (1979).
[20] *G. Giusti, B. Galanti, A. Mancini,* Serum Guanase Activity in Viral Hepatitis and Some Other Hepatic and Extra-hepatic Diseases, Enzymologia *38*, 373–382 (1970).
[21] *N. M. Quast, K. J. Clayson, P. E. Strandjord,* An Improved Method for the Assay of Serum Guanine Deaminase Activity, Am. J. Med. Tech. *34*, 513–521 (1968).
[22] *R. Kuzmits, H. Seyfried, A. Wolf, M. M. Müller,* Evaluation of Serum Guanase in Hepatic Diseases, Enzyme *25*, 148–152 (1980).
[23] *G. Giusti, B. Galanti,* Ricerche su Alcune Deaminasi del Siero in corso di Epatite Virale Umana e Sperimentale, Min. Med. *56-II*, 4448–4452 (1965).

3.3.2 UV-method, 339 nm

Fritz Heinz

General

The enzyme guanase catalyzes the hydrolytic deamination of guanine to xanthine and ammonia, and of 8-azaguanine to 8-azaxanthine and ammonia. Among human tissues, the highest activities are found in liver, kidney, and brain, low activity in intestinal mucosa and only traces in skeletal muscle, spleen and pancreas [1]. Guanase is absent from white and red blood cells [1]. The existence of isoenzymes was reported for brain and liver [2, 3]. Within the cell, guanase activity is found in the nucleus, the light mitochondrial fraction and the soluble fraction [4, 5].

Due to the high activities in the liver and kidney and the absence of guanase from blood cells, determination of the activity in sera may provide diagnostic information about kidney and liver lesions [5–9].

Application of method: in biochemistry and in clinical chemistry.

Enzyme properties relevant in analysis: the activity in sera is due to the existence of high guanase activities in liver. Substrate conversion proceeds linearly with time.

Methods of determination: the following methods have been described for the determination of guanase activity. The change of absorbance in the UV spectrum in the range of 240 – 254 nm [11, 12], due to the deamination of guanine to xanthine or 8-azaguanine to 8-azaxanthine, is measured [12]. In another method xanthine, the product of the guanase reaction, is oxidized to uric acid by xanthine oxidase and this reaction is followed by measuring the increase in optical density at 293 or 290 nm [13]. The other product of the guanase reaction, ammonia, can directly be measured colorimetrically [14] or by coupling with the glutamate dehydrogenase system [15]. In a very sensitive method [1], radioactively-labelled guanine is incubated with tissue extracts, cell extracts or sera for a definite time. Before counting the radioactivity, the product has to be separated chromatographically from the substrate [16].

International reference methods and standards: neither standardization at the international level nor the existence of reference standard materials is known so far.

Enzyme effectors: guanase from rabbit liver may be inhibited by sulphydryl-blocking agents such as PCMB. Strong competitive inhibitors are the purine precursor 5-aminoimidazole-4-carboxamide ($K_I = 3.05 \times 10^{-5}$ mol/l) and the corresponding ribonucleoside (5.68×10^{-4} mol/l). With tetrahydrofolic acid, a competitive inhibitor constant of 2.58×10^{-4} mol/l was determined [17].

Assay

Method Design [18]

Principle

(a) $$\text{Guanine} + H_2O \xrightarrow{\text{guanase}} \text{xanthine} + NH_3$$

(b) $$\text{Xanthine} + H_2O + O_2 \xrightarrow{\text{XOD*}} \text{urate} + H_2O_2$$

(c) $$H_2O_2 + \text{ethanol} \xrightarrow{\text{catalase**}} \text{acetaldehyde} + 2\,H_2O$$

(d) $$\text{Acetaldehyde} + NADP^+ + H_2O \xrightarrow{\text{AlDH***}} \text{acetate} + NADPH + H^+$$

Sum:

(e) $$\text{Guanine} + H_2O + NADP^+ + O_2 \longrightarrow \text{urate} + NH_4^+ + NADPH + \text{acetate}$$

* XOD, Xanthine oxidase, Xanthine: oxygen oxidoreductase, EC 1.2.3.2.
** Catalase, Hydrogenperoxide: hydrogen peroxide oxidoreductase, EC 1.11.1.6.
*** AlDH, Aldehyde dehydrogenase, Aldehyde: NAD(P)$^+$ oxidoreductase, EC 1.2.1.5.

The amount of guanine deaminated per unit time, measured by the increase in absorbance due to the formation of NADPH, is a measure of the catalytic activity of guanase. The equilibrium of the guanine deamination is far in favour of uric acid. Blanks are necessary for routine assays on sera. The following method can be easily adapted to automated equipment.

Optimized conditions for measurement: the coenzyme NADP is used in the following assay, because in sera the re-oxidation rate of NADPH is smaller than that of NADH and aldehyde dehydrogenase from yeast accepts either coenzyme. The addition of uricase to the reaction mixture further oxidizes urate to allantoin, producing an additional mole of H_2O_2 corresponding to another mole of NAD(P)H.

High ethanol concentration is necessary to prevent the normal catalase reaction with two moles of hydrogen peroxide.

Potassium ions are necessary to activate aldehyde dehydrogenase from yeast.

Temperature conversion factors: factors for the conversion of the results determined at 30 °C to another temperature are not reported.

Equipment

Spectrophotometer or spectral-line photometer capable of exact measurements at 339 nm, Hg 334 nm or Hg 365 nm, and with a thermostatted cuvette holder, recorder or stopwatch.

Reagents and Solutions

Purity of reagents: in the commercially available enzymes aldehyde dehydrogenase from yeast, catalase and if necessary uricase, the NADPH oxidation activities together should not exceed 0.01% of the activity of aldehyde dehydrogenase; NADP-dependent alcohol dehydrogenase should be absent. The quality of commercially available NADP and guanine is sufficient for good results. Other chemicals should be of analytical grade.

Preparation of solutions: all solutions in re-purified water (cf. Vol. II, chapter 2.1.3.2).

1. Tris buffer (Tris, 0.1 mol/l, KCl, 0.1 mol/l; pH 8.0):

   dissolve 12.11 g Tris base and 7.45 g KCl in approx. 700 ml water and adjust the pH to 8.0 by the addition of HCl, 2 mol/l, make up with water to 1000 ml.

2. Guanine solution (6.6 mmol/l):

   dissolve 1 mg guanine in NaOH, 0.1 mol/l.

3. Aldehyde dehydrogenase (50 kU/l):

   dissolve an equivalent to 50 U of the lyophilized powder (*Boehringer Mannheim* 171832) in 1 ml cold water.

4. Xanthine oxidase (4 kU/l):

   use a preparation of 10 mg/ml, ca. 0.4 U/mg (*Boehringer Mannheim* 110442) without further dilution.

5. Working solution (for about 20 determinations):

   to 18 ml Tris buffer (1) add 0.04 ml catalase (*Boehringer Mannheim* 106810; 20 mg/ml, ca. 52 kU), 10 mg disodium NADP and 2 ml ethanol. NADP can be replaced by NAD.

Stability of solutions: store all solutions at 4 °C. The working solution is stable for two days. Prepare the guanine solution freshly every week.

Procedure

Collection and treatment of specimen: collect blood from the vein. Addition of oxalate, citrate, fluoride or heparin is unobjectionable. Centrifuge for 10 min at about 3000 *g* in order to obtain plasma or serum. In sera of healthy persons the activity is low (0.5 – 1 U/l). In sera of patients suffering from acute hepatitis and chronic alcoholism elevated enzyme levels can be detected. Reduced activities are found in patients with chronic kidney diseases.

Stability of enzymes in the sample: sera may be stored for 3 days at 4 °C without change in activity.

Details for measurement in tissues: 1 g of tissue is added to 9 ml triethanolamine buffer (0.05 mol/l, pH 7.5). The mixture is homogenized and centrifuged at 100000 *g* for 30 min at 4 °C.

Assay conditions: wavelength 339 (Hg 334, Hg 365) nm; light path 10 mm; final volume 1.7 ml; 30 °C (thermostatted cuvette holder). Measure against air. A sample blank with water instead of guanine solution (2) is run simultaneously. Before starting the assay, adjust the temperature of the solutions to 30 °C.

Measurement

Pipette successively into a semi-micro cuvette:			concentration in the assay mixture	
working solution	(5)	1.00 ml	ethanol	1.01 mol/l
			Tris	53 mmol/l
			KCl	53 mmol/l
			NADP	0.37 mmol/l
			catalase	1530 kU/l
aldehyde dehydrogenase	(3)	0.05 ml	AlDH	1.47 kU/l
xanthine oxidase	(4)	0.05 ml	XOD	118 U/l
sample		0.5 ml	volume fraction	0.294
mix thoroughly with a plastic spatula; wait for 5 min, until a linear change in absorbance can be registered (blank),				
guanine solution	(2)	0.1 ml	guanine	0.388 mmol/l
mix, wait for 4 min till the change in absorbance becomes linear, then read the absorbance, repeat the reading after exactly 5 or 10 min or monitor the reaction on the recorder.				

The value $\Delta A/\Delta t$ must be <0.050/min at Hg 365 nm (<0.06/min at Hg 334 nm or 339 nm). Otherwise dilute samples with Tris buffer, 0.05 mol/l, pH 8.0.

Calculation: correct readings of sample for readings of blank, yielding $\Delta A/\Delta t$; use calculation formula (k) or (k_1) resp. (cf. "Formulae", Appendix 3). For values of ε cf. Appendix 4. The following relations are valid for the catalytic concentration of the sample (t in minutes):

	Hg 334 nm	339 nm	Hg 365 nm	
$b =$	$550 \times \Delta A/\Delta t$	$540 \times \Delta A/\Delta t$	$972 \times \Delta A/\Delta t$	U/l
$b =$	$9.17 \times 10^3 \times \Delta A/\Delta t$	$9.0 \times 10^3 \times \Delta A/\Delta t$	$16.2 \times 10^3 \times \Delta A/\Delta t$	nkat/l

Validation of Method

Precision, accuracy, detection limits and sensitivity: with an enzyme solution containing a mean of 54.3 U/l (n = 10) the standard deviation was calculated to be 0.683 and the coefficient of variation 1.26%. The lowest activity which could be detected at Hg 334 nm was approximately 0.2 U/l sample, when 0.5 ml of sample was used. Sensitivity was found to be $\Delta A/\Delta t = 0.01/10$ min (*Eppendorf* photometer, Hg 334 nm).

Sources of error: factors which influence the measured activities are unknown, as described by *Haeckel* [19] for the determination of uric acid, which uses the same indicator system.

Specificity: guanase activity is measured specifically since the reaction is started with the substrate and interfering NAD(P)H producing processes are eliminated by the blank.

Reference ranges: only few results have been published. Approx. 0.5 – 1 U per l sera were detected in healthy persons. In rat tissues, we found 1 U/g liver, 0.75 U/g brain, 0.42 U/g kidney, 0.36 U/g heart, and 0.11 U/g muscle.

References

[1] *R. Levine, T. C. Hall, C. A. Harris,* Guanase Activity in Normal and Neoplastic Human Tissue, Cancer *16*, 269 – 272 (1963).
[2] *K. S. Kumar, P. S. Krishnan,* Further Studies on the Isozymic Forms of Soluble Rat Liver Guanine Deaminase, Biochem. Biophys. Res. Commun. *39*, 600 – 608 (1970).
[3] *K. S. Kumar, A. Sitaramayya, P. S. Krishnan,* Guanine Deaminase in Rat Liver and Mouse Liver and Brain, Biochem. J. *128*, 1079 – 1088 (1972).
[4] *S. Kumar, K. K. Tewazi, P. S. Krishnan,* Solubilization and Partial Purification of Particulate Guanine Deaminase from Rat Brain, J. Neurochem. *13*, 1550 – 1552 (1966).
[5] *J. L. Withehouse, E. M. Knights jr., C. L. Santos, A. C. Hue,* Clinical Evaluation of Serum Guanase as a Liver Function Test, Clin. Chem. *10*, 632 (1964).
[6] *A. Bel, R. Dietsch, A. Alary, B. Savoye, R. Levart, J. Nesmoz, M. Nyssen,* La Guanine Déaminase Enzyme de Cytolyse Hépatique. Sa Valeur Diagnostique à Propos de 500 Malades, Presse Méd. *78*, 495 – 499 (1970).
[7] *G. Giusti, B. Galanti, A. Mancini,* Serum Guanase Activity in Viral Hepatitis and Some Other Hepatic and Extra-Hepatic Diseases, Enzymologia *38*, 373 – 382 (1970).
[8] *B. Kramp, W. P. Teichmann, W. Rehpenning,* Über die Bestimmung der Serumguanaseaktivität bei Leberparenchymerkrankungen, Dtsch. Z. Verdau.-Stoffwechs.-Krankh. *33*, 11 – 15 (1973).
[9] *G. Y. Kebejian, U. A. Al-Khalidi,* Serum Guanase in Kidney Diseases, Eur. J. Clin. Invest. *3*, 41 – 43 (1973).
[10] *A. C. Hue, A. H. Free,* An Improved Method for the Determination of Guanase in Serum or Plasma, Clin. Chem. *11*, 708 – 715 (1965).
[11] *H. M. Kalckar,* Differential Spectrophotometry of Purine Compounds by Means of Specific Enzymes. III. Studie of Enzymes of Purine Metabolism, J. Biol. Chem. *167*, 461 – 475 (1947).
[12] *A. Roush, E. R. Norris,* Deamination of Azaguanine by Guanase, Arch. Biochem. Biophys. *29*, 124 – 129 (1950).
[13] *H. M. Kalckar,* Differential Spectrophotometry of Purine Compounds by Means of Specific Enzymes. I. Determination of Hydroxypurine Compounds, J. Biol. Chem. *167*, 429 – 443 (1947).
[14] *W. T. Caraway,* Colorimetric Determination of Serum Guanase Activity, Clin. Chem. *12*, 187 – 193 (1966).
[15] *B. G. Wolters, B. J. Nusse, J. Bootsma, A. Groen,* Determination of Serum Guanase by Means of a NADH Linked Reaction, Clin. Chim. Acta *4*, 223 – 230 (1972).
[16] *U. A. Al-Khalidi, S. Aftimus, S. Musharrafieh, N. N. Khuri,* A Method for the Determination of Plasma Guanase in Finger-Tip Blood, Clin. Chim. Acta *29*, 381 – 384 (1970).
[17] *M. D. Glantz, A. S. Lewis,* Guanine Deaminase from Rabbit Liver, in: *P. A. Hoffee, M. E. Jones* (eds.), Methods in Enzymology, Vol. *LI*, Academic Press, New York 1978, pp. 512 – 517.
[18] *F. Heinz, S. Reckel, J. R. Kalden,* A New Spectrophotometric Assay for Enzymes of Purine Metabolism. II. Determination of Guanase Activity, Enzyme *24*, 247 – 254 (1979).
[19] *R. Haeckel,* The Use of Aldehyde Dehydrogenase to Determine H_2O_2-Producing Reactions. I. The Determination of the Uric Acid Concentration, J. Clin. Chem. Clin. Biochem. *14*, 101 – 107 (1976).

3.4 Adenosine Deaminase

Adenosine aminohydrolase, EC 3.5.4.4

3.4.1 UV-method

Fritz Heinz

General

Adenosine deaminase is an enzyme of the so-called purine salvage pathway that catalyzes the hydrolytic cleavage of adenosine to inosine and ammonia. Besides the naturally occurring 6-aminopurine, ribo- and 2′-deoxyribonucleosides, various synthetic derivates possessing a 6-aminopurine configuration are deaminated [1].

Adenosine deaminase is present in a great number of mammalian tissues, with the highest activities in the intestinal mucosa [2]. The enzyme is predominantly localized in the cytosolic fraction of the cells [3]. Among blood cells, adenosine deaminase activity is found in erythrocytes, lymphocytes and granulocytes [4].

Different isoenzyme patterns have been described for red blood cells and tissues [5]. Elevated enzyme levels can be detected in sera of patients suffering from liver cirrhosis [6]. Adenosine deaminase deficiency was found to be associated with severe impairment of immunity. Since the immune response is altered in many forms of neoplasia or in immune-suppressed patients [7–9], there is considerable interest in this enzyme at present.

Application of method: in biochemistry and clinical chemistry.

Enzyme properties relevant in analysis: the catalytic activity of serum is due to the existence of different isoenzymes [5] of different origins.

Methods of determination: the most common assay methods make use of the change in optical density in the UV range. At 265 nm adenosine and inosine possess different absorption coefficients, which may be used for the determination of adenosine deaminase activity [10]. However, the absorbance of adenosine at 265 nm is high and for this reason it is difficult to reach saturation conditions [11]. In combination with the nucleoside phosphorylase and xanthine oxidase reactions, the adenine part of adenosine can be converted to uric acid, which shows a maximum of absorbance at 293 nm [12]. However, high concentrations of protein, e.g. in serum, which absorb at this wavelength, may interfere.

In other assay systems the appearance of ammonia can be measured by means of the *Berthelot*-reaction [13], ammonia-sensitive electrodes [14] or the NADH-dependent glutamate dehydrogenase reaction [15].

Methods which utilize ^{14}C-labelled adenosine offer the greatest detectability. After a definite incubation time the reaction must be terminated and the substrate and the product, inosine, together with further degradation products e.g. xanthine, have to be separated by chromatography [3]. Although they might show lower detection limits, methods utilizing radioactive materials can only be applied in routine laboratories with radiological safety precautions and they are not easily automated. The conventional UV-method we use overcomes the difficulties associated with measurements at 293 and 265 nm.

International reference methods and standards: neither standardization at the international level nor the existence of reference standard materials is known so far.

Enzyme effectors: specific activators of adenosine deaminase which need to be added to the assay mixture are not known.

Assay

Method Design [5]

Principle

(a) $$\text{Adenosine} + H_2O \xrightarrow{\text{adenosine deaminase}} \text{inosine} + NH_3$$

(b) $$\text{Inosine} + P_i \xrightarrow{\text{NP*}} \text{hypoxanthine} + \text{ribose-1-P}$$

(c) $$\text{Hypoxanthine} + 2\,H_2O + 2\,O_2 \xrightarrow{\text{XOD**}} \text{urate} + 2\,H_2O_2$$

(d) $$2\,H_2O_2 + 2\,\text{ethanol} \xrightarrow{\text{catalase}^+} 2\,\text{acetaldehyde} + 4\,H_2O$$

(e) $$2\,\text{Acetaldehyde} + 2\,\text{NAD(P)}^+ + 2\,H_2O \xrightarrow{\text{AlDH}^{++}} 2\,\text{acetate} + 2\,\text{NAD(P)H} + 2\,H^+$$

Sum:

$$\text{Adenosine} + P_i + 2\,O_2 + 2\,\text{NAD(P)}^+ + 2\,\text{ethanol} + H_2O \longrightarrow \text{urate} + NH_4^+ + \text{ribose-1-P} + 2\,\text{acetate} + 2\,\text{NAD(P)H} + H^+$$

---

* NP, Nucleoside phosphorylase, EC 2.4.2.1.

** XOD, Xanthine oxidase, EC 1.2.3.2.

$^+$ EC 1.11.1.6.

$^{++}$ AlDH, Aldehyde dehydrogenase, EC 1.2.1.5.

The amount of adenosine hydrolyzed per min detected by the increase of absorbance due to the production of NAD(P)H is a measure of the catalytic activity of adenosine deaminase and corresponds to the formation of 2 moles of NAD(P)H. The equilibrium of the complete reaction sequence is in favour of acetate production, because the reactions catalyzed by adenosine deaminase, catalase, xanthine oxidase and aldehyde dehydrogenase are practically irreversible under the test conditions. Blanks are necessary for routine assays. The following method can be easily adapted to automated equipment.

Optimized conditions of measurement: NAD is used in the following assay; however, it can be replaced by NADP because aldehyde dehydrogenase from yeast accepts either coenzyme. The high ethanol concentration is necessary to prevent the normal catalase reaction with two moles of hydrogen peroxide. The K^+ ions are activators for the aldehyde dehydrogenase from yeast.

The sensitivity of the determination can be increased if uric acid is further oxidized to allantoin by uricase.

Temperature conversion factors: factors for the conversion of the results determined at 30 °C to another temperature have not been reported.

Equipment

Spectrophotometer or spectral-line photometer capable of exact measurement at 339 nm or Hg 334 nm, and with a thermostatted cuvette holder; recorder or stopwatch.

Reagents and Solutions

Purity of reagents: in the commercially-available enzymes aldehyde dehydrogenase from yeast, catalase and the specially prepared nucleoside phosphorylase, the NADH oxidation activities together should not exceed 0.01% of that of aldehyde dehydrogenase. NAD-dependent alcohol dehydrogenase must be absent. The nucleoside phosphorylase must be of the inosine-guanosine-specific type, because adenosine-specific enzyme preparations split adenosine to adenine and ribose 1-phosphate. Adenine is further oxidized by xanthine oxidase to 2,8-dihydroxyadenine, producing hydrogen peroxide. If a commercially-available nucleoside phosphorylase is used, it must be free from adenosine deaminase activity. In our hands commercially-available nucleoside phosphorylase preparations contain adenosine deaminase activity; therefore, we isolate the inosine-guanosine-specific enzyme from rabbit liver according to *Lewis & Glantz* [16] to step 5 (cf. Appendix, p. 313).

NAD and adenosine are of the highest commercially-available purities. The other chemicals are of analytical grade.

Preparation of solutions: all solutions in re-purified water (cf. Vol. II, chapter 2.1.3.2).

1. Phosphate buffer (0.1 mol/l, pH 7.5):

   dissolve 19.171 g $K_2HPO_4 \cdot 3\ H_2O$ and 2.177 g KH_2PO_4 in water to 1000 ml.

2. Nucleoside phosphorylase (10 U/ml):

   an inosine-guanosine-specific nucleoside phosphorylase isolated from rabbit liver to step 5 according to *Lewis & Glantz* [16]. Use dialyzed solution according to Appendix, p. 314.

3. Adenosine (18.7 mmol/l):

   dissolve 5 mg adenosine in 1 ml phosphate buffer (1).

4. Aldehyde dehydrogenase (ca. 50 U/ml):

   dissolve an equivalent to 50 U of the lyophilized powder (*Boehringer Mannheim* 171832) in 1 ml cold water.

5. Xanthine oxidase (10 mg/ml, ca. 4 U/ml):

   the commercially-available enzyme preparation (*Boehringer Mannheim* 110434) was used without further dilution.

6. Catalase (20 mg/ml, ca. 1.3×10^6 U/ml):

   a commercially-available enzyme preparation (*Boehringer Mannheim* 106810) was used without further dilution.

7. Working solution (for about 10 determinations):

   to 9 ml phosphate buffer (1) add 0.02 ml catalase (6), 1 ml ethanol and 5 mg NAD (*Boehringer Mannheim* 127990).

Stability for solutions: store all solutions at 0°C to 4°C. The working solution is stable at least for two days, the nucleoside phosphorylase preparation for 6 weeks and the aldehyde dehydrogenase solution for at least two days.

Procedure

Collection and treatment of specimen: collect blood from the vein. Addition of oxalate, citrate, fluoride or heparin is without effect.

Centrifuge for 10 min at about 3000 *g* in order to obtain plasma or serum.

For the determination of adenosine deaminase activity in patients suffering from immune deficiency, dilute 0.1 ml blood with 0.9 ml phosphate buffer (1) saturated with digitonin to lyse the cells. For the determination of adenosine deaminase in whole blood, the addition of nucleoside phosphorylase can be omitted, because red blood cells contain high activities (3000 U/l) of an inosine-guanosine-specific nucleoside phosphorylase.

Stability of the enzyme in the samples: the enzyme is stable in human serum for 24 h at 25 °C and at least for one week at 4 °C. Whole blood samples may be stored for 2 weeks at 4 °C. A loss in activity could not be detected during one week at 20 °C.

Details for measurement in tissues: in tissues, adenosine deaminase is localized in the soluble part of the cell and may be extracted as usual.

Assay conditions: wavelength 339 (Hg 334, Hg 365) nm; light path 10 mm; final volume 1.32 ml; 30 °C (thermostatted cuvette holder). Before starting the assay, adjust the temperature to 30 °C.

The values of $\Delta A/\Delta t$ must be <0.05/min at Hg 365 nm (<0.1/min at Hg 334 nm or 339 nm). Otherwise, dilute the samples with phosphate buffer (1) pH 7.5.

Measurement

Pipette successively into a semi-micro cuvette:			concentration in the assay mixture	
working solution	(7)	1.0 ml	phosphate ethanol NAD catalase	68.2 mmol/l 1.31 mol/l 0.57 mmol/l 1.97×10^6 U/l
AlDH		0.05 ml	AlDH	1890 U/l
XOD		0.02 ml	XOD	61 U/l
NP[a]		0.05 ml	NP	380 U/l
sample		0.1 ml	volume fraction	0.076
mix thoroughly with a plastic spatula; wait for ca. 5 min, until a linear change in absorbance is registered (blank);				
adenosine		0.1 ml	adenosine	1.42 mmol/l
mix, wait 2 min until the change in absorbance becomes linear, then read the absorbance, repeat the reading after exactly 5 and 10 min or monitor the reaction on the recorder.				

[a] For the whole lysed blood NP can be obmitted, because red blood cells contain high activities.

Calculation: use calculation formula (k) or (k_1), respectively (cf. "Formulae", Appendix 3). Values for ε are given in Appendix 4. The following relations are valid (Δt in minutes):

	Hg 334 nm	339 nm	Hg 365 nm	
$b =$	$1068 \times \Delta A/\Delta t$	$1048 \times \Delta A/\Delta t$	$1941 \times \Delta A/\Delta t$	U/l
$b =$	$1.79 \times 10^4 \Delta A/\Delta t$	$1.75 \times 10^4 \times \Delta A/\Delta t$	$3.23 \times 10^4 \times \Delta A/\Delta t$	nkat/l

Validation of Method

Precision, accuracy, detection limits and sensitivity: for an enzyme solution (cell extract) containing a mean of 35 U/l ($n = 10$) the standard deviation was 0.72 and the coefficient of variation was 1.5%. For assays at 334 nm Hg the lowest value which can be determined is approximately 0.2 U/l sample, if 0.5 ml of sample is used. Sensitivity is found to be $\Delta A/\Delta t = 0.001/10$ min (*Eppendorf* photometer, $\lambda = 334$ nm).

Source of error: influences of therapeutics on the measured activities are unknown, as described by *Haeckel* [17] for the determination of uric acid, which uses the same indicator system. Nucleoside phosphorylases, which split adenosine to adenine and ribose-1-phosphate, or nucleoside hydrolases, which split adenosine to adenine and ribose, may interfere with the method, because adenine is oxidized by xanthine oxidase to 2,8-dihydroxyadenine, producing hydrogen peroxide. The presence of the two enzymes, which we could not detect in animal tissues, can be demonstrated if the extracts are incubated for 24 h with adenosine in the presence of xanthine oxidase. After deproteinization the appearance of 2,8-dihydroxyadenine can be shown by its typical absorption spectrum.

Specificity: the adenosine deaminase activity is measured specifically since the reaction is started with the substrate and interfering NADH-producing processes are eliminated by the blank.

Reference ranges: only limited results have been published. Approx. 1 U per litre serum can be detected. For the whole blood of normal children, a mean value of 93.1 $\pm$ 20.1 ($n = 35$) U/l, corresponding to 0.695 $\pm$ 0.165 U per g haemoglobin, was found, principally due to the adenosine deaminase present in the red blood cells.

Appendix

Purification of purine nucleoside phosphorylase [16]

Homogenize 500 g thawed rabbit livers at 4°C in 2 l phosphate buffer (100 mmol/l; pH 7.0, containing 2-mercaptoethanol, 20 mmol/l). Centrifuge; add 209 g ammonium sulphate per litre supernatant. Clear the suspension by centrifugation, and precipitate the enzyme by addition of 238 g ammonium sulphate per litre. Dialyze against 2-mercaptoethanol (10 mmol/l), remove the precipitate by centrifugation and discard. Adjust pH of the supernatant to 7.5 with Tris base solution (2 mol/l), then adsorb the enzyme on a column of DE 52 (3.6 × 36 cm) equilibrated with phosphate buffer (50 mmol/l, pH 7.5, containing 2-mercaptoethanol, 10 mmol/l). Wash the column with the same buffer, eluate the nucleoside phosphorylase with citrate buffer (50 mmol/l, pH 6.8, containing 2-mercaptoethanol, 10 mmol/l). Pool the enzyme-containing fractions and precipitate the phosphorylase with ammonium sulphate as described above.

All phosphate buffers used in the following steps contain 2-mercaptoethanol, 10 mmol/l. pH is 7.5.

Dissolve the enzyme in phosphate buffer (50 mmol/l), dialyze against phosphate buffer (10 mmol/l). Centrifuge, adjust pH to 7.5. Adsorb the phosphorylase on a column of hydroxylapatite (2.6 × 30 cm) equilibrated with phosphate buffer (40 mmol/l). Wash the column and eluate the enzyme with phosphate buffer (100 mmol/l). Collect the fractions with phosphorylase activity and precipitate the enzyme by ammonium sulphate (561 g/l). Dissolve the precipitate in phosphate buffer (100 mmol/l), and dialyze against the same buffer.

The dialyzed solution can be stored at 4 °C for approx. 2 weeks without loss in activity.

References

[1] *C. L. Zielke, C. H. Suelter,* Purine, Purine Nucleoside, and Purine Nucleoside Aminohydrolases, in: *P. D. Boyer* (ed.), The Enzymes (third edition), Vol. *IV*, Academic Press, New York 1971, pp. 47–77.

[2] *J. E. Seegmiller, L. Thompson, H. Bluestein, R. Willis, S. Matsumoto, D. Carlson,* Nucleotide and Nucleoside Metabolism and Lymphocyte Function, in: *E. W. Gelfand, H. M. Dosch* (eds.), Biological Basis of Immunodeficiency, Raven Press, New York 1980, pp. 251–268.

[3] *M. B. van der Weyden, L. Bailey,* A Micromethod for Determinating Adenosine Deaminase and Purine Nucleoside Phosphorylase Activity in Cells from Human Peripheral Blood, Clin. Chim. Acta *82*, 179–184 (1978).

[4] *J. Meier, M. S. Coleman, J. J. Huttan,* Adenosine Deaminase Activity in Peripheral Blood Cells of Patients with Haematological Malignancies, Br. J. Cancer *33*, 312–319 (1976).

[5] *R. Hirschhorn,* Isozymes of Adenosine Deaminase, in: *C. L. Markert* (ed.), Isozymes, Vol. *II*, Academic Press, New York 1975, pp. 583–599.

[6] *G. C. Secchi, A. Rezzonico, N. Gervasini,* Adenosine Deaminase Activity in Liver Diseases, Enzymol. Biol. Clin. *8*, 67–72 (1967).

[7] *G. Dinescu-Romalo, C. Mihia, L. Vlad,* Decrease of Adenosine Activity in the Total Blood from Tumor Patients, Revue Roum. Biochem. *14*, 161–165 (1977).

[8] *B. Keogh, J. Pauly, G. Tritsch, A. Mittelman, G. P. Murphy,* Adenosine-Aminohydrolase Activity in the Erythrocytes, Lymphocytes, and Plasma of Healthy Subjects and Kidney Transplant Recipients, J. Surg. Oncol. *8*, 417–424 (1976).

[9] *H. M. Kalckar,* Differential Spectrophotometry of Purine Compounds by Means of Specific Enzymes. III. Studies on Enzymes of Purine Metabolism, J. Biol. Chem. *167*, 461–475 (1947).

[10] *H. Klenow,* The Enzymic Oxidation and Assay of Adenine, Biochem. J. *50*, 404–407 (1952).

[11] *W. Koerber, E. B. Meisterernst, G. Herrmann,* Quantitative Measurement of Adenosine Deaminase from Human Erythrocytes, Clin. Chim. Acta *63*, 323–333 (1975).
[12] *R. G. Martinek,* Micromethod for the Estimation of Serum Adenosine Deaminase, Clin. Chem. *9*, 620–625 (1963).
[13] *O. E. Hjemdahl-Monsen, D. S. Papastathopoulos, G. A. Rechnitz,* Automated Adenosine Deaminase Enzyme Determination with an Ammoniasensing Membrane Electrode, Anal. Chim. Acta *88*, 253–259 (1977).
[14] *G. Ellis, D. M. Goldberg,* A NADH Linked Kinetic Assay for Adenosine Deaminase EC 3.5.4.4. Activity, J. Lab. Clin. Med. *76*, 507–517 (1970).
[15] *F. Heinz, S. Reckel, R. Pilz, J. R. Kalden,* A New Spectrophotometric Assay for Enzymes of Purine Metabolism. IV. Determination of Adenosine Deaminase, Enzyme *25*, 50–55 (1980).
[16] *A. S. Lewis, M. D. Glantz,* Monomeric Purine Nucleoside Phosphorylase from Rabbit Liver. Purification and Characterization, J. Biol. Chem. *251*, 407–413 (1976).
[17] *R. Haeckel,* The Use of Aldehyde Dehydrogenase to Determine H_2O_2-Producing Reactions. I. The Determination of the Uric Acid Concentration, J. Clin. Chem. Clin. Biochem. *14*, 101–107 (1976).

3.4.2 Colorimetric Method

Giuseppe Giusti and Bruno Galanti

General

Adenosine deaminase, ADA, specifically reacts with adenosine and several adenine nucleoside analogues [1]. The enzyme is widely distributed in animal and human tissues. The highest activity is found in intestinal mucosa and spleen; lower activity is found in liver, skeletal muscle, kidney and serum [2, 3]. Activity is higher in lymphocytes and leukocytes than in erythrocytes [4]. The enzyme is present in the cytoplasmic fraction and a certain amount is located in the nucleus. ADA has been purified from calf intestine [1] and from human erythrocytes [5]. Commercial preparations are available (cf. Vol. II, Chapter 2.2.1). Human serum ADA increases in acute viral hepatitis and active cirrhosis and only to a much lesser extent in other hepatic diseases [6, 7]. Very high serum ADA levels have been observed in typhoid fever [3, 8]. Increased serum ADA has been found in tumour-bearing and leukaemic patients (cf. [3] for references). Lymphocyte ADA levels can be considered a parameter of immune response; in fact, lymphocyte ADA activity increases after *in vitro* and *in vivo* stimulation and kidney allograft rejection can be forecast by a lymphocyte ADA increase [9]. Lymphocyte ADA activity is decreased in tumour patients [10]. ADA is decreased in erythrocytes from patients with severe combined immunodeficiency, while heterozygous carriers of this autosomal recessive defect have half the normal enzyme activity (cf. [11] for references).

Application of method: in biochemistry and clinical chemistry.

Enzyme properties relevant in analysis: K_m values of human enzymes are: 0.039 mmol/l in erythrocytes, 0.044–0.046 mmol/l in leukocytes, liver and kidney, 0.060 mmol/l in spleen. A much higher value (2.031 mmol/l) is found in normal human serum [12]. Optimum pH ranges of human serum ADA and tissue ADA are 5.5 to 6.5 and 6.5 to 8.5, respectively. Human serum ADA also differs from human tissue enzymes in activation energy and relative substrate specificity [13]. Amphibians, birds and mammals have multiple forms of ADA. Some human tissues display as many as four molecular species of ADA (cf. [5] for references). All methods for measuring ADA activity indicate that deamination of substrate proceeds linearly with time.

Methods of determination: ADA activity can be determined by various methods. The reaction can be followed by measuring the fall in absorbance at 265 nm accompanying the conversion of adenosine to inosine [1]. Inosine may be converted by nucleoside phosphorylase and xanthine oxidase to uric acid with increased absorbance at 293 nm [14]. Since H_2O_2 is formed during the latter reaction, ADA activity may be determined by the assay of hydrogen peroxide by a NAD(P)H-coupled reaction by means of catalase and aldehyde dehydrogenase in the presence of ethanol [15].

The ammonia produced can be assayed by a colorimetric method [3, 16]. Alternatively, ammonia production can be coupled by means of glutamate dehydrogenase to NADH oxidation so that the resultant fall in absorbance at 339 nm can be monitored [12].

ADA activity can be also measured by radiometric assays with radiolabelled substrate; adenosine and the product inosine are separated by thin-layer, paper or column chromatography or other techniques (cf. [4, 17] for references).
Qualitative screening tests for measuring ADA activity in dried blood spots on filter paper have also been described [18].

The isoenzyme pattern in extracts of red cells, white cells, and lymphoid cells can be investigated using starch-gel or cellulose acetate electrophoresis, thin-layer or ion-exchange chromatography, gel filtration and other techniques (cf. [19] for references).

The simple, sensitive colorimetric method of *Galanti & Giusti* [3, 14] is described here. This method, which uses adenosine, 20 mmol/l, in the assay mixture ensures optimized conditions for measuring ADA activity from all known sources.

International reference method and standards: neither standardization at the international level nor the existence of reference standard materials is known.

Enzyme effectors: specific activators of adenosine deaminase required in the assay mixture are not known. Cu^{2+}, Ag^{+}, Cr^{3+} and to a lesser extent Zn^{2+}, Co^{2+} and Ni^{2+} are inhibitors. Several adenosine analogues inhibit ADA activity to various extents. ATP, ADP, AMP and cyclic AMP were found to be competitive inhibitors of ADA isozymes from normal human erythrocytes [20].

Assay

Method Design

Principle

(a) $\text{Adenosine} + H_2O \xrightarrow{\text{ADA}} \text{inosine} + NH_3$

(b) $NH_3 + ClO^- \longrightarrow NH_2Cl + OH^-$

(c) $NH_2Cl + C_6H_5\text{-}OH + OH^- \longrightarrow H_2N\text{-}C_6H_4\text{-}O^- + Cl^- + H_2O$

(d) $H_2N\text{-}C_6H_4\text{-}O^- + C_6H_5\text{-}OH + 1/2\ O_2 \longrightarrow O{=}C_6H_4{=}N\text{-}C_6H_4\text{-}O^- + H_2O$

The equilibrium of the reaction is far to the right. Ammonia is determined by the *Chaney & Marbach* modification of *Berthelot's* reaction. Ammonia forms an intensely blue indophenol with sodium hypochlorite and phenol in alkaline solution. Sodium nitroprusside is the catalyst. The ammonia concentration is directly proportional to the absorbance of the indophenol. The reaction catalyzed by ADA is stopped at the end of the incubation period by the addition of the phenol nitroprusside solution (16).

Optimized conditions for measurement: the optimum conditions for measurement [3, 12, 16] are adenosine, 20 mmol/l, phosphate buffer (ionic strength approx. 0.10) and a pH range between 6.2 and 6.8.

Temperature conversion factors: the temperature conversion factors for human serum and calf-intestine ADA as calculated by *Giusti & Gakis* [13] are reported in tables 1 and 2.

Table 1. Temperature conversion factors for human serum adenosine deaminase.

Reaction temperature, °C	Standard temperature		
	25 °C	30 °C	37 °C
25	1.000	1.305	1.856
30	0.766	1.000	1.422
35	0.595	0.776	1.104
37	0.538	0.702	1.000
45	0.367	0.479	0.681
50	0.292	0.381	0.542
55	0.233	0.304	0.432

Table 2. Temperature conversion factors for adenosine deaminase highly purified from calf intestine (suspension in ammonium sulphate, 2.8 mol/l, 2 mg/ml).

Reaction temperature, °C	Standard temperature 25 °C	30 °C	37 °C
25	1.000	1.181	1.468
30	0.847	1.000	1.243
35	0.726	0.857	1.066
37	0.681	0.804	1.000
45	0.535	0.632	0.785
50	0.463	0.547	0.680
55	0.402	0.475	0.590

Equipment

Spectrophotometer, spectral-line photometer or simple photometer (with tungsten lamp and filter) suitable for accurate measurements at wavelengths between 620 and 650 nm; water-bath (37 °C).

Reagents and Solutions

Purity of reagents: all reagents are of analytical grade. Solutions must be prepared with doubly distilled, ammonia-free water. Ammonia can be removed by addition of a little H_2SO_4 and $KMnO_4$ and a second distillation from a glass apparatus. This precaution is necessary if the ammonia content of the tap water is high.

Preparation of solutions (for about 20 determinations): prepare all solutions with re-purified water (cf. Vol. II, chapter 2.1.3.2).

1. Phosphate buffer (50 mmol/l; pH 6.5):

   dissolve 4.73 g $NaH_2PO_4 \cdot H_2O$ and 5.62 g $Na_2HPO_4 \cdot 12\ H_2O$ in water and dilute to 1000 ml with boiled water.

2. Buffered adenosine solution (adenosine, 21 mmol/l, phosphate, 50 mmol/l; pH 6.5):

   add ca. 30 ml phosphate buffer (1) to 280 mg adenosine in a 50 ml volumetric flask, warm in a hot water-bath and cool under running tap water. Dilute to 50 ml with phosphate buffer (1).

3. Ammonium sulphate stock solution (15 mmol/l):

   dissolve 1.982 g anhydrous ammonium sulphate in water, make up to 1000 ml and mix thoroughly.

4. Ammonium sulphate standard solution (75 µmol/l; NH_3, 0.15 mval/l):

   dilute 0.5 ml ammonium sulphate stock solution (3) (precision pipette) to 100 ml with phosphate buffer (1).

5. Phenol/nitroprusside solution (phenol, 106 mmol/l; nitroprusside, 0.17 mmol/l):

   dissolve 10 g phenol and 50 mg sodium nitroprusside in approx. 500 ml water and dilute to 1000 ml.

6. Alkaline hypochlorite solution (NaOCl, 11 mmol/l; NaOH, 125 mmol/l):

   mix 125 ml NaOH, 1 mol/l, and 16.4 ml Clorox* (contains 5% w/v NaOCl) dilute to 1000 ml with water.

Note: solution (5) corresponds to the diluted reagents of *Chaney & Marbach*. Solution (6) contains more hypochlorite than their diluted reagent. Both solutions are available from commercial sources, but the concentrations vary. Presumably all these solution are suitable for the ADA assay, because the concentrations of phenol, nitroprusside, NaOH and NaOCl in the reaction mixture are not critical for the formation of indophenol. We use the reagents from *Boehringer Mannheim GmbH,* whose composition is that of solutions (5) and (6).

Stability of solutions: store all solutions at 0°C to 4°C. Solutions (1), (3) – (5) (in dark bottles) and (6) are stable for at least 2 months. Adenosine crystallizes out from solution (2) at 4°C; it can be brought into solution again by warming the volumetric flask, but a little ammonia is released. Therefore, we prepare the daily requirement so as to obtain an ammonia-free adenosine solution. Solution (5) should be discarded if it becomes brown.

Procedure

Collection and treatment of specimen: use venous blood. Addition of oxalate (1 mg/ml), citrate (1 mg/ml), EDTA (1 mg/ml) or phenol-free heparin (0.2 mg/ml) does not interfere with the assay. Addition of fluoride gives unsatisfactory results. Use only serum or plasma free from haemolysis.

Stability of enzyme in sample: the samples should not be stored for longer than 48 – 72 h at 4°C. Storage of sera for more than 5 days at 4°C results in a release of ammonia, even if microbial contamination is avoided. This gives high blank values.

* Clorox is a commercial hypochlorite bleaching agent (*Clorox Co.* Oakland Calif.). Other similar preparations, such as sodium hypochlorite solution *Merck*, Darmstadt, Germany, are suitable.

Details for measurement in other samples: ADA activity can be measured in human erythrocytes lysed in distilled water. Haemoglobin can be removed according to *Ellis & Goldberg* [12]. ADA activity can be measured on human white blood cells, granulocytes and lymphocytes separated by centrifugation techniques and disrupted by sonication or by freezing and thawing (cf. [4, 10, 12] for references).

To measure activity in tissues, prepare homogenates in phosphate buffer, 50 mmol/l, pH 7.0 and centrifuge for 15–30 min at high speed (15000–20000 *g* at 0°C). If the protein content of the complete reaction mixture is below 3 mg/ml and the haemoglobin content is negligible, deproteinization can be omitted. With a higher protein content, deproteinize as described for the determination of guanase activity by *Caraway* [21]. In this case the concentration of the *Chaney & Marbach* reagents should be changed. It is always necessary to check whether, in the absence of adenosine, ammonia is released or bound (see under "Sources of error").

Assay conditions: wavelength 620–650 nm (optimum 620 nm); light path 10 mm; incubation volume 1.05 ml; incubation temperature 37°C; final volume 7.05 ml. Read against water. Also prepare a sample blank and a reagent blank.

Concentrations in incubation mixture:

phosphate 47.6 mmol/l
adenosine 20.0 mmol/l

Volume fraction of sample 0.048

Measurement

Pipette successively in test tubes:	reagent blank	standard	sample blank	sample
phosphate buffer (1)	1.00 ml	–	–	–
buffered adenosine solution (2)	–	–	1.00 ml	1.00 ml
ammonium sulphate standard solution (4)	–	1.00 ml	–	–
sample (serum)	–	–	–	0.05 ml
water*	0.05 ml	0.05 ml	–	–
mix and cap tubes with Parafilm, incubate for 60 min in a 37°C water-bath				
phenol/nitroprusside solution (5)	3.0 ml	3.0 ml	3.0 ml	3.0 ml
sample (serum)	–	–	0.05 ml	–
alkaline hypochlorite solution (6)	3.0 ml	3.0 ml	3.0 ml	3.0 ml
mix, incubate for 30 min in a 37°C water-bath, measure absorbances against water.				

* The addition of water can be omitted without causing any appreciable error.

If absorbance exceeds 1.000, dilute the sample 2–5 times with water and measure again. With this value as a guide, dilute serum accordingly with phosphate buffer and repeat the assay.

Calculation: readings for sample and standard are corrected by subtracting readings for sample blank and reagent blank, respectively, yielding ΔA_{sample} and $\Delta A_{standard}$. The following relation is valid for the catalytic concentration in serum:

$$b = \frac{\Delta A_{sample}}{\Delta A_{standard}} \times 50 \qquad \text{U/l}.$$

Validation of Method

Precision, accuracy, detection limits and sensitivity: when the ADA assay was repeated ten times on each of 5 different serum samples the coefficients of variation ranged from 1.88% to 2.50%. Ten simultaneous assays on a serum sample with a *Beckman* DU-2 spectrophotometer gave a mean value of 69.38 ± 1.46 U/l, while with a *Bausch & Lomb* spectrophotometer 10 assays gave a mean of 69.45 + 1.36 U/l.

Data on accuracy are not available since standard reference material has not yet been established.

The lowest value which can be measured at 628 nm with a sufficiently sensitive photometer is approximately 5 U/l with an imprecision $< \pm 20\%$. With a *Beckman* DU-2 spectrophotometer, first grade analytical reagents and ammonia-free water it is possible to measure an activity concentration of 1.7 U/l, corresponding to 5 nanomoles of ammonia released into the reaction mixture. It is advisable to make up an appropriate calibration curve of ammonium sulphate for assay of low activities.

Sources of error: effects of drugs and other therapeutic measures: A direct effect of therapeutics on the serum ADA level is not known. Treatment of typhoid fever patients with antibiotics results in a gradual decrease in the values.

Interference in the assay technique: with the ADA activity, release of ammonia from adenosine in the sample blank (or in the adenosine blank) is negligible. Neither release nor binding of ammonia occurs if serum and ammonia are incubated for 60 min at 37 °C. Purified ADA is competitively inhibited by urea and by several purine and pyrimidine compounds. However, the high substrate concentration used by us and the low concentration of these substances in serum prevents any interference.

With *Berthelot's* reaction it is essential that the procedure is carried out in a room which is free from traces of ammonia vapours. The absorbance of the reagent blank measured at 630–640 nm against water should be below 0.030.

The serum present in the reaction mixture interferes slightly with *Berthelot*'s reaction. The interference can be measured and corrected according to *Ellis & Goldberg* [22].

The following substances which normally or occasionally occur in serum cause no significant interference in the measurement of ammonia nitrogen with the reagents of *Chaney & Marbach*: glucosamine, citrulline, bilirubin, glutamine, haemoglobin, histidine, uracil, arginine, lysine, sulphadiazine, alanine, phenylalanine, uric acid, creatinine, salicylic acid.

If necessary, the following compounds can be added to the reaction mixture: citric acid, sodium citrate, sodium chloride. On the other hand, the following compounds must not be present in the assay system as they interfere with the formation of indophenol under the conditions described here: glycine buffer, glycylglycine buffer, triethanolamine buffer and, in particular, Tris buffer [23].

Specificity: under the conditions described above, the method is specific for the determination of adenosine deaminase activity.

Reference ranges: our reference interval at 95% for normal human serum is 10 – 25 U/l at 37 °C. These values derived from the results for two hundred normal subjects are in agreement with data of *Ellis & Goldberg* [12] obtained under optimized conditions. Several authors, not using optimized methods, report lower values (cf. [3] for references). In some diseases we found the following values of serum ADA activity (U/l at 37 °C): acute viral hepatitis, 79 ± 22; chronic active hepatitis, 81 ± 33; liver cirrhosis, 70 ± 21; obstructive jaundice, 43 ± 16; typhoid fever 112 ± 31.

With regard to the ADA content of human blood cells, mean normal activity content (nmol $\times$ h^{-1} $\times$ mg^{-1} protein at 37 °C) for human erythrocytes is 63 ± 24, for leukocytes it is 750 ± 280 and for lymphocytes it is 2105 ± 1170 [4]. *Orfanos et al.* [18], expressing values as mU/g Hb at 37 °C, reported normal levels of ADA activity in human erythrocytes of 1311 ± 301 for adults and of 1290 ± 268 for newborns; 96% of all observed values fell within the range 750 – 1850. Values for ADA activity in human tissue are reported in ref. [2].

Suitability of the method for assay of other deaminases: we have demonstrated that the present method is suitable for the assay of other nucleoside and nucleotide deaminases with the appropriate substitution of the buffered substrate solution (2). For serum guanase assay, use as substrate 8-azaguanine, 0.4 mmol/l, in sodium barbital-sodium buffer, pH 6.3; see [12] for details. For other nucleoside and nucleotide deaminases see [23] for details.

References

[1] *N. O. Kaplan,* Specific Adenosine Deaminase from Intestine, in: *S. P. Colowick, N. O. Kaplan* (eds.), Methods in Enzymology, Vol. *II*, Academic Press, New York 1955, pp. 473 – 475.

[2] *M. B. Van der Weyden, W. N. Kelley,* Adenosine Deaminase: Characterization of the Molecular Heterogeneity of the Enzyme in Human Tissue, Adv. Exp. Med. Biol. *76A*, 235 – 248 (1977).

[3] *G. Giusti,* Adenosine Deaminase, in: *H. U. Bergmeyer* (ed.), Methods of Enzymatic Analysis, 2nd edit., Verlag Chemie, Weinheim, and Academic Press, New York 1974, pp. 1092 – 1099.

[4] *M. B. Van der Weyden, L. Bailey,* A Micromethod for Determining Adenosine Deaminase and Purine Nucleoside Phosphorylase Activity in Cells from Human Peripheral Blood, Clin. Chim. Acta *82*, 179 – 184 (1978).
[5] *P. E. Daddona, W. N. Kelley,* Human Adenosine Deaminase: Purification and Subunit Structure, J. Biol. Chem. *252*, 110 – 115 (1977).
[6] *B. Galanti, G. Giusti,* Ricerche sull'Attività Adenosina-Aminoidrolasi del Siero in Diverse Condizioni di Patologia Umana, Minerva Medica *59II*, 5867 – 5874 (1968).
[7] *D. M. Goldberg, G. Ellis, A. M. Ward,* Diagnostic Triad for Portal Cirrhosis, Clin. Chim. Acta *72*, 379 – 382 (1976).
[8] *B. Galanti, G. Giusti,* Attività Adenosina Deaminasi del Siero nella Febbre Tifoidea, Boll. Soc. ital. Biol. sper. *45*, 327 – 330 (1968).
[9] *C. T. Lum, D. E. R. Sutherland, W. G. Yasmineh, J. S. Najarian,* Peripheral Blood Mononuclear Cell Adenosine Deaminase Activity in Renal Allograft Recipients, J. Surg. Res. *24*, 388 – 395 (1972).
[10] *M. Russo, R. Giancane, G. Apice, B. Galanti,* Adenosine Deaminase and Purine Nucleoside Phosphorylase Activities in Peripheral Lymphocytes from Patients with Solid Tumours, Br. J. Cancer *43*, 196 – 200 (1981).
[11] *A. M. Goldblum, F. C. Schmalstieg, J. A. Nelson, G. C. Mills,* Adenosine Deaminase and Other Enzyme Abnormalities in Immune Deficiency States. Birth Defects: Original Article Series, Vol. *XIV*, N 6A, pp. 73 – 84 (1978). The National Foundation.
[12] *G. Ellis, D. M. Goldberg,* A Reduced Nicotinamide Adenine Dinucleotide-linked Kinetic Assay for Adenosine Deaminase Activity, J. Lab. Clin. Med. *76*, 507 – 517 (1970).
[13] *G. Giusti, C. Gakis,* Temperature Conversion Factor, Activation Energy, Relative Substrate Specificity and Optimum pH of Adenosine Deaminase from Human Serum and Tissues, Enzyme *12*, 417 – 425 (1971).
[14] *D. A. H. Hopkinson, P. J. L. Cook, H. Harris,* Further Data on the Adenosine Deaminase Polymorphism and a Report of a New Phenotype, Ann. Hum. Genet. *32*, 361 – 367 (1969).
[15] *F. Heinz, S. Reckel, R. Pilz, J. R. Kalden,* A New Spectrophotometric Assay for Enzymes of Purine Metabolism: IV Determination of Adenosine Deaminase, Enzyme *25*, 50 – 55 (1980).
[16] *B. Galanti, G. Giusti,* Metodo Colorimetrico Rapido per la Determinazione dell'Attività Adenosina Deaminasi e 5-AMP Deaminasi del Siero, Boll. Soc. ital. Biol. sper. *42*, 1316 – 1320 (1966).
[17] *W. Meier, J. F. Conscience,* A Fast and Simple Radiometric Assay for Adenosine Deaminase Using Reversed-Phase Thin-Layer Chromatography, Anal. Biochem. *105*, 334 – 339 (1980).
[18] *A. P. Orfanos, E. W. Naylor, R. Guthrie,* Micromethod for Estimating Adenosine Deaminase Activity in Dried Blood Spots on Filter Paper, Clin.Chem. *24*, 591 – 594 (1978).
[19] *W. Martin, C. Ziegler,* The Determination of Red Cells Isoenzyme Systems ADA, AK and PGH, by means of Agarose Gel Thin-layer Electrophoresis, Aerztl. Lab. *24*, 125 – 129 (1978).
[20] *L. H. Siegenbeck van Henkelom, A. Boom, H. A. Barstra, G. E. J. Staal,* Characterization of Adenosine Deaminase Isozymes from Normal Human Erythrocytes, Clin. Chim. Acta *72*, 109 – 115 (1976).
[21] *W. T. Caraway,* Colorimetric Determination of Serum Guanase Activity, Clin. Chem. *12*, 187 – 193 (1966).
[22] *G. Ellis, D. M. Goldberg,* Assay of Human Serum and Liver Guanase Activity with 8-Azaguanine as Substrate, Clin. Chim. Acta *37*, 47 – 52 (1972).
[23] *G. Giusti, B. Galanti,* Metodo Colorimetrico Rapido per la Determinazione di Alcune Attività Deaminasi, Boll. Soc. ital. Biol. sper. *42*, 1312 – 1316 (1966).

3.5 Adenosinetriphosphatases

ATP phosphohydrolase, EC 3.6.1.3

Harvey S. Penefsky and Michael F. Bruist

General

The ATPase of beef-heart mitochondria was first isolated from the inner mitochondrial membrane in 1960 [1, 2]. The membrane-bound form of the enzyme is now considered to catalyze the terminal transphosphorylating reaction of oxidative phosphorylation: that is, the synthesis of ATP from ADP and P_i [3, 4]. Similar if not identical energy-transducing ATPases of this form have since been shown to be ubiquitous in nature. The ATPases, which are discrete enzyme proteins, are found in animal and plant mitochondria, on the chloroplast thylakoid membrane and on the bacterial plasma membrane and are part of a larger complex of proteins and phospholipid which couple an electrochemical potential gradient to ATP synthesis or ion transport [3].

Other types of ATPases of equal importance also participate in ion transport. These include the Ca^{2+} ATPase of the sarcoplasmic reticulum [5] and the Na^+, K^+ ATPase of the plasma membrane [6].

Application of method: in biochemistry.

Enzyme properties relevant in analysis: the formation of P_i and ADP from ATP in multi-enzyme systems does not necessarily result from the action of a discrete enzyme but could reflect ATP "utilization". The methods described in this chapter are best applied to purified or partially-purified preparations but can be used with proper precautions in more complex systems. For example, membrane-linked functions can be dissociated from ATPase activity by physical disruption or by uncoupling agents such as 2,4-dinitrophenol (oxidative phosphorylation) or A-23187 (the sarcoplasmic reticulum Ca^{2+}-ATPase).

The enzymatic activity of many ATPases may be partially or completely masked. The chloroplast ATPase, including the soluble, homogeneous enzyme, must first be activated and mitochondrial preparations may be masked by an inhibitory protein [3].

Product inhibition (by ADP) of many ATPases must be carefully considered in the application of any assay system.

Methods of determination: a variety of methods has been described. The most important are regenerating systems for ATP, measurement of the release of $^{32}P_i$ from [γ-^{32}P]ATP and a pH-metric procedure.

International reference methods and standards: neither reference methods nor standards are currently available.

Enzyme effectors: a divalent metal ion is required for most conditions of assay. Mitochondrial and bacterial ATPases and Na^+, K^+-ATPase require Mg^{2+} while the enzymes of chloroplast (but see ref. [7]) and the sarcoplasmic reticulum require Ca^{2+}. Oxy-anions such as sulphite, chromate and bicarbonate stimulate the mitochondrial enzyme [7]. Vanadate is a remarkably potent inhibitor of the Na^+, K^+-ATPase [6].

3.5.1 UV-method, Regenerating System for ATP

Assay

Method Design

Principle

(a) $$ATP + H_2O \underset{Mg^{2+}}{\overset{ATPase}{\rightleftharpoons}} ADP + P_i$$

(b) $$ADP + PEP \underset{K^+}{\overset{PK^*}{\rightleftharpoons}} ATP + pyruvate$$

(c) $$Pyruvate + NADH + H^+ \overset{LDH^*}{\rightleftharpoons} lactate + NAD^+$$

Reactions (a), (b) and (c) are employed (cf. ref. [1, 8]). Each mole of ATP hydrolyzed results in the oxidation of 1 mol of NADH. The method is adaptable to automated equipment.

Optimized conditions for measurement: a pH of 7.5 ensures good activity of the lactate dehydrogenase. The initial concentration of NADH is varied in response to the sensitivity desired.

Equipment

A recording spectrophotometer with thermostatted cuvette holders. Good stability on the 0 to 0.1 absorbance scale is especially useful for kinetics studies.

---

* Abbreviations:
PK pyruvate kinase; ATP: pyruvate 2-*O*-phosphotransferase, EC 2.7.1.40;
LDH lactate dehydrogenase, L-lactate: NAD^+ oxidoreductase, EC 1.1.1.27.

Reagents and Solutions

Purity of reagents: contamination of ATP solutions by ADP will result in an initial oxidation of NADH. Commercially available phosphoenol pyruvate, lactate dehydrogenase and pyruvate kinase are suitable.

Preparation of solutions (for about 20 determinations): all solutions in re-purified water (cf. Vol. II, chapter 2.1.3.1).

1. Tris buffer (1 mol/l, pH 7.5):

   dissolve 121.1 g Tris base in 800 ml water. Adjust pH to 7.5 with glacial acetic acid and bring volume to 1 litre with water.

2. $MgSO_4$ (0.1 mol/l):

   dissolve 2.46 mg $MgSO_4 \cdot 7\ H_2O$ in 100 ml water.

3. Phosphoenolpyruvate (0.1 mol/l):

   dissolve 234 mg PEP, trisodium salt, in 10 ml water.

4. NADH (2 mmol/l):

   dissolve 3.1 mg NADH, disodium salt, in 2 ml twice-diluted Tris buffer (1).

5. ATP (0.2 mol/l):

   dissolve 1.228 g mg $ATP\text{-}Na_2 \cdot 3.5\ H_2O$ in 8 ml water. Adjust pH to 7.4 with NaOH, 3 mol/l. Add water to 10 ml.

6. Pyruvate kinase, PK (160 kU/l):

   centrifuge 0.08 ml enzyme suspension in ammonium sulphate (10 mg protein/ml) for 5 min at 12000 *g* at 5 °C. Remove supernatant and dissolve the sediment in 0.92 ml Tris buffer (1).

7. Lactate dehydrogenase, LDH (62 kU/l):

   centrifuge 0.05 ml enzyme suspension in ammonium sulphate solution (5 mg protein/ml) for 5 min at 12000 *g*, remove the supernatant and dissolve the sediment with 1 ml twice-diluted Tris buffer (1).

Stability of solutions: NADH solution (2) prepared freshly each week and stored at −20 °C is stable for two months. Lactate dehydrogenase solution (3) is prepared freshly each day.

Procedure

Collection and treatment of specimen: if large sample volumes are used (greater than 0.05 ml) consideration should be given to possible inhibition by anions [9].

Stability of the enzyme in specimen or sample: aqueous solutions of the purified enzyme are unstable below about 15 °C, but are stable indefinitely if quick-frozen in liquid nitrogen [9]. Membrane-bound ATPase, e.g., sub-mitochondrial particles, is stable at 5 °C and is stored at −70 °C.

Assay conditions: wavelength 339 nm; light path 10 mm; final volume 2.5 ml; 30 °C. Measure against air. Prepare a blank which contains water instead of sample.

Measurement

Pipette successively into the cuvette:			concentration in assay mixture	
Tris buffer	(1)	0.125 ml	Tris	50 mmol/l
$MgSO_4$ solution	(2)	0.100 ml	Mg^{2+}	4.0 mmol/l
PEP solution	(3)	0.050 ml	PEP	2.0 mmol/l
LDH solution	(7)	0.040 ml	LDH	1 kU/l
NADH solution	(4)	0.060 ml	NADH	48 µmol/l
ATP solution	(5)	0.025 ml	ATP	2.0 mmol/l
PK solution	(6)	0.040 ml	PK	2.56 kU/l
water		2.0 − x ml		
mix, record absorbance on the 0 to 0.1 scale until a constant absorbance is reached,				
sample		x ml	volume fraction	variable
mix, record absorbance with time.				

Calculation: use eqns. (k) or (k_1) from "Formulae", Appendix 3. For values of ε: cf. Appendix 4. The catalytic concentration in the sample is (Δt in minutes):

$$b = \frac{2.5}{\varepsilon \times v} \times 1000 \times \Delta A/\Delta t = \frac{396.8}{v} \times \Delta A/\Delta t \quad \text{U/l}$$

$$b = \frac{6613}{v} \times \Delta A/\Delta t \quad \text{nkat/l}$$

where v is sample volume in the assay mixture.

Validation of Method

Precision, accuracy, detection limits and sensitivity: the repeatability of the rates of hydrolysis in replicate assays is ± 5%. Data on accuracy are not available. Detection limits are determined by the drift and noise in the spectrophotometer. A usable sensitivity is 0.1 nmol ATP hydrolyzed per min.

Sources of error: if the enzyme sample contains ADP an initial rapid oxidation of NADH will be superimposed on the apparent enzyme-catalyzed rate. KCN (final concentration 1 mmol/l) must be added to the cuvette to inhibit cytochrome oxidase when the ATPase of mitochondrial fractions is analyzed. Rates are corrected for cyanide-insensitive oxidation.

3.5.2 Radiometric Method

This method is preferred for the determination of Ca^{2+}-dependent ATPases which are inhibited by Mg^{2+} present in the PK/LDH system.

Assay

Principle: $^{32}P_i$ is released from [γ-^{32}P]ATP and precipitated from unreacted radioactive substrate [10]. Because of the high sensitivity of the method, only small amounts of product need accumulate and inhibition by ADP is avoided.

Optimal conditions for measurement: saturating concentrations of ATP may be used to obtain maximum velocity. The amount of product formed is limited to 1 to 5 µmol/l (formation of product should in no case exceed 10% of the initial substrate) by limiting the amount of enzyme added.

Equipment

A scintillation counter. Radioactivity is quantitated either with a scintillation cocktail or via *Cerenkov* counting.

Reagents and Solutions

Purity of reagents: [γ-^{32}P]ATP should be low in $^{32}P_i$ (less than 2% of the total radioactivity) and should be free from ^{32}P in the α and β positions of ATP. Radiochemical purity is determined by chromatography on polyethyleneimine-cellulose [11].

Preparation of solutions (for about 80 determinations):

1. Tris-sulphate buffer (Tris, 1 mol/l; pH 8.0):

   dissolve 121.1 g Tris in 800 ml water, adjust pH with sulphuric acid, 10 mol/l and make up with water to 1000 ml.

2. ATP (0.2 mol/l):

   dissolve 1.25 g ATP-$Na_2 \cdot 3.5\ H_2O$ with 7 ml water. Adjust pH to 7.4 with NaOH, 1 mol/l. Add water to 10 ml.

3. $MgSO_4$ (0.2 mol/l):

   dissolve 2.469 g $MgSO_4 \cdot 7\ H_2O$ with 100 ml water.

4. [γ-^{32}P]ATP (ATP, 0.1 mmol/l; 10^3 cpm/nmol):

   to 2 ml ATP solution, 0.1 mmol/l, add 2 μl [γ-^{32}P]ATP solution [12], 10^8 cpm/ml, 1.7×10^7 cpm/nmol.

5. Perchloric acid (60% w/v; P_i, 10 mmol/l):

   dilute 60 ml $HClO_4$, 70%, sp. gr. 1.67, to 100 ml with water.

6. P_i precipitating reagent:

   to 80 ml HCl, 0.067 mol/l, add 1 ml triethanolamine, 6.5 mol/l, 1 ml bromine water, 1%, 3 g ammonium molybdate, make up with HCl, 0.067 mol/l, to a final volume of 100 ml.

7. HCl (1 mol/l):

   dilute 10 ml HCl, 12 mol/l, with 110 ml water.

8. NaOH (1 mol/l):

   dissolve 40 g NaOH with 100 ml water.

9. Acetic acid (3 mol/l):

   dilute 17.6 ml acetic acid, 17 mol/l, with 82.4 ml water.

Stability of solutions: store [γ-^{32}P]ATP, solution (4), at neutral pH at −20°C. The ATP solution (2) should be kept at 4°C and can be used for up to 90 days. The Tris buffer (1) as well as $MgSO_4$ solution (3) are stable for months at room temperature as long as no growth of micro-organisms occurs. The P_i precipitating reagent is stored at room temperature and is usable for about 1 month. The solutions (5), (7), (8) and (9) are stable for many months at room temperature.

Procedure

Assay conditions: final volume 1.0 ml; 30°C (thermostatted water-bath). Reaction is started by adding enzyme sample to a reaction mixture equilibrated at 30°C. Prepare a reagent blank by adding 0.1 ml 60% perchloric acid-P_i solution (5), before the ATPase sample. Prepare a control which is identical with the reaction mixture above except that a known amount of $^{32}P_i$ replaces the [γ-^{32}P]ATP. The control is used to determine the loss of $^{32}P_i$ during washing of the phosphomolybdate pellet.

Measurement

Incubation and precipitation

Pipette into a 15 ml test tube:			concentration of assay mixture	
Tris buffer	(1)	0.05 ml	Tris	50 mmol/l
ATP solution	(2)	0.03 ml	ATP	6 mmol/l
$MgSO_4$ solution	(3)	0.03 ml	$MgSO_4$	6 mmol/l
[γ-^{32}P]ATP solution	(4)	0.03 ml	[γ-^{32}P]ATP	2.3×10^{-6} Ci/l
water		0.86 – x ml		
mix, adjust temperature of the reaction mixture to 30°C; start reaction by adding				
sample		x ml		
incubate for 1 to 5 min (depending on the activity of the sample), stop the reaction by adding				
perchloric acid	(5)	0.10 ml	P_i	1 µmol/l
centrifuge to remove protein precipitates if any; add to the quenched reaction mixture				
P_i precipitating reagent	(6)	1.10 ml		
allow 10 min at room temperature for the dense phosphomolybdate precipitate to form; centrifuge for 10 min at 200 *g*.				

Radiometry

The supernatant, which is virtually free from $^{32}P_i$, may be counted as a measure of the amount of [γ-^{32}P]ATP remaining. Add 0.3 ml supernatant to 3 ml of a scintillation fluid which is compatible with aqueous samples (e.g. Liquiscint). Samples also may be counted via the *Cerenkow* method in the ^{32}P channel.

Wash the pellet once by re-suspension in 1 ml HCl solution (7) followed by centrifugation. Remove the supernatant, dissolve the pellet with 150 µl NaOH solution (8) and transfer to a scintillation vial. Wash the walls of the test tube with 150 µl acetic acid solution (9). Combine the acetic acid wash and the NaOH solution, add 3 ml scintillation fluid. Count in the ^{32}P channel.

Calculation: the blank and the control have to be accounted for (for abbreviations used here cf. "Formulae", Appendix 3). Radioactivity of P_i formed in the presence of added ATPase

$$C = C_{sample} - C_{blank} \qquad \text{cpm};$$

accounting for the recovery of $^{32}P_i$ after washing of the phosphomolybdate pellet yields

$$C_{ATPase} = \frac{C_{^{32}P_{i\,(added)}}}{C_{^{32}P_{i\,(found)}}} \times C \qquad \text{cpm}$$

the catalytic concentration of the sample is

$$b = \frac{C_{ATPase}}{X \times \Delta t \times v} \qquad \text{U/l}$$

X is specific radioactivity of [γ-^{32}P]ATP in cpm/nmole.

Validation of Method

Precision, accuracy, detection limits and sensitivity: ATPase activity of replicate samples is repeatable to within ± 5%. Data on accuracy are not available. Detection limits and sensitivity are determined by losses of $^{32}P_i$ during washing of the phosphomolybdate pellet. Recovery of $^{32}P_i$ is between 85 and 93%. Hydrolysis of 0.01 nmol [γ-^{32}P]ATP is readily detected and sensitivity is 0.05 nmol/min.

Sources of error: failure to control loss of $^{32}P_i$ during washing of the pellet.

3.5.3 pH-Metric Method

Assay

Method Design

Principle

(a) $$\mathrm{ATP + H_2O \xrightleftharpoons[\text{ATPase}]{Mg^{2+}} ATP + P_i + H^+}$$

The hydrolysis of ATP is accompanied by the net production of one proton at pH 8. The reaction is carried out in a lightly buffered medium in a pH-stat. This instrument maintains a constant pH in the reaction mixture by delivering to it precisely controlled amounts of alkali (titrant).

Optimized conditions for measurement: the chloroplast ATPase (CF_1) is used as an example. Standard conditions are ATP, 0.5 mmol/l; $CaCl_2$, 6 mmol/l; NaCl, 50 mmol/l and EDTA, 1 mmol/l.

Equipment

The instrument used, obtained from *Radiometer,* Copenhagen, consists of a reaction cell, titration assembly, pH meter (PHM64) autotitrator (TTT60) and recorder. Separate hydrogen and calomel electrodes (these provide a faster response time than a combined electrode) are mounted in the reaction cell. Nitrogen is flushed over the reaction mixture to exclude CO_2. The pH meter and autotitrator and associated microprocessor control the delivery of titrant from a 0.25 ml syringe mounted in the titration assembly. The volume of NaOH delivered is recorded versus time. Directions for construction of a suitable apparatus have been published [13].

Reagents and Solutions

Purity of reagents: commercial preparations of ATP are freed from ADP as described [14] and stored at pH 5 at $-20\,°C$. All solutions are prepared with re-purified and freshly boiled water cooled under nitrogen (cf. Vol. II, chapter 2.1.3.1). Special precautions are taken to exclude CO_2 from water. CO_2 is removed from frozen solutions by thawing under vacuum.

Preparation of solutions (for about 20 determinations):

1. Concentrate (NaCl, 250 mmol/l; $CaCl_2$, 30 mmol/l; EDTA, 5 mmol/l):

   dissolve 14.61 g NaCl, 5.41 g $CaCl_2 \cdot 2\,H_2O$ and 1.86 g Na_2EDTA in a final volume of 1 litre using CO_2-free (boiled) water. Store under N_2 in a stoppered flask.

2. Titrant (NaOH, 1 mmol/l):

   dissolve 0.04 g NaOH (free from carbonate) in 1 litre water.

3. NaCl/Ca^{2+}/EDTA:

   add 70 ml water to 20.0 ml concentrate (1). Adjust pH to 7.9 to 8.0 initially with NaOH, 10 mmol/l, and then with NaOH, 1 mmol/l. Add water to a final volume of 100.0 ml.

4. Substrate (ATP, 10 mmol/l):

   add 6.21 mg ATP-$Na_2 \cdot 3.5\,H_2O$ to 0.2 ml concentrate (1). Adjust pH to about 7 with NaOH, 100 mmol/l, then to pH 7.9 to 8.0 with NaOH, 1 mmol/l. Freeze the solution and thaw under vacuum. Adjust volume to 1.0 ml with water. Keep on ice under N_2. The concentration of ATP is confirmed by checking the absorbance at 259 nm. It should be 0.154 for a 1:1000 dilution in water at pH 7.

Stability of solutions: prepare solution (4) daily. Other solutions will be re-usable for some weeks if stored under N_2 in glass-stoppered bottles sealed with Parafilm.

Procedure

Collection, treatment and stability of specimen and sample: Activated CF_1 (chloroplast ATPase) [14] is dissolved in a buffer containing NaCl, 50 mmol/l, Tris chloride, 10 mmol/l, pH 8.0 and EDTA, 1 mmol/l. The final protein concentration is 0.3 mg/ml. Keep the sample at room temperature. Prepare fresh samples daily.

Assay conditions: the speed of titrant delivery should be about twice the expected reaction velocity (for 0.02 units of enzyme, deliver 0.04 ml/min of NaOH, 1 mmol/l). The recorder chart speed should be sufficient to display the rate accurately after 30 s.

Measurement: soak the hydrogen electrode in HCl, 0.1 mol/l and the calomel electrode in saturated KCl overnight before use. Fill and empty the delivery syringe several times. Rinse electrodes, stirrer and delivery tip with water and dab dry with a paper towel.

Pipette into the reaction cell:			concentration in assay mixture	
$NaCl/Ca^{2+}/EDTA$ solution	(3)	5 ml	NaCl Ca^{2+} EDTA	50 mmol/l 6 mmol/l 1 mmol/l
substrate solution	(4)	10 μl	ATP	1 mmol/l
flush water-saturated N_2 stream (0.5 l/min) over the surface of solution, stir at the fastest speed that does not produce a vortex or bubbles; bring pH to 8.0 with titrant (2) if necessary; when pH reading is stable, start recorder;				
sample		5 μl		
pH should not be altered. Allow 10 s for the instrument to reach a proper steady state.				

If reaction conditions are changed, the amount of enzyme added should be adjusted so that the rate of titrant addition is approximately the same.

Calculation: the catalytic content of the sample, related to mg protein is given by eqns. (l) and (l_1) in "Formulae", Appendix 3:

$$z_c/m_c = \frac{1000 \times R \times c}{n \times m_s} \quad \text{U/mg}$$

$$z_c/m_c = \frac{16667 \times R \times c}{n \times m_s} \quad \text{nkat/mg}$$

where is

R rate of titrant addition, ml/min
c titrant concentration, mmol/l
m_s mass of sample protein, mg
n number of protons liberated per mole ATP hydrolyzed.

At pH 8.0 the ratio* H^+: ATP is 1.

* Consult ref. [15], Fig. 1b for the ratio at other pH values.

Validation of Method

Precision, accuracy, detection limits and sensitivity: repeated determinations show a reproducibility of 5%. The lower limit of detection is determined by the titrant concentration and the quality of the instrument. With the instrument from *Radiometer* and NaOH 0.3 nmol/l titrant, rates of 10 nmol/min are routinely measured.

Sources of error: slow electrode response times and changes in CO_2 concentration of the reaction mixture during the analysis constitute the largest sources of error. Inadvertent addition of a buffer to the reaction mixture will decrease response time and cause apparent slowing of the rate.

References

[1] *M. E. Pullman, H. S. Penefsky, A. Datta, E. Racker,* Partial Resolution of the Enzymes Catalyzing Oxidative Phosphorylation. 1. Purification and Properties of Soluble, Dinitrophenol-stimulated Adenosine Triphosphatase, J. Biol. Chem. *235*, 3322 – 3329 (1960).

[2] *H. S. Penefsky, M. E. Pullman, A. Datta, E. Racker,* Partial Resolution of the Enzymes Catalyzing Oxidative Phosphorylation. 2. Participation of a Soluble Adenosine Triphosphatase in Oxidative Phosphorylation, J. Biol. Chem. *235*, 3330 – 3336 (1960).

[3] *H. S. Penefsky,* Mitochondrial ATPase, in: *A. Meister* (ed.), Advances in Enzymology and Related Areas of Molecular Biology, Vol. *49*, John Wiley and Sons, New York 1979, pp. 223 – 280.

[4] *R. L. Cross,* The Mechanism and Regulation of ATP Synthesis by F_1-ATPase, in: *E. S. Snell, P. D. Boyer, A. Meister, C. C. Richardson* (eds.), Annu. Rev. Biochem. *46*, Annual Reviews, Inc., Palo Alto 1977, pp. 688 – 714.

[5] *L. deMeis, A. L. Vianna,* Energy Interconversion by the Ca^{++}-Dependent ATPase of the Sarcoplasmic Reticulum, in: *E. S. Snell, P. D. Boyer, A. Meister, C. C. Richardson* (eds.), Annu. Rev. Biochem. *48*, Annual Reviews Inc., Palo Alto 1977, pp. 275 – 292.

[6] *L. C. Cantley,* Structure and Mechanism of the (Na, K)-ATPase, in: *D. R. Sanadi* (ed.), Current Topics in Bioenergetics *11*, Academic Press, New York 1981, pp. 201 – 237.

[7] *H. S. Penefsky,* Mitochondrial and Chloroplast ATPases, in: *P. D. Boyer* (ed.), The Enzymes, Vol. *X*, Academic Press, New York 1974, pp. 375 – 394.

[8] *H. S. Penefsky,* Differential Effects of Adenylyl Imidodiphosphate on Adenosine Triphosphate Synthesis and the Partial Reactions of Oxidative Phosphorylation, J. Biol. Chem. *249*, 3579 – 3585 (1974).

[9] *H. S. Penefsky, R. C. Warner,* Partial Resolution of the Enzymes Catalyzing Oxidative Phosphorylation. VI Studies on the Mechanism of Cold Inactivation of Mitochondrial Adenosine Triphosphatase, J. Biol. Chem. *240*, 4694 – 4702 (1965).

[10] *C. Grubmeyer, R. L. Cross, H. S. Penefsky,* Mechanism of ATP Hydrolysis by Beef Heart Mitochondrial ATPase. Rate Constants for Elementary Steps in Catalysis at a Single Site, J. Biol. Chem. *257*, 17092 – 12100 (1982).

[11] *C. Grubmeyer, H. S. Penefsky,* The Presence of Two Hydrolytic Sites on Beef Heart Mitochondrial Adenosine Triphosphatase, J. Biol. Chem. *256*, 3718 – 3727 (1981).

[12] *I. M. Glynn, J. B. Chappell,* A Simple Method for the Preparation of ^{32}P-Labelled Adenosine Triphosphate of High Specific Activity, Biochem. J. *90*, 147 – 149 (1964).

[13] *B. D. Werner, G. Boehme, M. S. Urdea, K. H. Pool, J. I. Legg,* Construction and Evaluation of a Sensitive, Inexpensive pH Stat/Titrator System, Analyt. Biochem. *106*, 175 – 185 (1980).

[14] *M. F. Bruist, G. G. Hammes,* Further Characterization of Nucleotide Binding Sites on Chloroplast Coupling Factor One., Biochemistry *20*, 6298 – 6305 (1980).

[15] *R. A. Alberty,* Standard Gibbs Free Energy, Enthalpy and Entropy Changes as a Function of pH and pMg for Several Reactions Involving Adenosine Phosphates, J. Biol. Chem. *244*, 3290 – 3302 (1969).

4 Lyases, Ligases

4.1 Orotidine-5′-phosphate Decarboxylase

Orotidine-5′-phosphate carboxy-lyase, EC 4.1.1.23

Sebastian Reiter and Wolfgang Gröbner

General

Orotidine-5′-phosphate decarboxylase (OMP-DC) catalyzes the irreversible conversion of orotidine 5′-phosphate (OMP) to uridine 5′-phosphate (UMP) and CO_2 in the pathway of pyrimidine nucleotide biosynthesis. The enzyme was first demonstrated in yeast [1], later in mammalian tissues and higher plants. The subcellular and tissue distribution in animals (rats) was studied by *Pausch et al.* (1972) [2]; OMP-DC is exclusively located in the cytosol. Pure preparations have been obtained from yeast [3] and *Ehrlich* ascites carcinoma [4]. The polypeptide isolated from mammalian cells contains two enzymatic activities: besides OMP-DC, it also exhibits activity of orotate phosphoribosyltransferase (OPRT, EC 2.4.2.10), the preceding enzyme in the pathway; it is therefore able to catalyze the two-step conversion of orotic acid to UMP via OMP. A hereditary deficiency of OPRT and OMP-DC causes a rare disorder characterized by excessive oroticaciduria, retarded growth and development and megaloblastic anaemia [5].

Application of method: in biochemistry and in clinical chemistry. OMP-DC may also be used in a coupled assay with OPRT to determine either OPRT activity itself or orotic acid and 5-phosphoribosyl 1-pyrophosphate concentrations.

Enzyme properties relevant in analysis: the occurrence of two isoenzymes was demonstrated by *McClard et al.* (1980) [4]. Linearity of substrate conversion with respect to time depends on the enzyme source and the assay conditions; linearity over a period of 2 hours has been reported for OMP-DC from human erythrocytes [5].

The reaction rate is proportional to concentration of protein except at low protein concentrations [5, 6].

Methods of determination: two methods for determination of catalytic activity are used:

- a direct spectrophotometric assay is based on the different absorption coefficients of substrate OMP and product UMP at 285 nm [1]. It can be used only with more purified enzyme preparations and substrate concentrations greater than 0.015 mmol/l. OMP-DC in erythrocyte lysate can be determined by this method after removal of haemoglobin [7]; a reference value for OMP-DC activity related to haemoglobin-free erythrocyte protein is not available.

- the micro-radiochemical method described here for the determination of OMP-DC in erythrocyte lysates is simple, highly sensitive and specific. It can be used for crude enzyme preparations with high absorption of UV-light and low OMP-DC activity, e.g. erythrocyte lysates, the most suitable enzyme source for clinical purposes.

International reference method and standards: do not exist so far.

Enzyme effectors: OMP-DC from yeast is not fully active in the absence of agents which maintain the protein sulphydryl groups in a reduced state, especially at advanced stages of purification [3].

OMP-DC from mammalian tissues has no effector requirements. OMP-DC is inhibited by several purine and pyrimidine nucleotides including nucleotide derivatives of anti-metabolites, barbituric acid, allopurinol and oxipurinol which are formed *in vivo* after administration of these drugs [8, 9].

Assay

Method Design

Principle

$$\text{Orotidine 5'-phosphate} \xrightarrow{\text{OMP-DC}} \text{uridine 5'-phosphate} + CO_2 .$$

The reaction is irreversible and proceeds quantitatively. Most authors determine the catalytic activity of OMP-DC by measuring the amount of $[^{14}C]CO_2$ liberated per unit time from $[^{14}C]$orotidine 5′-phosphate labelled at the carboxyl moiety. CO_2 is trapped in an alkaline solution. Radioactivity is determined in a liquid scintillation counter (cf. Fig. 1, [10, 11]). Automated equipment for the micro-radiochemical method is not available.

Optimized conditions for measurement: the pH curves for OMP-DC are flat and broad. The optimal pH for the enzyme of human erythrocytes was found to lie within the range from 6.2 to 6.8 [5] and for the enzyme of beef erythrocytes from 7.2 to 7.8 [12]. In all recent studies the activity of the human erythrocyte enzyme has been determined at pH 7.4; therefore this pH should be used in order to obtain activities comparable with the reference values. Either phosphate or Tris-HCl can be used as the incubation buffer. The substrate concentration curve of OMP-DC from human erythrocytes is triphasic, with K_m values of 33, 1.7 and 0.082 mol/l [13]. This phenomenon is due to the existence of the OPRT/OMP-DC polypeptide in different molecular-weight forms which exhibit different values of K_m, V [6] and specific activity [14]. Saturation of the enzyme is reached at substrate concentrations between 0.05 mmol/l [5] and 0.1 mmol/l [15].

Temperature conversion factors: the ratio of activities of rat liver OMP-DC at 25 °C and 37 °C is 1:2.4 [2]. No more data are available yet.

Equipment

Shaking water-bath; glass test tubes (10 ml); rubber caps with disposable plastic centre wells (*Kontes Glass,* Vineland, New Jersey, Fig. 1); pipettes; stopwatch; disposable syringe (10 ml) and injection needle; scissors; scintillation vials; liquid scintillation counter; spectrophotometer (for protein determination).

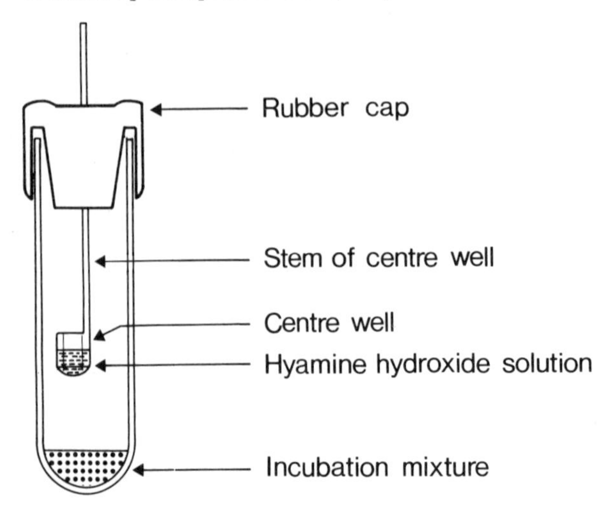

Fig. 1. Test tube with disposable plastic centre well containing hyamine hydroxide.

Reagents and Solutions

Purity of reagents: [carboxyl-^{14}C]orotidine 5′-phosphate and unlabelled orotidine 5′-phosphate are commercially available more than 98% pure.

Preparation of solutions (for about 100 determinations): all solutions in re-purified water (cf. Vol. II, chapter 2.1.3.2).

1. NaCl solution (0.9% w/v):

   dissolve 9 g NaCl in 1 l water.

2. Phosphoric acid (1 mol/l):

   add 6.74 ml ortho-phosphoric acid 85% (1 l = 1.71 kg) to 80 ml water and dilute to 100 ml with water.

3. Phosphate buffer (50 mmol/l; pH 7.4):

   dissolve 4.45 g $Na_2HPO_4 \cdot 2\,H_2O$ in 400 ml water, adjust to pH 7.4 with H_3PO_4 (2), and dilute to 500 ml with water.

4. [carboxyl-^{14}C]Orotidine 5′-phosphate (0.1 mmol/l; 0.2 mCi/mmol):

   dissolve 2.17 mg unlabelled orotidine 5′-phosphate, trisodium salt, in 40 ml phosphate buffer (3), add 0.05 ml radioactive stock solution of [carboxyl-^{14}C]orotidine 5′-phosphate (0.02 mCi/ml ethanol: water 1 : 1; specific activity 20 – 40 mCi mmol) and dilute to 50 ml with phosphate buffer (3).

5. Perchloric acid (4 mol/l):

   add 34.4 ml perchloric acid 70% (1 l = 1.67 kg) to 50 ml water and dilute to 100 ml with water.

6. Scintillation fluid:

   dissolve 9.45 g 2,5-diphenyloxazole(PPO) and 0.2 g 2,2′-p-phenylenbis-(5-phenyloxazole)(POPOP) in 2.5 l toluene.

7. Hyamine hydroxide (1 mol/l):

   commercial solution of methylbenzethonium hydroxide in methanol.

Stability of solutions: phosphate buffer (3) and NaCl solution (1) are stable at 0 – 4 °C as long as no microbial contamination occurs. The substrate solution (4) should not be stored since decomposition of [^{14}C]OMP occurs in aqueous solutions; the stock solution contains 50% ethanol to reduce the rate of decomposition to 1% per 6 months when stored at – 10 °C.

Procedure

Collection and treatment of specimen: collect venous blood in a heparinized tube. Centrifuge for 10 min at 600 *g*. Remove plasma and buffy coat by suction and wash the erythrocytes twice with cold isotonic NaCl solution (1). Lyse the washed erythrocytes by freezing (– 20 °C) and thawing twice. Dilute the lysate 1 + 9 with phosphate buffer (3).

Stability of the enzyme in the sample: the stability of OMP-DC in haemolysates incubated at 37 °C was studied by *Tax et al.* (1976b) [16]: after 5 hours, 30% of starting activity was left when haemolysate was incubated with Tris-HCl buffer as compared to 81% with phosphate buffer. The effect of pH on the stability of OMP-DC from yeast and rat liver was studied by *Creasey & Handschumacher* (1961) [17]. *Gröbner & Kelley* (1975) [14] showed that the different molecular forms of OPRT/OMP-DC exhibit different stability on storage at 4°C: while the larger forms of both enzymes are relatively stable, the small form loses about 25% of activity in 48 hours at 4°C. At −10°C no loss of OMP-DC activity was observed over a period of 1 week [5].

Details for measurement in tissues: in all tissues studied the activity of OMP-DC was localized entirely in the cytosol [2]. The 100000 *g* supernatant of tissue homogenates in isotonic sucrose or buffer is therefore used for measurement of OMP-DC in tissues. The enzyme from animal tissues shows characteristics with respect to K_m, pH optimum and different molecular-weight forms similar to those of the erythrocyte enzyme. Since many tissues have much higher nucleotidase and phosphatase activities than erythrocytes, addition of EDTA to the incubation mixture is recommended to prevent dephosphorylation of the substrate OMP [2].

Assay conditions: incubation volume 1 ml; 37°C (water-bath); reaction time 60 min. Duplicate assays. For blanks use 0.5 ml phosphate buffer (3) instead of diluted erythrocyte lysate. Scintillation countings at 4°C for 5 min.

Measurement

Incubation

Pipette into each centre well 0.2 ml hyamine hydroxide solution (7).

<table>
<tr><td colspan="2">Pipette successively into each test tube (kept in ice-water):</td><td>concentration in assay mixture</td></tr>
<tr><td>substrate solution (4)

diluted erythrocyte lysate</td><td>0.5 ml

0.5 ml</td><td rowspan="2">[^{14}C]OMP 0.05 mmol/l
phosphate 47.5 mmol/l
volume fraction 0.5</td></tr>
<tr><td colspan="2">mix thoroughly, then immediately close the test tube with the air-tight rubber cap holding the prepared plastic centre well (Fig. 1); incubate exactly 60 min at 37°C (water-bath), terminate the reaction by injection of perchloric acid (5) through the rubber cap into the reaction mixture</td></tr>
<tr><td>perchloric acid (5)</td><td>0.2 ml</td><td rowspan="2">$HClO_4$ 0.67 mol/l</td></tr>
<tr><td colspan="2">shake the tubes for another 60 min at 37°C to trap all the [^{14}C]CO_2 evolved.</td></tr>
</table>

Scintillation counting

1. Since hyamine causes quenching the overall counting efficiency η has to be determined:

   place into each of 5 scintillation vials

substrate solution (4)	0.01 ml (0.2 nCi)
hyamine hydroxide (7)	0.20 ml
scintillation fluid (6)	10.00 ml

   one centre well.

   In 5 blanks replace substrate solution (4) by phosphate buffer (3). Measure average counting rates C (cpm) for both series; the difference is the net counting rate C_n.

2. For measurement of incubation mixtures pipette into each 20 ml-scintillation vial 10 ml scintillation fluid (6). Remove the rubber caps with the centre well from the test tubes, wipe the outside of the centre well and transfer the centre well into a scintillation vial. (Rubber cap and centre well are separated from each other by cutting the stem of the centre well with scissors.) Measure radioactivity in a liquid scintillation counter; calculate mean values for the counting rates of sample and blank; the difference between C_n for sample and blank is ΔC_n for the incubation assay.

Attempts have been made to modify this convenient method in order to diminish the consumption of scintillation fluid and the volume of radioactive waste: the alkaline solution in the centre well can be replaced by a NaOH-soaked paper strip which is dried before being transferred into the scintillation fluid allowing repeated use of the scintillation vials and fluid [18].

Protein determination

For calculation of specific activity the protein concentration of the diluted erythrocyte lysate has to be determined. A suitable method is described in Vol. II, chapter 1.3, p. 84.

Calculation

1. *Counting efficiency η*

$$\eta = \frac{C_n}{0.2 \times 2220} \quad \text{cpm/dpm}$$

0.2 radioactivity of 0.01 ml substrate solution in the scintillation vial, nCi,
2220 disintegrations per min per nCi, dpm.

The overall counting efficiency usually ranges from 0.6 to 0.75 cpm/dpm.

2. *Specific activity of OMP-DC*

In the samples of erythrocyte lysate the specific activity (catalytic activity content z_c/m_s) is expressed in the literature as nmol of CO_2 formed per hour per mg protein. It is calculated as follows:

$$z_c/m_s = \frac{\Delta C_n}{\eta \times 2220 \times 0.2 \times \rho_{protein}} \quad \text{arbitr. units/mg} .$$

In this assay
0.2 is the specific radioactivity of substrate solution (mCi/mmol)
$\rho_{protein}$ is mass concentration of protein in the sample (mg/ml).
0.2 specific radioactivity of substrate, nCi/nmol

Note: these units are not international units. For conversion to U/g multiply by 1/60000.

Validation of Method

Precision, accuracy, detection limits and sensitivity: measurements of specific activity show very good reproducibility. To our knowledge the only data available concerning the precision and accuracy of the method are reported by *Fox* (1971) [11]: OMP-DC activity was assayed in triplicate in haemolysates from 9 individuals; determination of the relative standard deviation (SD/mean) for each individual gave a range of 0.42 to 5.06% with a mean of 2.45%.

Routine determinations of OMP-DC activity should be performed in duplicate. Detection limits and sensitivity can be increased by using a higher specific activity of the substrate.

Sources of error: several nucleotide derivatives of anti-metabolites, barbituric acid, allopurinol and oxipurinol formed *in vivo* after administration of these drugs inhibit OMP-DC [8, 9]. Continuous intake of these drugs induces a slow increase of OMP-DC activity, especially in erythrocytes, where eight-fold levels of activity can be reached [19]. This increase of enzyme activity seems to be due to enzyme stabilization *in vivo* [20, 16].

When higher activities of pyrimidine 5′-nucleotidase or other phosphatases are present in the incubation mixture irregular kinetics may arise due to rapid degradation of the substrate OMP to orotidine which is not decarboxylated by OMP-DC.

Specificity: OMP is the only existing substrate for OMP-DC. Therefore the reaction is absolutely substrate specific.

Reference ranges: there are different values for normal OMP-DC activity in erythrocyte lysates of adults given in the literature depending on the assay conditions used.

Fox et al. (1973) [21] have studied the activity of OMP-DC in 145 normal blood donors: the substrate concentration used was 0.05 mmol/l, the buffer sodium phosphate 50 mmol/l, pH 7.4; the activity content ranged from 0.2 to 2.2 nmol/h and mg protein; the activity distribution was non-*Gaussian*: the peak activity of the curve was about 0.5 nmol/h and mg protein, whereas the arithmetic mean amounted to 0.755 nmol/h and mg protein. Using the same assay conditions *Foster et al.* (1973) [22] found for 11 normal subjects an average activity content of 0.6 ± 0.23 nmol/h and mg haemoglobin.

Beardmore et al. (1972) [19] determined the activity of OMP-DC at a substrate concentration of 0.005 mmol/l in Tris buffer, 10 mmol/l, pH 7.4. The average value for 37 control subjects was 0.128 ± 0.068 nmol/h and mg protein.

Tax et al. (1976a) [15] used a OMP concentration of 0.1 mmol/l and Tris buffer, 50 mmol/l, pH 7.4; they found in 26 normal subjects an activity content of 0.52 ± 0.2 nmol/h and mg protein.

From these data we conclude that OMP-DC activity of erythrocytes should be measured at substrate concentrations of at least 0.05 mmol/l and that normal activity content amounts to about 0.5 nmol/h and mg protein.

Patients with hereditary oroticaciduria have about 1% of normal OMP-DC activity [21].

References

[1] *I. Lieberman, A. Kornberg, E. S. Simms,* Enzymatic Synthesis of Pyrimidine Nucleotides. Orotidine-5′-phosphate and Uridine-5′-phosphate, J. Biol. Chem. *215,* 403–415 (1955).

[2] *J. Pausch, D. Keppler, K. Decker,* Activity and Distribution of the Enzymes of Uridylate Synthesis from Orotate in Animal Tissues, Biochim. Biophys. Acta *258,* 395–403 (1972).

[3] *R. S. Brody, F. H. Westheimer,* The Purification of Orotidine-5′-phosphate Decarboxylase from Yeast by Affinity Chromatography, J. Biol. Chem. *254,* 4238–4244 (1979).

[4] *R. W. McClard, M. J. Black, L. R. Livingstone, M. E. Jones,* Isolation and Initial Characterization of the Single Polypeptide that Synthesizes Uridine-5′-monophosphate from Orotate in *Ehrlich* Ascites Carcinoma. Purification by Tandem Affinity Chromatography of Uridine-5′-monophosphate synthase, Biochemistry *19,* 4699–4706 (1980).

[5] *L. H. Smith jr., M. Sullivan, C. M. Huguley jr.,* Pyrimidine Metabolism in Man. IV. The Enzymatic Defect of Orotic Aciduria, J. Clin. Invest. *40,* 656–664 (1961).

[6] *G. K. Brown, R. M. Fox, W. J. O'Sullivan,* Interconversion of Different Molecular Weight Forms of Human Erythrocyte Orotidylate Decarboxylase, J. Biol. Chem. *250,* 7352–7358 (1975).

[7] *R. F. Silva, D. Hatfield,* Orotate Phosphoribosyltransferase: Orotidylate Decarboxylase (Erythrocyte), in: *S. P. Colowick, N. O. Kaplan,* (eds.), Methods in Enzymology, Vol. *LI,* Academic Press, New York 1978, pp. 143–154.

[8] *G. K. Brown, W. J. O'Sullivan,* Inhibition of Human Erythrocyte Orotidylate Decarboxylase, Biochem. Pharmacol. *26,* 1947–1950 (1977).

[9] *B. W. Potvin, H. J. Stern, S. R. May, G. F. Lam, R. S. Krooth,* Inhibition by Barbituric Acid and its Derivatives of the Enzymes in Rat Brain which Participate in the Synthesis of Pyrimidine Ribotides, Biochem. Pharmacol. *27,* 655–665 (1978).

[10] *S. H. Appel,* Purification and Kinetic Properties of Brain Orotidine 5′-phosphate Decarboxylase, J. Biol. Chem. *243,* 3924–3929 (1968).

[11] *R. M. Fox,* A Simple Incubation Flask for $^{14}CO_2$ Collection, Anal. Biochem. *41*, 578–580 (1971).
[12] *D. Hatfield, J. B. Wyngaarden,* 3-Ribosylpurines. I. Synthesis of (3-Ribosyluric Acid)5′-Phosphate and (3-Ribosylxanthine)5′-Phosphate by a Pyrimidine Ribonucleotide Pyrophosphorylase of Beef Erythrocytes, J. Biol. Chem. *239*, 2580–2586 (1964).
[13] *W. J. M. Tax, J. H. Veerkamp,* Inhibition of Orotate Phosphoribosyltransferase and Orotidine-5′-phosphate Decarboxylase of Human Erythrocytes by Purine and Pyrimidine Nucleotides, Biochem. Pharmacol. *28*, 829–831 (1979).
[14] *W. Gröbner, W. N. Kelley,* Effect of Allopurinol and its Metabolic Derivatives on the Configuration of Human Orotate Phosphoribosyltransferase and Orotidine 5′-phosphate Decarboxylase, Biochem. Pharmacol. *24*, 379–384 (1975).
[15] *W. J. M. Tax, J. H. Veerkamp, J. M. F. Trijbels,* Activity of Purine Phosphoribosyltransferases and of Two Enzymes of Pyrimidine Biosynthesis in Erythrocytes of ten Mammalian Species, Comp. Biochem. Physiol. *54B*, 209–212 (1976a).
[16] *W. J. M. Tax, J. H. Veerkamp, F. J. M. Trijbels, E. D. A. M. Schretlen,* Mechanism of Allopurinol-mediated Inhibition and Stabilization of Human Orotate Phosphoribosyltransferase and Orotidine Phosphate Decarboxylase, Biochem. Pharmacol. *25*, 2025–2032 (1976b).
[17] *W. A. Creasey, R. E. Handschumacher,* Purification and Properties of Orotidylate Decarboxylases from Yeast and Rat Liver, J. Biol. Chem. *236*, 2058–2063 (1961).
[18] *K. Prabhakararao, M. E. Jones,* Radioassay of Orotic Acid Phosphoribosyltransferase and Orotidylate Decarboxylase Utilizing a High-voltage Paper Electrophoresis Technique or an Improved $^{14}CO_2$-Release Method, Anal. Biochem. *69*, 451–467 (1975).
[19] *T. D. Beardmore, J. S. Cashman, W. N. Kelley,* Mechanism of Allopurinol-mediated Increase in Enzyme Activity in Man, J. Clin. Invest. *51*, 1823–1832 (1972).
[20] *R. M. Fox, M. H. Wood, W. J. O'Sullivan,* Studies on the Coordinate Activity and Lability of Orotidylate Phosphoribosyltransferase and Decarboxylase in Human Erythrocytes, and the Effects of Allopurinol Administration, J. Clin. Invest. *50*, 1050–1060 (1971).
[21] *R. M. Fox, M. H. Wood, D. Royse-Smith, W. J. O'Sullivan,* Hereditary Orotic Aciduria: Types I and II, Am. J. Med. *55*, 791–798 (1973).
[22] *D. M. Foster, C. Soong Lee, W. J. O'Sullivan,* Allopurinol and Enzymes of de Novo Pyrimidine Biosynthesis, Biochem. Med. *7*, 61–67 (1973).

4.2 Fructose-1,6-bisphosphate Aldolase

D-Fructose-1,6-bisphosphate D-glyceraldehyde-3-phosphate-lyase, EC 4.1.2.13

Peter Willnow

General

Fructose-1,6-bisphosphate aldolase (ALD) is widely distributed, due to its central position in the carbohydrate metabolic pathway. It was discovered in yeast in 1934 [1] and, in 1943, was isolated from rat muscle by *Warburg & Christian.* All the enzymes investigated so far can be classified in two groups.

Group 1: enzymes from all animals, higher plants and green algae. There are small differences with respect to substrate specificity and behaviour towards inhibitors.

Group 2: enzymes from bacteria, yeast and blue-green algae. These mainly differ from the members of group 1 in having lower pH optima, bivalent metal ions in the active centre and a requirement for potassium ions for activity.

The enzyme is found in all tissues but in different quantities.

Table 1. D-Fructose-1,6-bisphosphate aldolase activity content in human tissues (2). (U/g wet weight at 25 °C)

Tissue	Activity content U/g	Tissue	Activity content U/g
Liver	4.8	heart	4.2
Muscle	40.2	spleen	0.03
Uterus	0.78	kidney	
Stomach		cortex	1.5
Musc.	2.2	medulla	0.62
Mucosa	0.93	cerebrum	
Lymphgang.	2.9	cortex	4.5
Fat tissue	1.4	medulla	2.1
Erythrocytes	0.86	cerebellum	3.9
		lung	0.40

In normal human sera the measured activity is relatively low. The enzyme is also found in cerebrospinal fluid, saliva, gastric juice and bile. The half-life in serum is 21 ± 2 hours.

Application of method: the determination of ALD activity has a role in clinical chemistry for the diagnosis of liver diseases and myocardial infarction. However, the greatest elevations in activity in serum are observed in primary myopathies, especially *Duchenne* muscular dystrophy.

Enzyme properties relevant in analysis: ALD is an enzyme built up from four sub-units. There are 3 different polypeptide chains (α, β and γ chains), each of molecular weight 40000. The three genetic loci which determine the structures of the three chains are expressed to different extents in various mature tissues, and at various stages of normal or abnormal development. Isoenzymes with all possible combinations have been found. The complete enzyme has a molecular weight of 150000 to 160000. In the procedure described there is a lag-phase of approx. 3 minutes; thereafter, substrate conversion proceeds linearly with time. Samples with an activity within the linear range show a linear substrate conversion for at least 10 minutes.

Methods of determination: several methods have been described for the determination of activity, all based on substrate conversion according to equation (a). There are differences regarding the assay of the reaction products; e.g. determination of triose phosphates hydrolyzable in alkali [3], or assay of triose phosphate by different colorimetric methods [4, 5]. Alternatively, activity can be determined with an optical test based on coupled enzymatic reactions, according to *Racker* [6].

Only the last of these procedures has gained practical importance. The following method is based on the same principle. The method can be used in biochemistry and clinical chemistry.

International reference methods and standards: neither an international reference method nor a standard material exists so far.

Enzyme effectors: the enzyme isolated from muscle is completely inhibited by heavy metals. A competitive inhibition is possible by glucose, fructose and fructose 6-phosphate [1]. Enzymes belonging to group 2 (from bacteria, yeast and blue-green algae) are inhibited by chelating agents. They need potassium ions for activation. Specific inhibitors for isoenzymes have not been described so far.

Assay

Method Design

Principle

(a) D-Fructose 1,6-bisphosphate $\xrightarrow{\text{ALD}}$ dihydroxyacetone phosphate + D-glyceraldehyde 3-phosphate

(b) D-Glyceraldehyde 3-phosphate $\xrightleftharpoons{\text{TIM*}}$ dihydroxyacetone phosphate

(c) Dihydroxyacetone phosphate + NADH + H^+ $\xrightleftharpoons{\text{GDH**}}$ glycerol 3-phosphate + NAD^+

As the equilibrium of reaction (b) lies 96% and that of reaction (c) totally towards the right of the equations, it is possible to convert F-1,6-P_2 completely. At appropriate optimization (excess of TIM and GDH), reaction (a) is rate-limiting and the decrease in absorbance measured at 339 (or Hg 334 or Hg 365) nm is directly proportional to the activity of ALD. Two moles of NADH are oxidized per mole of F-1,6-P_2 cleaved. The method is applicable to automated analyzers, which can measure at 339 nm (Hg 334 or 365 nm) and which can incubate for 3 minutes before the measurements are taken.

* TIM D-glyceraldehyde 3-phosphate ketolisomerase, EC 5.3.1.1.
** GDH *sn*-glycerol-3-phosphate: NAD^+ 2-oxidoreductase, EC 1.1.1.8.

Optimized conditions of measurement: the optimization has been carried out for the measurement of ALD activity in human sera: different conditions may apply to the analysis of other types of sample.

Buffer and pH: the rate of conversion is equal in Tris or DEA buffers. The pH optimum lies between 7.5 and 8.5 for the stated conditions.

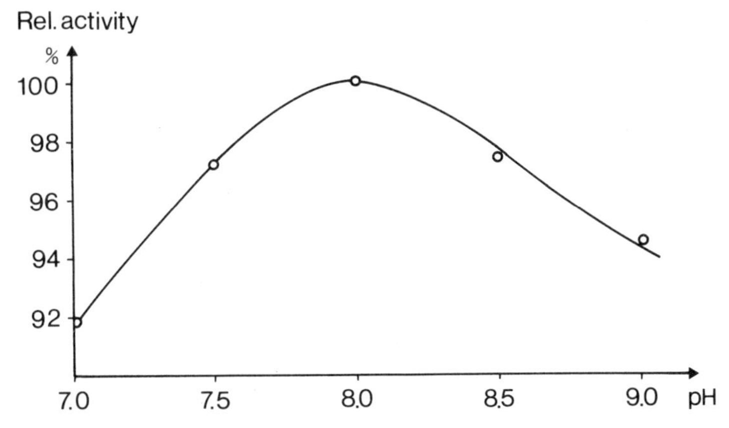

Fig. 1. Dependence of aldolase activity on buffer and pH; measurement as described under assay conditions. Sample: human serum

Substrate: the highest activity of the serum enzyme is measured with between 4 and 5 mmole F-1,6-P_2 per litre. Higher substrate concentrations inhibit the activity.

An activity at the dilution limit (approx. 100 U/l) shows a linear decrease of absorbance for at least 10 minutes.

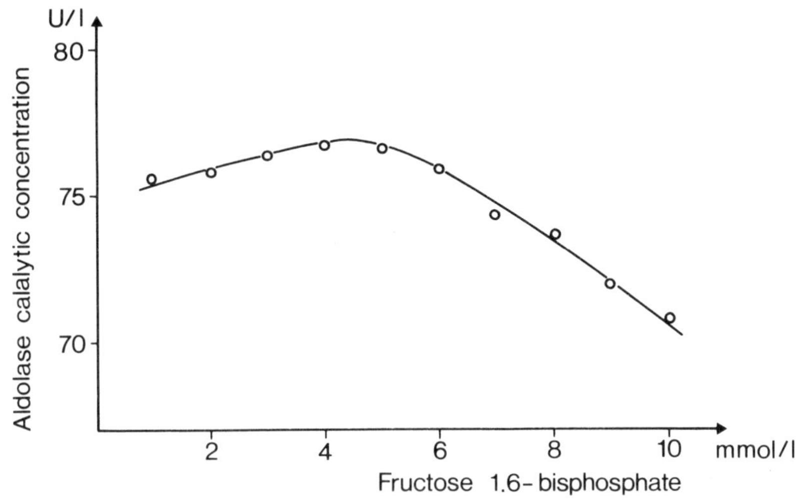

Fig. 2. Dependence of catalytic concentration of aldolase on fructose 1,6-bisphosphate concentration. All other concentrations as described under assay conditions. Sample: human serum

Temperature conversion factors: for 43 human serum samples with different activities a factor of 2.5 for the conversion from 25 °C to 37 °C has been found. The influence of variation in isoenzyme composition has not been investigated.

Equipment

Spectrophotometer or spectral-line photometer capable of exact measurements at 339 nm, or Hg 334 nm or Hg 365 nm; water-bath (30 °C); stopwatch or recorder attached to the photometer. The method is also applicable to automated systems which fulfill the above conditions.

Reagents and Solutions

Purity of reagents: specific activity of GDH 150 U/mg; TIM 5000 U/mg and LDH 500 U/mg, content of aldolase relative to auxiliary and indicator enzyme activities less than 0.001%. Suitable commercial preparations are available. F-1,6-P_2-$Na_3 \cdot 8\ H_2O$ should contain $\geqslant$60% F-1,6-P_2, F-1,6-P_2-$(CHA)_4 \cdot 10\ H_2O \geqslant$ 36%.

Preparation of solutions (for about 40 determinations): all solutions in re-purified water (cf. Vol. II, chapter 2.1.3.2).

1. Buffer/substrate solution (Tris, 100 mmol/l; pH 8.0; F-1,6-P_2, 5 mmol/l):

   dissolve 1.21 g Tris in 80 ml water and adjust to pH 8.0 with HCl, 5 mol/l. Dissolve 458.5 mg F-1,6-P_2-$(CHA)_4 \cdot 10\ H_2O$ or 275.1 mg F-1,6-$P_2Na_3 \cdot 8\ H_2O$ in the buffer solution and fill up to 100 ml with water.

2. NADH (β-NADH, 8.5 mmol/l):

   dissolve 12.06 mg NADH-Na_2 and 20 mg $NaHCO_3$ in 2 ml water.

3. GDH/TIM/LDH (30 kU/l, 500 kU/l, 250 kU/l):

   add 0.2 mg GDH, 0.1 mg TIM and 0.5 mg LDH to ammonium sulphate solution, 3.2 mol/l, make up to 1 ml, mix.

Stability of solutions: keep all solutions in a refrigerator at +2 to +8 °C.
Buffer/substrate solution (1) and NADH solution (2) are stable for at least 4 weeks, enzyme suspension (3) for one year.

Procedure

Collection and treatment of specimen: blood is collected from the vein without stasis. Citrate (5 mg/ml), oxalate (2 mg/ml), fluoride (2 mg/ml) or EDTA (1 mg/ml) can be added without interference. To obtain plasma or serum centrifuge 10 min at approx. 3000 *g*. Only samples without any haemolysis can be used.

Stability of the enzyme in the sample: aldolase is relatively stable in the presence of larger amounts of other proteins, even at slightly higher temperatures. Therefore it is possible to obtain plasma or serum at ambient temperature. The following losses of activity have been found during storage in a sealed tube [9].

At + 20 °C to + 25 °C	minus 15% in 5 days
+ 4 °C	minus 8% in 5 days
− 20 °C	minus 0% in 15 days.

Details for measurement in other samples: *Kellen et al.* [10] describe a procedure to determine human isoenzymes in tissues.

Assay conditions: wavelength 339 (Hg 334, Hg 365) nm, light path 10 mm, final volume 2.61 ml, 30 °C (thermostatted cuvette holder). Measure against air. Before starting the assay, adjust temperature of solutions to 30 °C.

Measurement

Pipette successively into the cuvette:			concentration in the assay	
buffer/substrate	(1)	2.5 ml	Tris	96 mmol/l
			F-1,6-P_2	4.8 mmol/l
NADH solution	(2)	0.05 ml	β-NADH	0.16 mmol/l
enzyme suspension	(3)	0.01 ml	GDH	⩾0.1 kU/l
			TIM	⩾1.5 kU/l
			LDH	⩾1.0 kU/l
sample		0.05 ml	volume fraction	0.019
mix, incubate for 3 min, read absorbance and start stopwatch. Repeat measurement after exactly 1, 2, 3 min or monitor the reaction on a recorder.				

Calculation: calculate from the measured absorbances a mean value for $\Delta A/\Delta t$ (t in minutes). Use calculation formula (k) or (k_1) (cf. "Formulae", Appendix 3). Values for ε: cf. Appendix 4.

The following relations are valid for the catalytic concentration in the sample:

	Hg 334 nm	339 nm	Hg 365 nm	
b =	$4223 \times \Delta A/\Delta t$	$4143 \times \Delta A/\Delta t$	$7676 \times \Delta A/\Delta t$	U/l
b =	$70386 \times \Delta A/\Delta t$	$69047 \times \Delta A/\Delta t$	$127928 \times \Delta A/\Delta t$	nkat/l

Validation of Method

Precision, accuracy, detection limits and sensitivity: the imprecision for within-run and day-to-day performance is given in Table 2.

Table 2. Precision of ALD activity determination; sample: human sera with added ALD; number of determinations: within-run $n = 20$
day-to-day $n = 10$

Activity concentration U/l	Imprecision, within-run		Activity concentration U/l	Imprecision, day-to-day	
	SD U/l	RSD %		SD U/l	RSD %
10.7	0.25	2.3	10.4	0.32	3.1
17.6	0.29	1.6	16.8	0.60	3.6
64.6	1.14	1.8	60.0	2.64	4.4

Data about accuracy are not available since standard reference material is not established yet.

The lowest measurable activity depends on the quality of the equipment used. It is approx. 0.5 U/l with an estimated error of ±25%. The upper limit of the measuring range is 100 U/l. Sensitivity is $\Delta A/\Delta t = 0.005/10$ min.

Sources of error: similar values are found in plasma and serum. Due to the high activity of aldolase in erythrocytes it is necessary to use only samples free from haemolysis. The relation between activity in human serum and erythrocytes is approx. 1 to 150 [8]. Lipaemic sera can be used, but grossly lipaemic specimens should be diluted to avoid too high initial absorbance values. Interference by therapeutics has not been found so far.

Specificity: the method is specific for aldolase. Interference by endogenous substrates in the sample (e.g. pyruvate) are removed by the presence of LDH in the reagent during the pre-incubation period.

After start of the reaction there is a lag-phase, so an incubation period of about 3 minutes is necessary. During that period equilibrium is established; the decrease in absorbance measured after that period is directly proportional to the aldolase activity in the sample.

Reference ranges: due to the lack of an accepted and generally used reference method many different reference values are found in the literature.

Sitzmann [7] found the following values with a similar method in human sera at 25 °C:

adults	2.8 – 3.5 U/l
newborn	3.8 – 9.5 U/l
1. – 6. month	2.0 – 6.2 U/l
7. – 12. month	2.1 – 4.5 U/l
infants	1.8 – 3.8 U/l

References

[1] *J. F. Taylor,* Aldolase from Muscle, in: *S. P. Colowick, N. O. Kaplan* (eds.), Methods in Enzymology, Vol. *I*, Academic Press Inc., New York 1955, pp. 310 – 315.
[2] *E. Schmidt, F. W. Schmidt,* Enzym-Muster menschlicher Gewebe, Klin. Wochenschr. *38*, 957 – 962 (1960).
[3] *O. Mayerhof, K. Lohmann,* Über die enzymatische Gleichgewichtsreaktion zwischen Hexosediphosphorsäure und Dioxyacetonphosphorsäure, Biochem. Z. *273*, 73 – 79 (1934).
[4] *S. B. Barker, W. H. Summerson,* The Colorimetric Determination of Lactic Acid in Biological Material, J. Biol. Chem. *138*, 535 – 554 (1941).
[5] *J. A.Sibley, A. L. Lehninger,* Determination of Aldolase in Animal Tissues, J. Biol. Chem. *177*, 859 – 872 (1949).
[6] *E. Racker,* Spectrophotometric Measurement of Hexokinase and Phosphohexosekinase Activity, J. Biol. Chem. *167*, 843 – 854 (1947).
[7] *F. C. Sitzmann,* Normalwerte, Hans Marseille Verlag, München 1956, p. 12.
[8] *F. Bruns, Chr. Kirchner,* Über die Aldolaseaktivität von Serum und roten Blutzellen verschiedener Spezies, Naturwiss. *41*, 141 – 142 (1954).
[9] *S. Feissli, G. Forster, G. Laudahn, E. Schmidt, F. W. Schmidt,* Normal-Werte und Alterung von Hauptketten-Enzymen im Serum, Klin. Wochenschr. *44*, 390 – 396 (1966).
[10] *J. A. Kellen, A. Chan, B. Caplan, A. Malkin,* Aldolase Isoenzym Patterns during Human Ontogeny and in Lung, Kidney and Breast Cancer, Enzyme *25*, 228 – 235 (1980).

4.3 Citrate Synthase (Condensing Enzyme)

Citrate oxaloacetate-lyase (*pro*-3*S*-$CH_2COO \rightarrow$ acetyl-CoA), EC 4.1.3.7

Mark Stitt

General

Citrate synthase catalyzes the reaction by which acetyl-CoA condenses with oxaloacetate and enters the *Krebs* cycle, and is located in the matrix of the mitochondria. Its high activity and the availability of sensitive assay methods make it a suitable enzyme for use as a marker for mitochondria during subcellular fractionation procedures. It has been used during non-aqueous fractionation of liver [1] and heart [2], as well as wheat protoplasts [3] and spinach leaf (*Gerhardt,* unpubl.). It should be noted that citrate synthase takes part in the glyoxalate cycle and can be present in the glyoxysomes [4, 5] in tissues carrying out gluconeogenesis from fat. It is therefore worthwhile to check the validity of the use of citrate synthase as an exclusive marker for mitochondria by comparison with the distribution of other mitochondrial matrix enzymes, such as fumarase.

Application of method: in biochemical studies of animal and plant tissue.

Enzyme properties relevant in analysis: the reaction proceeds linearly with time.

Methods of determination: a variety of methods for measurement of citrate synthase are described in [6, 7] but all suffer from lack of linearity or sensitivity.

International reference methods and standards: neither standardization at the international level nor the existence of reference standard materials is known so far.

Enzyme effectors: citrate synthase from animal [6] and plant [5] sources is influenced by a range of metabolites including ATP, NADH and succinyl-CoA.

Assay

Method Design

Principle

(a) $$\text{L-Malate} + \text{APAD}^{+} \xrightleftharpoons{\text{MDH}} \text{oxaloacetate} + \text{APADH} + \text{H}^{+}$$

(b) $$\text{Oxaloacetate} + \text{acetyl-CoA} + \text{H}_2\text{O} \xrightarrow{\text{CS}} \text{citrate} + \text{CoA}$$

MDH = malate dehydrogenase (EC 1.1.1.37), APAD = acetylpyridine-adenine dinucleotide.

The amount of oxaloacetate used per unit time is a measure of the catalytic activity of citrate synthase. In the absence of other reactions using oxaloacetate, the use of oxaloacetate can be monitored by the rate of APAD reduction. In the presence of excess malate dehydrogenase and malate, the removal of oxaloacetate is balanced by a regeneration of oxaloacetate and a concomitant reduction of APAD.

It might be noted that the removal of oxaloacetate does not exactly balance the generation of oxaloacetate and reduction of APAD (cf. "Preceeding Indicator Reactions", Vol. I, chapter 2.4.2.2, p. 175) and use of the absorption change, ΔA, can lead to an underestimation of the total enzyme activity in this assay. The relation between oxaloacetate and reduced APAD is defined by the equilibrium constant for malate dehydrogenase, in the presence of a large excess of malate and oxidized APAD. During the assay, as oxaloacetate is consumed by citrate synthase, the reduced APAD concentration increases and the oxaloacetate concentration decreases. This means that the generation of oxaloacetate (and concomitant reduction of APAD) actually underestimates the rate of consumption of oxaloacetate by citrate synthase. This consumption is given by the sum of the generation of oxaloacetate from malate plus the rate of decrease of oxaloacetate concentration. The decrease in oxaloacetate concentration is proportional to the increase in the ratio of reduced : oxidized APAD in the presence of an excess of malate. The following expression is correct for the consumption of oxaloacetate and acetyl-CoA by citrate synthase

$$\Delta A_{\mathrm{AcCoA}} = A_2 - A_1 \times \frac{(A_1)^2}{A_2}$$

where A_1 is the absorbance at the start of the period Δt and A_2 is the absorbance at the end.

Optimized conditions for measurement: citrate synthase is best assayed under alkaline conditions. Addition of cations is not necessary.

Temperature conversion factors: not available.

Equipment

Spectrophotometer or spectral-line photometer with the possibility of exact measurement at Hg 365 nm, preferably with a thermostatted cuvette holder and water-bath, and a recorder. For maximal detectability a dual wavelength spectrophotometer can be used, with measurement at 365 nm against a reference wavelength of 440 nm.

Reagents and Solutions

Purity of reagents: malate dehydrogenase should be free from enzyme activities consuming oxaloacetate.

Preparation of solutions (for about 100 determinations): all solutions in re-purified water (cf. Vol. II, chapter 2.1.3.2).

1. Triethanolamine buffer (0.7 mol/l, pH 8.5):

   dissolve 1.86 g triethanolamine in 75 ml water, adjust to pH 8.5 with HCl, 1 mol/l, and dilute to 100 ml with water.

2. L-Malate (140 mol/l, pH 7.5):

   dissolve 37.52 mg L-malic acid in 1.0 ml water, adjust pH to 7.5 with KOH, 1 mol/l, and adjust volume to 2 ml with water.

3. Acetylpyridine-adenine dinucleotide (10 mmol/l):

   dissolve 12.5 mg APAD in 1.9 ml water.

4. Malate dehydrogenase (MDH, 600 kU/l):

   centrifuge malate dehydrogenase suspension (600 U/mg) for 2 min, remove the $(NH_4)_2SO_4$ supernatant and re-dissolve the sediment in 1 ml trithanolamine buffer (1).

5. Acetyl-coenzyme A (8.4 mmol/l):

   dissolve 14 mg acetyl-CoA in 2 ml water.

Stability of solutions: solutions of malate, APAD and acetyl-CoA can be stored at −40 °C. APAD should be stored in weakly acidic conditions. Triethanolamine buffer can be stored at 4 °C.

Procedure

Collection and treatment of specimens: the exact procedure varies according to the tissue and the nature of the fractionation [1 − 3]. In general, after carrying out a sub-cellular fractionation, the fractions obtained can be subdivided into 50 µl aliquots which are then rapidly frozen. This can be achieved by pipetting the aliquots into small containers placed in a tight-fitting aluminium block pre-cooled in liquid N_2. The samples can then be stored at −85 °C and thawed immediately before assay. This allows a large number of enzymes to be assayed, and if necessary the analyses repeated, without undue experimental haste.

Stability of the enzyme in the specimen: in an unfrozen sample the stability depends on levels of phenols and proteases in the tissue. At −85 °C samples can be stored for 2 − 3 weeks without loss of activity.

Details for measurement in tissues: it is important to ensure that the mitochondria have been fully disrupted by sonication or detergent. The detergent concentration depends on the protein and lipid content of the tissue and can be determined in preliminary experiments (see e.g. [3]). Excessive levels of detergent, or long sonication with inadequate cooling or pauses, also lead to loss of enzyme activity.

Assay conditions: wavelength Hg 365 nm (360 and 440 nm in a dual wavelength spectrophotometer); light path 10 mm; final volume 930 µl in a half-micro cuvette; 30 °C. The assay is carried out in less than 1 ml to decrease the cost of the reagents as well as the amount of extract required.

Measurement

Pipette successively into the cuvette:			concentration in assay mixture	
triethanolamine buffer	(1)	750 µl	triethanolamine	80.6 mol/l
L-malate solution	(2)	20 µl	L-malate	3 mmol/l
APAD solution	(3)	20 µl	APAD	0.22 mmol/l
malate dehydrogenase solution	(4)	20 µl	MDH	12.9 kU/l
mix thoroughly with a plastic spatula, wait 5 min or until the equilibrium between malate and oxaloacetate is reached;				
sample		100 µl	volume fraction	0.108
mix, wait until steady reading is achieved;				
acetyl-coenzyme A solution	(5)	20 µl	AcCoA	0.18 mmol/l
mix, record increase in absorbance A_1 to A_2 per unit time.				

Calculation: for calculation use eqn. (k) or (k_1), cf. "Formulae", Appendix 3; however, see p. 355. The absorption coefficient of APADH at Hg 365 nm is 0.91 $l \times mmol^{-1} \times mm^{-1}$.

The following relations are valid for the catalytic concentration in the sample (extract) (t in minutes):

$$b = 1.022 \times 10^3 \times [A_2 - (A_1)^2/A_2]/\Delta t \quad \text{U/l}$$

$$b = 1.703 \times 10^4 \times [A_2 - (A_1)^2/A_2]/\Delta t \quad \text{nkat/l}$$

To calculate the catalytic content related to mg chlorophyll, b must be divided by the mass concentration of chlorophyll of the extract (mg/l).

Both A_1 and A_2, refer to the absorbance which is dependent upon reduced APAD, and do not include other absorbance due to other components in the cuvette absorbing at Hg 365 nm.

Validation of Method

Precision, accuracy, detection limits and sensitivity: for values of about 10 U/l an imprecision of ±0.5 U/l has been found using a dual wavelength photometer and spinach leaf extract. Data about accuracy are not available due to the lack of standard reference material. The detection limits with a dual wavelength photometer are about $\Delta A/\Delta t = 0.0001\ min^{-1}$. In practice, the sensitivity will depend upon the blank values for APAD reduction in the absence of acetyl-CoA. With extracts of spinach or wheat leaves, these are about $\Delta A/\Delta t = 0.0005\ min^{-1}$ (about 0.05 $nmol/min^{-1}$) so that activities of $\Delta A/\Delta t = 0.001\ min^{-1}$ (about 0.1 nmol/min) can be measured well above the blank rate.

Sources of error: the reaction of acetyl-coenzyme A with oxaloacetate is specific but the measurement of this reaction by reduction of APAD is only valid if the malate-oxaloacetate equilibrium has been established. The reaction loses linearity when a significant proportion of the malate has been consumed. Inhibition of the enzyme will occur if incorrectly stored APAD is used (see above) or if the acetyl-coenzyme A has decayed. After adding the extract, a temporary rapid alteration in absorbance can occur in the absence of acetyl-coenzyme A due to the presence in the extract of malate, as well as amino acids which can donate amino groups to oxaloacetate. In this case, acetyl-coenzyme A should not be added until a steady, low rate of APAD reduction is achieved.

Specificity: citrate synthase in highly specific for its substrates acetyl-CoA and oxalo-acetate. Fluoroacetyl-CoA, fluorooxaloacetate, citryl-CoA and malyl-CoA are also substrates [7].

Reference ranges: wheat leaves contain approximately 0.35 U/mg chlorophyll [3] and liver contains approximately 0.6 U/mg protein [1].

References

[1] *R. Elbers, H. W. Heldt, P. Schmucker, S. Soboll, H. Wiese,* Measurement of ATP/ADP Ratios in Mitochondria and in the Extramitochondrial Compartment by Fractionation of Freeze-stopped Liver Tissue in Non-aqueous Media, Hoppe-Seyler's Z. physiol. Chem. *355*, 378–393 (1974).

[2] *S. Soboll, K. Werdan, M. Bozsik, M. Müller, E. Erdman, H. W. Heldt,* Distribution of Metabolites between Mitochondria and Cytosol of Cultured Fibroplastoid Rat Heart Cells, FEBS Lett. *100,* 125–128 (1980).

[3] *M. Stitt, R. McC. Lilley, H. W. Heldt,* Adenine Nucleotide Levels in the Cytosol, Chloroplasts and Mitochondria of Wheat Leaf Protoplasts, Plant Physiol. *70,* 971–977 (1982).

[4] *B. Axelrod, H. Beevers,* Differential Response of Mitochondrial and Glyoxysomal Citrate Synthase to ATP, Biochim. Biophys. Acta *256,* 175–178 (1974).

[5] *J. T. Wiskich,* Control of the Krebs Cycle, in: *D. D. Davies* (ed.), The Biochemistry of Plants, Vol. *2,* Metabolism and Respiration, Academic Press, New York 1980, pp. 243–278.

[6] *D. Shephard, P. B. Garland,* Citrate Synthase from Rat Liver, in: *S. P. Colowick, N. O. Kaplan* (eds.), Methods in Enzymology, Vol. *XIII,* Academic Press, New York 1969, pp. 11–16.

[7] *P. A. Srere,* Citrate Synthase, in: *S. P. Colowick, N. O. Kaplan* (eds.), Methods in Enzymology, Vol. *XIII,* Academic Press, New York 1969, pp. 3–11.

4.4 Fumarase

L-Malate hydro-lyase, EC 4.2.1.2

Mark Stitt

General

Fumarase catalyzes a reaction of the *Krebs* cycle and is found in the matrix of mitochondria. It is commonly used as a marker during subcellular fractionation [1, 2]. The high activity of fumarase, and the absence of marked regulatory properties, make it suitable for use as a marker enzyme. A simple assay procedure is described here, which allows assay of fumarase with relatively high sensitivity.

Application of method: in analysis of plant and animal tissues.

Enzyme properties relevant in analysis: substrate conversion proceeds linearly with time but there may be an initial lag in crude extracts with low activities. Fumarase is inhibited by anions, the concentrations of which should be standardized.

Methods of determination: various methods have previously been described [3]. The method described here is a modification of these techniques, which has been applied during subcellular fractionation of spinach (*Gerhardt,* unpubl.) and wheat [2] leaf material. *Hatch* has developed [4] an assay in which fumarase activity is measured in the direction of malate formation, and the malate is detected by coupling its production to NADPH reduction by malic enzyme. A potential problem with this assay when applied to plant extracts is that malate levels in such extracts can be very high and this interferes with the accuracy of the assay unless the samples are first desalted.

International reference methods and standards: neither standardization at the international level nor the existence of reference standard materials is known so far.

Enzyme effectors: fumarase is particularly sensitive to inhibition by a number of anions (e.g. chloride, malonate) so the anion concentration in the assay should be standardized.

Assay

Method Design

Principle

$$\text{L-Malate} \xrightleftharpoons{\text{fumarase}} \text{fumarate} + H_2O$$

The accumulation of fumarate can be measured by the increase in absorbance at 240 nm. In the presence of high malate concentrations the reaction proceeds with a linear rate for up to one hour.

Optimal conditions of measurement: pH 7.5, with high malate concentration in the absence of other organic anions.

Temperature conversion factors: not available.

Equipment

Spectrophotometer with ultraviolet light source, temperature controlled cuvette-holder, cuvette storer, water-bath, recorder and quartz cuvettes. For sensitive measurements a dual wavelength spectrophotometer is required.

Reagents and Solutions

Purity of reagents: the solutions should be free from chloride and organic anions.

Preparation of solutions (for about 200 determinations): all solutions in re-purified water (cf. Vol. II, chapter 2.1.3.2).

1. Potassium phosphate (0.1 mol/l, pH 7.5):

   dissolve 0.34 g KH_2PO_4 in 80 ml H_2O, adjust pH to 7.5 with KOH, 3 mol/l, and dilute to 100 ml with water.

2. L-Malate (0.8 mol/l, pH 7.5):

   dissolve 107 mg L-malic acid in 0.6 ml water, adjust pH to 7.5 with KOH, 3 mol/l, and dilute to 1 ml with water.

Stability of solutions: phosphate buffer is best stored frozen; prolonged storage at 4°C leads to microbial contamination, resulting in interfering activity as well as a large increase in absorbance at 240 nm which seriously impairs the assay sensitivitiy. A high blank absorption is an indication of microbial contamination. L-Malate can be stored frozen.

Procedure

Collection and treatment of specimens: see chapter 4.3 “citrate synthase”, p. 356.

Stability of the enzyme in the specimen: fumarase activity in an extract of wheat protoplasts decreased 4- to 5-fold within 15 min, even in the presence of bovine serum albumin and polyvinyl pyrrolidine (*Lilley, R. McC.* pers. commun.). The stability of the enzyme in extracts should be checked and the enzyme assayed immediately after preparation of the extracts; alternatively extracts should be frozen rapidly with liquid N_2.

Details for measurement in tissues: fumarase is located in the matrix of mitochondria and the activity can be underestimated if the mitochondria are not fully disrupted (cf. chapter 4.3 "citrate synthase").

Occasionally an extract may contain high levels of compounds absorbing at ultraviolet wavelengths, so that the background signal is very high. In this case the extracts may be first dialyzed, or de-salted by passage through Sephadex G25 (coarse) (*Pharmacia,* Uppsala).

Assay conditions: wavelength 240 nm; light path 10 mm; 30 °C; final volume 595 µl; thermostatted cuvette holder and pre-equilibrated solutions. With a dual wavelength photometer, use 320 nm as the reference wavelength.

Measurement

Pipette successively into the cuvette:			concentration in assay mixture	
phosphate solution	(1)	500 µl	phosphate	84 mmol/l
malate solution	(2)	35 µl	L-malate	46 mmol/l
extract		60 µl	volume fraction	0.101
mix and record increase of absorbance at 240 nm				

Calculation: for calculation use eqn. (k) or (k_1), cf. "Formulae", Appendix 3. The absorption coefficient at 240 nm, pH 7.5, is $\varepsilon = 0.244\ l \times mmol^{-1} \times mm^{-1}$. The following relations are valid for the catalytic concentration in the sample (extract) (t in minutes):

$$b = 4064 \times \Delta A/\Delta t \quad \text{U/l}$$

$$b = 6.773 \times 10^4 \times \Delta A/\Delta t \quad \text{nkat/l.}$$

To calculate the catalytic content related to mg chlorophyll, b must be divided by the mass concentration of chlorophyll of the extract (mg/l).

Validation of Method

Precision, accuracy, detection limits and sensitivity: for values of 10 U/l an imprecision of ± 0.8 U/l has been found for manual performance. Data about inaccuracy are not available due to the lack of standard reference material. The detection limit is about $\Delta A/\Delta t = 0.001\ \mathrm{min}^{-1}$ and the rate is linear for about 1 hour. In extracts of wheat protoplasts using a dual wavelength photometer, alterations of $\Delta A/\Delta t = 0.001$ min^{-1} (0.4 nmol/min) can be registered with an imprecision of $< \pm 25\%$. The measurement at high sensitivity depend greatly upon the stability of the background signal.

Sources of error: error can arise due to sedimentation of protein during the assay. A blank may be carried out without malate as a control. Variable levels of anions in different extracts can lead to error. The fumarase assay is sensitive to temperature and solutions and cuvettes should be pre-equilibrated at the same temperature as the cuvette holder.

Although specific and easy to carry out, the direct measurement of fumarase by monitoring fumarate formation at 240 nm is limited by the low absorption coefficient, especially when a dual wavelength spectrophotometer is not available and the interference from non-specific alterations of absorption can be significant. In this case, the alternative assay by *Hatch* [4] may be preferred (cf. p. 359).

Specificity: the formation of fumarate from malate is specific.

Reference ranges: wheat protoplasts contain approximately 0.25 U/mg chlorophyll [2].

References

[1] *R. Hampp,* Rapid Separation of the Plastid, Mitochondrial and Cytoplasmic Fractions from Intact Leaf Protoplasts of *Avena,* Planta *150,* 291 – 293 (1980).

[2] *R. McC. Lilley, M. Stitt, G. Mader, H. W. Heldt,* Rapid Fractionation of Wheat Leaf Protoplasts Using Membrane Filtration, Plant Physiol. *70,* 965 – 970 (1982).

[3] *R. L. Hill, R. A. Bradshaw,* Fumarase, in: *S. P. Colowick, N. O. Kaplan* (eds.), Methods in Enzymology, Vol. *XIII,* Academic Press, New York 1969, pp. 91 – 99.

[4] *H. D. Hatch,* A Simple Spectrophotometric Assay for Fumarate Hydratase in Crude Tissue Extracts, Anal. Biochem. *85,* 271 – 278 (1978).

4.5 δ-Aminolaevulinate Dehydratase

Porphobilinogen synthase, EC 4.2.1.24

Karl-Heinz Schaller and Alexander Berlin

General

The enzyme δ-aminolaevulinate dehydratase (AL-D) in erythrocytes catalyzes the condensation of 2 moles of δ-aminolaevulinic acid to form 1 mole of porphobilinogen. During erythropoiesis in chordates the enzyme functions as part of the haem synthesizing machinery, but survives attrition during erythrocyte maturation. Activity of the enzyme can easily be assayed in samples of peripheral blood [1]. It is essential not only for haemoglobin formation, but also for the synthesis of haem-containing respiratory enzymes, among them catalase and cytochromes P-450, b, c and c_1, as well as cytochromes a and a_3 which contain modified haem as a prosthetic group [2]. AL-D activity has been found in many organisms and tissues. According to *Gibson et al.* [3], this enzyme should be present in the cytoplasm of all cells possessing aerobic metabolism. AL-D is not a mitochondrial enzyme. In animals, the liver has the highest activity, but the kidneys and the bone marrow are also rich in this enzyme. AL-D has also been isolated from various plants and micro-organisms [3].

The extreme sensitivity of AL-D to divalent lead ions has led to the use of such activity measurements as an indirect measure of blood lead in human subjects [4, 5].

In recent years considerable interest and practical importance has been associated with the use of the enzyme activity as a bioanalytical measure of environmental lead exposure. This has been aided by the development of a standardized method for the measurement of this enzyme for its proposed use as indicated in the council directives of the Commission of the European Communities [6].

Hernberg [4] has given an extensive report on the biological significance of AL-D inhibition and the use of AL-D as a test of exposure.

AL-D has also been used as a measure of ethanol consumption in alcoholics [7]. *In vivo* and *in vitro*, exposure to carbon monoxide (smoking) causes a consistent, small but significant diminution of AL-D activity [8].

Application of method: in biochemistry and toxicology, in environmental and occupational medicine.

Methods of determination: several methods proposed for measuring AL-D activity in erythrocytes are based upon the enzyme's conversion of two molecules of aminolaevulinic acid to porphobilinogen, which is then measured spectrophotometrically by reaction with *Erlich's* reagent. In general, these have involved modifications of the original *Bonsignore* method [9] and usually suggest a change in the buffer system, for better control of the reaction pH. Assay methods vary from study to study, making comparison of results almost impossible. In an effort to achieve some uniformity in the assay of the enzyme, *Berlin & Schaller* [10] in 1974 proposed the "European Standardized Method", which has since become widely accepted.

In recent years estimation of the restored activity as an index of blood lead content has been proposed by several authors. The treatment of blood samples from lead-exposed mammals with glutathione [11], dithiothreitol (DTT) [12, 13], zinc [14], or zinc and glutathione at 50 °C [1], restores the activity of AL-D to normal levels, and the extent of reactivation correlates with the blood lead level.

The amount of AL-D in human erythrocytes can now be determined directly by radioimmunoassay [13].

International reference methods and standards: a standardization exists at the European level [10]. The existence of reference standard material is not known so far.

Assay

Method Design

Principle

$$2\ \delta\text{-Aminolaevulinate} \xrightarrow{\text{AL-D}} \text{porphobilinogen} + 2\,H_2O$$

The quantity of porphobilinogen produced in 60 min is a measurement of the δ-aminolaevulinic acid dehydratase activity. Porphobilinogen is measured photometrically at 555 nm after reaction with modified *Ehrlich's* reagent.

Optimal conditions for measurement: the catalytic activity of δ-aminolaevulinate dehydratase is measured at 37 °C in presence of excess δ-aminolaevulinic acid (4 mmol/l). The pH optimum of 6.4 is most important. This has been specially studied by *Hernberg* [4]. The European standardized method takes the aspects of optimization into consideration [10].

Equipment

Spectrophotometer or spectral-line photometer capable of exact measurement at 555 nm.

Reagents and Solutions

Purity of reagent: the reagents must be free from lead and zinc.

Preparation of solutions (for about 40 determinations): all solutions in re-purified water (cf. Vol. II, chapter 2.1.3.2).

1. Phosphate buffer (200 mmol/l; pH 6.4)

   1a: dissolve 1.78 g disodium hydrogen phosphate 2-hydrate in 100 ml water;
   1b: dissolve 1.38 g sodium dihydrogen phosphate 1-hydrate in 100 ml water;
   mix 29 ml solution (1a) and 71 ml solution (1b).

2. δ-Aminolaevulinate (10 mmol/l; pH 6.4):

   dissolve 167.6 mg δ-aminolaevulinic acid hydrochloride in approx. 50 ml solution (1b); the pH is adjusted to 6.4 with solution (1a). The volume is then adjusted to 100 ml with phosphate buffer (1).

3. Mercury(II) chloride/trichloroacetic acid ($HgCl_2$, 5 mmol/l; TCA, 612 mmol/l):

   dissolve 1.358 g $HgCl_2$ in 100 ml trichloroacetic acid (100 g in 1000 ml water).

4. *Ehrlich's* Reagent:

   dissolve 2.5 g 4-dimethylamino benzaldehyde in 50 ml glacial acetic acid, add 24.5 ml perchloric acid (spec. gr. 1.7), and 4 ml mercury (II) chloride solution (0.25 g $HgCl_2$ in 10 ml glacial acetic acid). Mix, cool, and make up to 100 ml with glacial acetic acid in a volumetric flask. Store in a dark bottle. If at this stage any brown colouration appears, the reagent must be discarded.

Stability of solutions: store all solutions at 0 °C to 4 °C. The phosphate solutions (1a) and (1b) are stable as long as no microbial contamination occurs. δ-Aminolaevulinic acid is unstable in alkaline solution.

Procedure

Collection and treatment of specimen: 2 ml venous blood is sampled with a plastic syringe (not lead stabilized) over dried heparin (5 mg). Prepare immediately four 0.2 ml aliquots in plastic tubes (not lead stabilized), cool at 4 °C. Three of the four blood aliquots are used for triplicate measurement and one for the blank. Pipette with *Marburg* type plunger pipettes (cf. Vol. I, chapter 3.1.2.2) previously calibrated.

Stability of the enzyme in the specimen: if the analysis is carried out within 3 h after sampling, the aliquots may be kept without cooling. The maximum permissible period for the conservation of samples at 4 °C is 24 h. Immediately prior to analysis all samples should be placed in an ice-water-bath for 10 min.

Assay conditions: wavelength 555 nm; light path 10 mm (or 20 mm if the absorbance is very low); volume for colour reaction 2.0 ml; incubation volume 2.5 ml; 37 °C ± 0.2 °C. Measure absorbance of samples against blank. Triplicate assays. Adjust temperature of solutions (not of samples) to 37 °C.

Measurement

Incubation

Pipette successively into plastic tubes:	samples	blank	concentration in assay mixture
sample (blood) water	0.20 ml 1.30 ml	0.20 ml 1.30 ml	volume fraction 0.08
mix thoroughly			
$HgCl_2$ (3) buffer/substrate (2)	 1.00 ml	1.00 ml 1.00 ml	aminolaevulinate 4 mmol/l phosphate 80 mmol/l
mix, keep at 47 °C for exactly 60 min			
$HgCl_2$ (3)	1.00 ml	–	
mix, enzyme reaction is stopped. Centrifuge for 10 min at 20000 *g*, filter supernatant through acid-resistant filter paper (e.g. Whatman No. 54). Use filtrate for colour reaction.			

Colour reaction

Mix with vortex mixer 1 ml of incubation solution (filtrate) with 1 ml modified *Ehrlich's* reagent [4]. Let stand for 5 min. Read absorbances at room temperature.

Calculation: for calculation use eqn. (k) or (k_1), cf. "Formulae", Appendix 3. $v_i = 0.5$. The catalytic activity is referred to litre erythrocytes using hematocrit value (%). For Δt = 60 min the following relation is valid:

$$b = \frac{100 \times 35 \times 2}{\mathrm{h} \times 0.0062 \times 10} \times \Delta A/\Delta t = \frac{1.13 \times 10^5}{\mathrm{h}} \times \Delta A \quad \mathrm{U/l}$$

h is haematocrit value, %
100 converts h into litre
35 is dilution factor of sample
2 converts 1 mol porphobilinogen into 2 mol δ-aminolaevulinate
0.0062 is absorbance coefficient, $l \times \mu mol^{-1} \times mm^{-1}$

Validation of Method

Precision, accuracy and detection limit: the "European Standardized Method" was tested by personnel from 16 different European laboratories [15]. Precision within series ranged from ±0.6–7.2% (mean: ±3.0%). Overall mean activity was 23.6 ± 3.8 U/l. This gives a coefficient of variation of ± 16% for interlaboratory measurements.

The evaluation of the present method determined by 16 laboratories on at least five individuals with four samplings over a period of three weeks gave an overall mean coefficient of variation of ±9.15%.

Data about accuracy are not available since standard reference material is not established yet. The simultaneously determined activity in the blood of 24 workers by the "European Standardized Method" and another micro-scale method shows a good correlation, a negligible y-intercept and a proportionality error (slope) in favour of the micro-scale method [16]. The detection limit of the method is 0.5 U/l [17].

Sources of error: the contamination of blood samples by metals such as zinc and chelating substances in an evacuated collection tube can severely affect the AL-D activity of the blood being investigated.

Experience has shown that, in particular, porphobilinogen is very sensitive to light. The whole analysis should be performed in the absence of direct sunlight in the laboratory. The enzyme activity is influenced strongly by the incubation temperature and the pH. The variance of the water-bath temperature from 37°C should not be greater than ±0.2°C. The pH of the substrate solution must be checked for each run.

Reference ranges: nineteen European laboratories carried out a restricted population survey of 50 men in each country. The results of the blood AL-D activity obtained for the individual groups showed a skewed distribution. Results are, therefore, given as the median and the ninety-five percentile (Fig. 1).

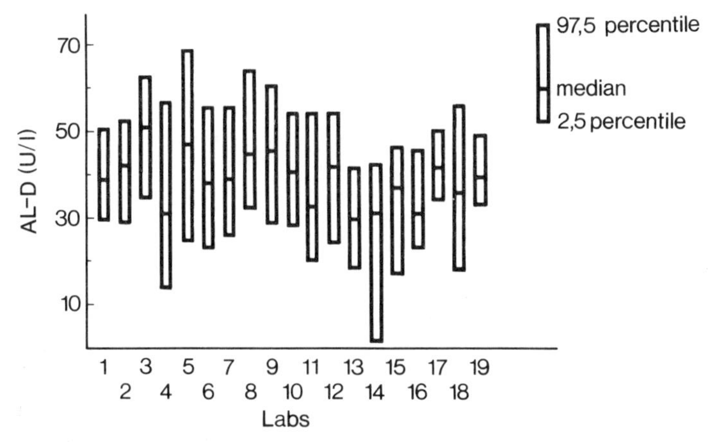

Fig. 1. Median, 2.5 and 97.5 percentiles for the AL-D determination in blood samples from 50 persons in 19 European cities [17].

The Council directive on biological screening of the population for lead [6] gives the following data for the relationship between blood levels and AL-D activity:

Blood lead level (μg/dl)	AL-D (U/l)
35 (upper normal limit)	20
30	25
20	35

References

[1] *R. A. Mitchel, J. E. Drake, L. A. Wittlin, T. A. Rejent,* Erythrocyte Porphobilinogen Synthase (Delta-Aminolaevulinate Dehydratase) Activity: A Reliable and Quantitative Indicator of Lead Exposure in Humans, Clin. Chem. *23*, 105 – 111 (1977).

[2] *F. De Matteis,* Disturbances of Liver Porphyrin Metabolism Caused by Drugs, Pharmacol. Rev. *19*, 523 – 557 (1967).

[3] *K. D. Gibson, A. Neuberger, J. J. Scott,* The Purification and Properties of delta-Aminolaevulinic Acid Dehydrase. Biochem. J. *61*, 618 – 629 (1955).

[4] *S. Hernberg, J. Nikkanen,* Effect of Lead on delta-Aminolaevulinic Acid Dehydrase – A Selective Review –, Pravoc. Lék. *24*, 2 – 3 (1972).

[5] *B. Haeger-Aronson, M. Abdulla, B. Fristedt,* Effect of Lead on delta-Aminolaevulinic Acid Dehydratase Activity in Red Blood Cells, Arch. Environ. Health *23*, 440 – 445 (1971).

[6] *Council Directive of 29 March 1977 on Biological Screening of the Population for Lead:* Official Journal of the European Communities 20 (no. L 105), 10 (28. 4. 1977).

[7] *N. Krasner, M. R. Moore, G. G. Thompson, W. Mc Intosh, A. Goldberg,* Depression of Erythrocyte delta-Aminolaevulinic Acid Dehydratase Activity in Alcoholics, Clin. Sci. Mol. Med. *46*, 415 – 418 (1974).

[8] *M. R. Moore, P. A. Merdedith,* The Effect of Carbon Monoxide upon Erythrocyte delta-Aminolevulinic Acid Dehydratase Activity, Arch. of Environ. Health, 158 – 161 (1979).

[9] *D. Bonsignore, P. Calissano, C. Cartasegna,* Un semplice metodo per la determinanzione della delta-amino-levulinico-deidratase nel sangue: Comportamanto dell'enzima nell' intossicazione saturnina, Med. Lav. *56*, 199 – 205 (1965).

[10] *A. Berlin, K. H. Schaller,* European Standardized Method for the Determination of delta-Aminolevulinic Acid Dehydratase Activity in Blood, Z. Klin. Chem. Klin. Biochem. *12*, 389 – 390 (1974).

[11] *Th. Haas, W. Mache, K. H. Schaller, K. Mache, G. Klavis, R. Stumpf,* Zur Bestimmung der Delta-Aminolävulinsäure-Dehydratase und ihrer diagnostischen Wertigkeit, Int. Arch. Arbeitsmed. *30*, 87 – 104 (1972).

[12] *J. L. Granick, S. Sassa, S. Granick, R. C. Levere, A. Kappas,* Studies in Lead Poisoning. II. Correlation between the Ratio of Activated to Inactivated delta-Aminolevulinic Acid Dehydratase of Whole Blood Level, Biochem. Med. *8*, 149 – 159 (1973).

[13] *H. Fujita, K. Sato, S. Sato,* Increase in the Amount of Erythrocyte delta-Aminolevulinic Acid Dehydratase in Workers with Moderate Lead Exposure, Int. Arch. Occup. Environ. Health *50*, 287 – 297 (1982).

[14] *V. N. Finelli, D. S. Klauder, M. A. Karaffa, H. G. Petering,* Interaction of Zinc and Lead of delta-Aminolevulinate Dehydratase, Biochem. Biophys. Res. Commun. *65*, 303 – 311 (1975).

[15] *A. Berlin, K. H. Schaller, H. Grimes, M. Langevin, J. Trotter,* Environmental Exposure to Lead: Analytical and Epidemiological Investigations Using the European Standardised Method for Blood delta-Aminolevulinic Acid Dehydratase Activity Determination, Int. Arch. Occup. Environ. Health *39*, 135 – 141 (1977).

[16] *D. C. Wigfield, J.-P. Farant,* Assay of delta-Aminolaevulinate Dehydratase in 10 μl of Blood, Clin. Chem. *27*, 100 – 103 (1981).

[17] *K. H. Schaller,* delta-Aminolaevulinsäure-Dehydratase, Analytische Methoden, Bd. 2, Kommission zur Prüfung gesundheitsschädlicher Arbeitsstoffe der Deutschen Forschungsgemeinschaft – Arbeitsgruppe "Analytische Chemie", Verlag Chemie, Weinheim (1975).

4.6 Adenylate Cyclase

ATP pyrophosphate-lyase (cyclizing), EC 4.6.1.1

Günter Schultz and Karl H. Jakobs

General

Adenylate cyclase, which was discovered by *Sutherland & Rall* in 1957, occurs in most types of mammalian cell, in lower animals, in unicellular organisms, in bacteria and probably in plants. In animal tissues, the majority of the enzyme is bound to the plasma membrane; only in the testis is most of the enzyme activity found in the 100000 *g* supernatant fraction. Occurrence, properties and regulation of adenylate cyclase have recently been reviewed [1]. Adenylate cyclases have been purified from bacteria [2] and, most recently, from rabbit myocardium [3].

The membrane-bound adenylate cyclase of most mammalian tissues underlies the bi-directional control of stimulatory and inhibitory hormones, acting via different stimulatory and inhibitory receptors, respectively [1, 4]. Hormone-binding to these receptors affects adenylate cyclase activity through distinct stimulatory and inhibitory regulatory proteins (N_s and N_i) which bind guanine nucleotides [5]. These regulatory components also mediate stimulatory and inhibitory effects of GTP, stable GTP analogues and fluoride. In addition, N_s is the target of cholera toxin (ADP-ribosylation of N_s) [1], while *B. pertussis* toxin interferes with hormonal inhibition by ADP-ribosylation of N_i [6].

Stimulatory and inhibitory hormonal effects on adenylate cyclase activity are easily observed in membrane preparations, provided that GTP (10^{-8} to 10^{-5} mol/l) is present. Usually, GTP itself has little or no effect on adenylate cyclase activity, whereas stable GTP analogues such as guanosine 5′-(γ-thio)triphosphate or guanosine 5′-(β,γ-imido)triphosphate cause stimulation or inhibition of the enzyme, depending on the conditions and the cell type used. Sodium (and lithium) ions have been shown to favour or even to permit inhibitory hormone effects, at least in some systems [4]. Na^+ and Li^+ also affect hormonal stimulation, but their exact sites of action have not been defined.

Ca^{2+} stimulates adenylate cyclase through calmodulin in some tissues, e.g. tissue from some parts of the central nervous system. It is not clear, however, whether the enzyme controlled by Ca^{2+}-calmodulin is the same as that controlled by hormones and guanine nucleotides via N_s and N_i. Soluble adenylate cyclase from bacteria,

---

Abbreviations:

N_s stimulatory guanine nucleotide-binding, regulatory component;

N_i inhibitory guanine nucleotide-binding, regulatory component.

unicellular organisms and mammalian testis is not influenced by agents which affect the mammalian enzyme via N_s or N_i. Forskolin, a diterpene compound which stimulates mammalian adenylate cyclase in membranes from any type of cell except for sperm by an apparent direct effect on the catalytic component, also has no effect on these forms of adenylate cyclase [7].

Application of method: determinations of adenylate cyclase activity are performed in biochemical and pharmacological studies on the control of cellular functions, mainly in studies on hormone-receptor functions. The assay conditions suggested below are suitable for studies of adenylate cyclase stimulation and inhibition in most mammalian cell types.

Methods of determination: principles and problems of determinations of adenylate cyclase activity have been reviewed [8]. The activity of adenylate cyclase is very low as compared to those of other enzymes that use ATP as a substrate. Therefore, the conversion of ATP to cAMP can only be followed by sensitive assays such as radioimmunoassay, or a binding assay with a cAMP-dependent protein kinase as a binding protein or, more easily, by determination of radioactively-labelled cAMP formed from labelled ATP.

The method described here is based on the determination of [^{32}P]cAMP formed from [α-^{32}P]ATP, using an easy but effective isolation procedure for the cyclic nucleotide [9]. This method provides high sensitivity in combination with high simplicity, particularly when the labelled ATP is prepared as described [10, 11]. If commercially-labelled ATP is used under conditions of very low substrate conversion, an alternative method for isolation of the product based on a double column procedure [12, 13] is recommended. The method described allows the assay of 100 tubes within 3 to 4 hours and can be applied to any system studied so far.

International reference methods and standards: the assay conditions have not been standardized, nor is standard reference material available.

Assay

Method Design

Principle

$$\text{ATP} \cdot \text{Me} \xrightarrow[\text{Me}^{2+}]{\text{adenylate cyclase}} \text{cAMP} + \text{PP}_i$$

where Me can be Mg or Mn.

In the assay described, the rate of conversion is determined by isolating and measuring the amount of [^{32}P]cAMP formed from [α-^{32}P]ATP. As the assay is generally performed with an unpurified enzyme, using more or less purified membranes, several side-reactions occurring simultaneously at much higher rates than the adenylate cyclase reaction itself have to be considered and overcome.

Cyclic AMP formed in the assay has to be separated from the remaining substrate and from other products formed from ATP. A two-step separation is necessary for the isolation of cAMP, conversion of which from ATP is generally not more than a few percent but often even less than 0.1%. We describe a procedure using $ZnCO_3$ co-precipitation of 5′-nucleotides, P_i and PP_i, and subsequent chromatography on a short alumina column [9]. In cases of very low conversion of ATP to cAMP, the two-column procedure of *Salomon* [12, 13] is recommended, which should be preceded by a $ZnCO_3$ precipitation step, mainly in order to reduce the volume of radioactive waste.

Optimized conditions for measurement

a) In order to exclude interference due to degradation of ATP by various ATPases and nucleotidases, an ATP-regenerating system is included in the assay. Creatine kinase and creatine phosphate are preferred to pyruvate kinase and phosphoenol pyruvate because phosphoenol pyruvate is more likely to interfere with the adenylate cyclase system itself than the first system.

b) Nucleotide pyrophosphatase action can be overcome by addition of myokinase. This addition may be necessary in some tissues, e.g. liver.

c) Cyclic nucleotide phosphodiesterase-induced degradation of labelled cAMP formed by the adenylate cyclase is reduced by including unlabelled cAMP and/or a phosphodiesterase inhibitor, e.g. IBMX*. A non-methylxanthine phosphodiesterase inhibitor, e.g. Ro 20-1724, is used when adenosine effects on purinergic receptors are studied.

d) On the other hand, adenosine formed in the assay can interfere by stimulatory or inhibitory effects via purinergic receptors, or by an inhibitory effect via the "P-site" of the adenylate cyclase system. Such adenosine effects can be prevented by addition of adenosine deaminase, by using dATP as the substrate and/or by including a methylxanthine as a purinergic antagonist [14].

In any system in which adenylate cyclase is studied without previous experience within the laboratory, conditions have to be established under which product formation reflects initial velocity, with cAMP formation being proportional to the incubation time and the amount of membrane used. In most cases, special conditions are required to achieve optimal substrate conversion that is linear with respect to time and amount of membrane protein. The "basal" rate of cAMP formation is high in a few tissues, e.g. adipocytes, resulting in a measurable adenylate cyclase activity under standard conditions as described on p. 375. In other tissues, e.g. liver, the adenylate cyclase activity is so low that the assay conditions have to be optimized very carefully in order

* IBMX 3-isobutyl-1-methylxanthine.

to achieve reliable determinations of enzyme activity. Preliminary studies aimed at optimizing the conditions may include increasing the amount of creatine phosphate and creatine kinase, adding myokinase, adjusting the pH of the reaction to a more alkaline value, increasing the specific radioactivity of the [α-^{32}P]ATP by lowering the ATP concentration and increasing the amount of [α-^{32}P]ATP per tube, and adjusting the concentrations of free divalent metal ions or membrane protein, temperature, and incubation time. On the other hand, the standard conditions have to be changed and adapted to the question asked in the study. Whereas hormonal stimulation is easily seen in most systems, there may be problems in demonstrating hormonal inhibition of the adenylate cyclase. Inhibition can be seen under the standard conditions described below but are more pronounced when the divalent metal ion concentration is low (free Mg^{2+}, 1 – 2 mmol/l, or free Mn^{2+}, 0.05 – 0.1 mmol/l) and when the incubation temperature is low (25 °C); stimulation of the enzyme by a hormone or forskolin may help the demonstration of inhibition.

Regulation of adenylate cyclase activity is mostly studied in membranes possessing the entire complex of hormone receptors, N_s, N_i and catalytic subunit. Purification of membranes can result in increased specific catalytic activity and hormonal stimulation of adenylate cyclase but may also result in decreased hormonal inhibition. When membrane preparation involves time-consuming steps, thermal inactivation of the catalytic component and proteolytic destruction of the inhibitory system should be kept small by keeping the temperature low (0 – 4 °C) at all steps and by inclusion of inhibitors of proteolytic enzymes. When membrane pre-incubation is required, ATP and/or forskolin should be included for protection of catalytic activity. For continuous studies in a system, membranes can be stored frozen (preferably at – 70 °C or lower temperatures) at least for several months.

Temperature conversion factors: have not been determined because of the complexity of the system.

Equipment

Refrigerated centrifuge for membrane preparation; centrifuge for $ZnCO_3$-pelleting (10000 *g* if possible); water-bath incubator for incubations; dispensers for multiple additions (e.g. in the $ZnCO_3$ step); liquid scintillation spectrometer; Plexiglas (Lucite) shield and radiation monitor. Short plastic columns of 7 – 8 mm diameter with reservoir (about 15 ml) are used. Such columns are available from *Evergreen Scientific,* Los Angeles, California (no. 3030 – 31), from *BioRad Laboratories* (no. 731 – 1550) or from *Sarstedt* (no. 91787, in combination with a polyethylene centrifuge tube, no. 55520, tip of which is cut off). A fritted plastic disk punched out of porous polyethylene sheet material (BelArt) is fitted into these columns.

Reagents and Solutions

Purity of reagents: all reagents (except for those used in the product-isolation procedure) should be of highest purity available. ATP should be of a GTP-free quality. [α-^{32}P]ATP should be prepared as described [10, 11]. If commercial preparations have to be used, ethanol-free, low specific-activity material must be ordered.

Preparation of alumina columns: pour dry neutral alumina (*Merck* aluminum oxide 90, *Woelm* N or similar material) into the columns to 2 cm length. Wash the columns with 10 ml Tris-HCl buffer, 0.1 mol/l, pH 7.5 (see below). Regenerate used columns by washing with 10 ml of the same Tris buffer. The columns can be re-used repeatedly for several months until an increase in the blank value is observed.

Preparation of solutions: re-purified water is used for all solutions (cf. Vol. II, chapter 2.1.3.2).

Stock solutions (frozen stock solutions should be used whenever possible)

1. Triethanolamine buffer (1 mol/l; pH 7.4 – 8.0; consider $\mathrm{d}pK_a/\mathrm{d}T = -0.020$):

   dissolve 18.57 g triethanolamine-HCl in 50 ml water, adjust to the desired pH value, e.g. 7.5, with NaOH, 1 mol/l, and dilute to 100 ml with water.

2. $MgCl_2$ (100 mmol/l):

   dissolve 2.033 g $MgCl_2 \cdot 6\,H_2O$ in 100 ml water.

3. EGTA (10 mmol/l):

   dissolve 380.4 mg EGTA in 100 ml water and adjust to neutral pH.

4. IBMX (10 mmol/l):

   dissolve 44.4 mg IBMX (3-isobutyl-1-methylxanthine) in 20 ml water; heating is required to re-dissolve the compound after freezing and thawing.

5. Forskolin (10 mmol/l):

   dissolve 4.11 mg forskolin in 1 ml ethanol.

6. cAMP (10 mmol/l):

   dissolve 34.7 mg cAMP $\cdot$ H_2O in 10 ml water.

7. ATP (10 mmol/l):

   dissolve 60.6 mg ATP-$Na_2H_2 \cdot 3\,H_2O$ in 10 ml water.

8. GTP (10 mmol/l):

   dissolve 56.7 mg GTP, disodium salt, in 10 ml water.
   Determine the actual concentrations of the nucleotide solutions by the UV absorbance at 260 nm, using absorption coefficients of 1.50 and 1.18 l $\times$ mmol^{-1} $\times$

mm^{-1} for adenine and guanine nucleotides, respectively; adjust the concentrations accordingly. Also adjust the pH values of these nucleotide solutions to 7.0 for determinations of A_{260} and for storage.

Other solutions

9. $ZnCl_2$ (125 mmol/l):

   dissolve 17.04 g $ZnCl_2$ in 1 l water.

10. Na_2CO_3 (125 mmol/l):

    dissolve 13.25 g anhydrous Na_2CO_3 in 1 l water.

11. Tris-HCl buffer (100 mmol/l):

    dissolve 12.11 g Tris in 0.9 l water, adjust the pH value to 7.5 by HCl, 1 mol/l, and make up to 1 l with water.

Working solutions (for 100 determinations, prepared fresh for every assay)

12. Standard component mixture*:

    thaw the required frozen stock solutions and mix the following components to give a final volume sufficient for 100 tubes (or for the expected tube number with an excess volume for 3 to 5 tubes); add solid substances of bovine serum albumin, creatine-P-$Na_2 \cdot 4\ H_2O$ and creatine kinase, lyophilized, ca. 380 U/mg (37°C).

Triethanolamine buffer (1)	500 μl
$MgCl_2$ solution (2)	500 μl
IBMX solution (4)	1000 μl
cAMP solution (6)	100 μl
ATP solution (7)	100 μl
Creatine-P-$Na_2 \cdot 4\ H_2O$	16.4 mg
Creatine kinase	4 mg (1.52 kU, 37°C)
Bovine serum albumin	10 mg
Water	800 μl
Total volume	3000 μl.

* Inclusion of ATP in this mixture is preferred to combination with the labelled ATP when membranes are pre-incubated (protection of the catalytic component).
In studies on the effect of Na^+, the buffer has to be adjusted by a base other than NaOH, and creatine phosphate-Na_2 has to be replaced by the corresponding Tris salt.
A heavy metal chelator (e.g. EDTA or EGTA) and a thiol (e.g. dithiothreitol) can be included in the assay but are not essential. Similarly, forskolin (10 – 30 μmol/l) can be included for adenylate cyclase stimulation, especially in studies on inhibition with enzymes of low activity.

13. [α-^{32}P]ATP (10 – 25 μCi/ml)*:

dilute labelled ATP from frozen stock solution (consider decay!) so that addition of 20 μl to each tube produces a radioactivity of 200000 – 500000 cpm per tube (this requires a solution with 20 – 50 μCi in 2 ml for 100 assay tubes).

Stability of solutions: store all solutions at 0 – 4 °C during use to avoid microbial growth, decomposition of creatine phosphate and nucleoside triphosphates, and thermal inactivation of adenylate cyclase. The working solutions can be used on the day of preparation.

Procedure

Preparation of specimen and sample, stability of adenylate cyclase in the sample: prepare plasma membranes according to standard procedures. Perform all steps of membrane preparation in the cold as the catalytic subunit of the adenylate cyclase system is very sensitive to elevated temperature and proteolytic attack eliminates hormonal inhibitions of the enzyme. After centrifugation, take the pellet up in cold diluted buffer and store on ice until use. EDTA, EGTA or a thiol can be added but are not required. Avoid vesicle formation as the membranes have to be accessible from either side. Freezing of membrane aliquots (at – 70 °C) for storage is generally possible.

For the determination of the blank value, prepare a small aliquot (sufficient for about 10 determinations) containing denatured membranes (i.e. heated for 5 min at 100 °C).

Assay conditions

Constant time assay should be carried out in a total volume of 100 μl in conical, 1.5 ml plastic tubes (*Eppendorf* type). Prepare the following samples (preferably in triplicates):

(1) samples containing [α-^{32}P]ATP and membranes, for the determination of [^{32}P]cAMP formed during the incubation under the various assay conditions tested.
(2) samples containing [α-^{32}P]ATP and denatured adenylate cyclase, for the determination of the blank value of the labelled ATP used,
(3) for the determination of the amount of [α-^{32}P]ATP added per assay tube, pipette 20 μl of the [α-^{32}P]ATP solution (13) into scintillation vials.

Time course assays: use 3.5 ml plastic tubes (12 × 55 mm). For a given number (n) of time points, pipette the (n + 1)-fold amounts each of

* As the ATP concentration is very low in this solution, it can be neglected for calculations. Unlabelled ATP, the concentration of which should be 0.05 to 0.2 mmol/l in the assay (the % substrate conversion will be too low with higher ATP concentrations) (9), can be combined with the tracer or can be included in the mixture containing the standard components.

(1) mixture of standard components (12),
(2) variable additions,
(3) [α-^{32}P]ATP solution (13), equilibrate-thermally,
(4) start by the addition of membrane suspension (15 s intervals between tubes); keep the volume in which the membranes are added small in order to minimize lowering of the assay temperature,
(5) stop the reaction at the desired time-points by transferring a 100 µl aliquot into a 1.5 ml conical plastic tube and immediately adding 400 µl $ZnCl_2$ solution (9),
(6) add 500 µl sodium carbonate solution (10).

Compounds added to all assay tubes, such as substrate, cofactors, inhibitors, ATP-regenerating system and buffer, should be included in the mixture. If more than 2 variable additions are required, another 10 or 20 µl of free volume can be gained by reducing the volume in which the membranes or the labelled substrate are added.

Measurement

Pipette successively into 1.5 ml tubes (on ice):			concentration in assay	
standard component mixture	(12)	30 µl	TEA	50 mmol/l
			$MgCl_2$	5 mmol/l
			IBMX	1 mmol/l
			cAMP	0.1 mmol/l
			ATP	0.1 mmol/l
			creatine-P	5 mmol/l
			CK	152 U/l
			albumin	1 mg/ml
variable addition A (e.g. a hormone)		10 µl		as required
variable addition B (e.g. GTP)		10 µl		as required
[α-^{32}P]ATP	(13)	20 µl		0.1 mmol/l
mix thoroughly, transfer vials at 15 s intervals to water-bath (25, 30 or 37 °C), allow to equilibrate thermally for 5 min				
membrane suspension, exactly on time (15 s intervals)		30 µl	volume fraction	0.3
mix, incubate for an exact period of time (2 to 30 min), stop by adding				
$ZnCl_2$	(9)	400 µl		50 mmol/l
Na_2CO_3	(10)	500 µl		62 mmol/l

Product isolation: labelled cAMP formed by the adenylate cyclase is separated from the remaining substrate and from side products such as 5′-nucleotides, P_i and PP_i by $ZnCO_3$ precipitation of these side products; this step is followed by chromatography on a short column of neutral alumina.

Centrifuge the samples for 3 – 5 min at about 10000 *g*. If only a low speed centrifuge (1000 – 2000 *g*) is available, freeze the samples prior to centrifugation, in order to form a coarser precipitate.

After centrifugation, transfer 800 µl of the supernatant fluid to the alumina columns. After draining of the sample, apply 2 ml Tris-HCl buffer (11), to be followed (after draining) by another 2 ml of this buffer. Collect the entire eluate in a liquid scintillation counting vial (20 or 5 ml capacity). ANDA (7-amino-1,3-naphthalene disulphonate in HCl, 0.1 mol/l, to give an acid pH value) can be added to a final concentration of 0.01 – 0.1% (w/v), in order to increase the counting efficiency in liquid scintillation counters with old photomultiplier tubes. Determine the amount of ^{32}P present in the eluate by measuring the *Čerenkov* radiation (counter settings as for ^{3}H).

Calculation: the catalytic activity related to the mass of protein (specific activity) is expressed as the formation of 1 pmol cAMP per min and mg protein. The amount of cyclic AMP formed is calculated from the fraction of [α-^{32}P]ATP converted to the cyclic nucleotide, considering the specific activity of ATP and the product recovery. The catalytic activity content is

$$z_c/m_s = \frac{C_p - C_{bl}}{R \times X \times m \times t} \text{ pmol} \times \text{min}^{-1} \times \text{mg}^{-1} (\text{mU/g})$$

where

C_p counting rate of the individual sample measured, cpm
C_{bl} counting rate of samples with denatured adenylate cyclase, cpm
R recovered fraction of [^{32}P]cAMP in the $ZnCO_3$ supernatant applied to the alumina column
X specific radioactivity of total ATP added per tube, cpm/pmol
m protein added per tube, mg
t incubation time, min.

Determination of blank values (C_{bl}) is necessary for every assay, as their magnitude varies with the batch and age of the [α-^{32}P]ATP used. In addition, non-enzymatic cAMP formation may occur. The determination of the blank values should be performed in separate samples containing all assay components including denatured adenylate cyclase. Blanks should be 0.02% or less with labelled ATP prepared as described [10, 11] and with commercial preparations.

Product recovery (R) is mainly the result of the fraction of the $ZnCO_3$ supernatant applied to the alumina column. Adsorption of cyclic AMP to $ZnCO_3$ and neutral alumina can be neglected; thus R will be 0.8 for the procedure suggested above.

Validation of Method

Precision, accuracy, detection limit and sensitivity: the standard deviation of adenylate cyclase determinations performed in triplicates is generally within a few %. On the other hand, the accuracy of adenylate cyclase determinations is poor as the enzyme is part of a membranous multi-component system which is affected by a variety of factors (cf. "General", pp. 369, 370). The detection limit has not been determined; it is possible to increase the sensitivity of the present method significantly by reducing the assay volume, increasing the specific activity of the substrate and other measures.

Sources of error: the complexity of the multi-component system makes the enzyme sensitive to various factors which are probably only partially known.

Specificity: the specificity of the assay depends largely on the labelled substrate used. Adenylate cyclase will also convert dATP and adenosine 5′-(β,γ-imido)triphosphate to the respective cyclic nucleotides.

References

[1] *E. M. Ross, A. G. Gilman,* Biochemical Properties of Hormone-sensitive Adenylate Cyclase, Annu. Rev. Biochem. *49*, 533 – 564 (1980).
[2] *M. Hirata, O. Hayaishi,* Adenyl Cyclase of Brevibacterium Liquefaciens, Biochim. Biophys. Acta *149*, 1 – 11 (1967).
[3] *T. Pfeuffer, H. Metzger,* 7-O-Hemisuccinyl-deacetyl Forskolin-Sepharose: a Novel Affinity Support for Purification of Adenylate Cyclase, FEBS Lett. *146*, 369 – 375 (1982).
[4] *K. H. Jakobs, K. Aktories, G. Schultz,* Inhibition of Adenylate Cyclase by Hormones and Neurotransmitters, Adv. Cyclic Nucleotide Res. *14*, 173 – 187 (1981).
[5] *M. Rodbell,* The Role of Hormone Receptors and GTP-regulatory Proteins in Membrane Transduction, Nature *284*, 17 – 22 (1980).
[6] *T. Katada, M. Ui,* Direct Modification of the Membrane Adenylate Cyclase System by Islet-activating Protein due to ADP-ribosylation of a Membrane Protein, Proc. Natl. Acad. Sci. USA *79*, 3129 – 3133 (1982).
[7] *K. B. Seamon, J. W. Daly,* Forskolin, Cyclic AMP and Cellular Physiology, Trends Pharmacol. Sci. *4*, 120 – 123 (1983).
[8] *G. Schultz,* General Principles of Assays for Adenylate Cyclase and Guanylate Cyclase Activity, in: *S. P. Colowick, N. O. Kaplan* (eds.), Methods in Enzymology, Vol. *XXXVIII*, Academic Press, New York 1974, pp. 115 – 125.
[9] *K. H. Jakobs, W. Saur, G. Schultz,* Reduction of Adenylate Cyclase Activity in Lysates of Human Platelets by the Alpha-adrenergic Component of Epinephrine, J. Cyclic Nucleotide Res. *2*, 381 – 392 (1976).
[10] *T. F. Walseth, R. A. Johnson,* The Enzymatic Preparation of (α-^{32}P)-Nucleoside Triphosphates, Cyclic (^{32}P)AMP, and Cyclic (^{32}P)GMP, Biochim. Biophys. Acta *562*, 11 – 31 (1979).
[11] *R. A. Johnson, T. F. Walseth,* The Enzymatic Preparation of (α-^{32}P)ATP, (α-^{32}P)GTP, (^{32}P)cAMP, and (^{32}P)cGMP, and their Use in the Assay of Adenylate and Guanylate Cyclases and Cyclic Nucleotide Phosphodiesterases, Adv. Cyclic Nucleotide Res. *10*, 135 – 167 (1979).
[12] *Y. Salomon, C. Londos, M. Rodbell,* A Highly Sensitive Adenylate Cyclase Assay, Anal. Biochem. *58*, 541 – 548 (1974).
[13] *Y. Salomon,* Adenylate Cyclase Assay, Adv. Cyclic Nucleotide Res. *10*, 35 – 55 (1979).
[14] *J. Wolff, C. Londos, D. M. F. Cooper,* Adenosine Receptors and the Regulation of Adenylate Cyclase, Adv. Cyclic Nucleotide Res. *14*, 199 – 214 (1981).

4.7 Guanylate Cyclase

GTP pyrophosphate-lyase (cyclizing), EC 4.6.1.2

Günter Schultz and Eycke Böhme

General

Guanylate cyclase, which was first described in 1969, has been detected in most types of mammalian cell, in lower animals, in unicellular organisms and in bacteria. In many mammalian tissues the enzyme occurs in two forms, one bound to membranes and the other found in the 100000 *g* supernatant fluid. Most tissues contain both forms, but the relative distribution varies from tissue to tissue, with extreme situations observed in intestinal mucosal cells, where guanylate cyclase is largely bound to plasma membranes, and in platelets, where the enzyme is almost completely soluble. Occurrence, properties and regulation of guanylate cyclase have recently been reviewed [1, 2]. Soluble and particulate forms of guanylate cyclase have been purified to apparent homogeneity from various mammalian tissues [2, 3] and from sea urchin sperm [4], respectively. Purified soluble guanylate cyclase has been obtained in forms that contain [5] or do not contain [6] a haem component.

The regulation of guanylate cyclase activity is poorly understood. Whereas the particulate enzyme from *Paramecium* underlies the control of calmodulin [7], the control of mammalian membrane-bound guanylate cyclase is not known. The soluble guanylate cyclase from mammalian tissues is controlled by redox mechanisms (for review see [1, 2]). Poly-unsaturated fatty acids, e.g. arachidonic acid, and possibly their hydroperoxides and oxygen radicals appear to act as physiological regulators in the course of hormonal regulations of cGMP* formation. In addition, various substances which inhibit smooth muscle contraction and platelet aggregation, and which contain NO or form NO-containing intermediates, can stimulate soluble guanylate cyclase (for review see [2]). The exact mechanism of this stimulation is not clear; some of these drugs may act as direct stimulants, whereas formation of active S-nitrosothiol intermediates may be required with others [8], and release of NO may also be involved.

Stimulation of soluble guanylate cyclase by NO-containing drugs is observed in crude enzyme preparations or with purified guanylate cyclase if the preparation is performed under reducing conditions preserving a haem component of the soluble enzyme; oxidation of crude or purified soluble guanylate cyclase reduces stimulation by NO-containing drugs. Stimulation by these drugs and by unsaturated fatty acids is

* Abbreviations: cGMP, cyclic GMP, guanosine 3′:5′-monophosphate; IBMX, 3-isobutyl-1-methylxanthine.

more pronounced when the enzyme activity is determined with Mg^{2+} rather than with Mn^{2+}, although the use of the latter cation results in much higher basal activities than are measured with Mg^{2+}, possibly as the result of a redox mechanism.

Application of method: determinations of guanylate cyclase activity are performed in biochemical and pharmacological studies on the control of cellular functions, especially in studies on drugs affecting vascular smooth muscle and platelet function. The assay conditions suggested below allow studies on guanylate cyclase regulation, at least by some cellular components and drugs.

Methods of determination: principles and problems of guanylate cyclase determination have been reviewed [9, 10]. Guanylate cyclase activity is generally very small; the activity of the membrane-bound form, especially, is lower than that of other particulate enzymes using GTP as a substrate. Whereas a sensitive assay (in general a radioimmunoassay) of cGMP formed is required for determination of particulate guanylate cyclase activity from most tissues [10], the activities of the particulate enzyme from intestinal mucosa and of soluble guanylate cyclases from most tissues can be assayed by use of [α-^{32}P]GTP or [^{3}H]GTP as a substrate. The method described below is based on the determination of [^{32}P]cGMP formed from [α-^{32}P]GTP, using an easy but effective isolation procedure for cGMP. The assay is similar to that described for adenylate cyclase activity in the preceding chapter [11]. This method provides high sensitivity and simplicity with ^{32}P-labelled GTP prepared as described [12, 13] and with some batches of commercial [α-^{32}P]GTP. The method allows 100 or more determinations within about half a working day.

International reference methods and standards: reference material is not available. Assay conditions have not been standardized but rather have to be adapted to the question under investigation.

Assay

Method Design

Principle

$$\text{GTP} \cdot \text{Me} \xrightarrow[\text{Me}^{2+}]{\text{guanylate cyclase}} \text{cGMP} + \text{PP}_i$$

where Me can be Mg or Mn.

For the assay described, the principles of determination and problems involved [9] are very similar to those described for the adenylate cyclase assay [11]. The rate of the guanylate cyclase-catalyzed reaction is determined by isolating cGMP and measuring the amount of [^{32}P]cGMP formed from [α-^{32}P]GTP.

The cGMP formed in the assay is separated from the remaining substrate and other ^{32}P-labelled products formed during the incubation period by a two-step procedure. After co-precipitation of 5′-nucleotides, of inorganic phosphate and pyrophosphate by zinc carbonate, the supernatant fluid is applied to an alumina column under neutral or acid conditions; acid conditions [14] generally provide a lower blank value, especially if commercial [α-^{32}P]GTP is used.

For every assay, the blank value has to be determined in separate samples; the magnitude of this blank depends on the batch and age of the labelled substrate used, on possible non-enzymatic formation of cGMP [15] and other factors. In addition, a few separate samples are included in every assay for determination of product recovery throughout the cGMP isolation procedure.

Optimized conditions for measurement: when the assay is performed with membranes or unpurified soluble guanylate cyclase, several side-reactions occurring simultaneously at much higher rates than the guanylate cyclase reaction have to be considered and overcome. These interfering reactions are in principle the same as those discussed for adenylate cyclase [11].

a) Degradation of GTP by GTPases and 5′-nucleotidases is overcome by inclusion of a GTP-regenerating system; creatine phosphate/creatine kinase are preferred to phosphoenol pyruvate/pyruvate kinase as phosphoenol pyruvate is a strong inhibitor of guanylate cyclase.

b) Degradation of labelled cGMP by cyclic nucleotide phosphodiesterases is reduced by a phosphodiesterase inhibitor, e.g. IBMX, and by unlabelled cGMP. When the assay is performed with purified enzyme preparations, addition of another protein, e.g. albumin, may be necessary for enzyme protection.

In any system in which guanylate cyclase is studied without previous experience within the laboratory, conditions have to be established under which product formation reflects initial velocity, with cGMP formation being proportional to the incubation time and the amount of protein used. As degradation of GTP is small with soluble guanylate cyclase preparations, cGMP formation will be linear over long periods of time provided that a thiol is added, which prevents time- and oxygen-dependent auto-activation of crude soluble guanylate cyclases (for review cf. [2]) and reduces enzyme inactivation by oxidation. With purified guanylate cyclase preparations, the addition of a protein will generally be necessary for enzyme protection. Addition of albumin results in guanylate cyclase activities higher than those seen with most other proteins, but the possibility of contamination of albumin with enzyme effectors, e.g. fatty acids, should be kept in mind.

Temperature conversion factors: have not been determined.

Equipment

The equipment required is in principle the same as that described for the adenylate cyclase assay [11] (cf. p. 372).

Reagents and Solutions

Purity of reagents: all reagents (except for those used in the product isolation procedure) should be of highest purity available. [α-^{32}P]GTP should be prepared as described [12, 13]. If commercial preparations have to be used, ethanol-free, low specific-activity material has to be ordered. [^{32}P]cGMP should be prepared from [α-^{32}P]GTP by use of a stimulated, purified or crude soluble guanylate cyclase, e.g. from lung or platelets, and should be purified as described below or elsewhere [16]. [^{3}H]cGMP used instead of [^{32}P]cGMP should also be purified and can be stored at −20°C with addition of ethanol to a final concentration of 50% (v/v) without significant decomposition.

Preparation of alumina columns: short columns of neutral alumina are prepared as described for the adenylate cyclase assay [11]. Pour dry neutral alumina (*Merck* aluminum oxide 90, *Woelm* N or similar material) into the columns (7−8 mm diameter) to 2 cm length. Wash the columns with 10 ml Tris-HCl buffer, 0.1 mol/l, pH 7.5 (see below). Because guanylate cyclase activity can vary considerably from sample to sample, especially if stimulatory drugs are being tested, variable amounts of radioactivity are retained; this results in considerable and variable blank values when the columns are re-used. Therefore, prepare columns generally from fresh alumina for every guanylate cyclase assay.

If acid conditions for the cyclic GMP isolation are used, pre-treat alumina columns with 2−5 ml perchloric acid, 0.2 mol/l (see below) instead of the Tris-HCl buffer.

Preparation of solutions: all solutions in re-purified water (cf. Vol. II, chapter 2.1.3.2).

Stock solutions (frozen stock solutions should be used whenever possible).

1. Triethanolamine-HCl buffer (1 mol/l; pH 7.4−8.0; consider dpK_a/dT = −0.020):

   dissolve 18.57 g triethanolamine-HCl in 50 ml water, adjust to the desired pH value of, e.g., 7.5, with NaOH, 1 mol/l, dilute to 100 ml with water.

2. $MgCl_2$ (100 mmol/l):

   dissolve 2.033 g $MgCl_2 \cdot 6\,H_2O$ in 100 ml water.

3. $MnCl_2$ (100 mmol/l):

   dissolve 1.979 g $MnCl_2 \cdot 4\,H_2O$ in 100 ml water.

4. IBMX (10 mmol/l):

   dissolve 44.4 mg IBMX in 20 ml water; heating is required to re-dissolve the compound after freezing and thawing.

5. GTP (10 mmol/l):

   dissolve 56.7 mg GTP, disodium salt, in 10 ml water.

6. Cyclic GMP (100 mmol/l):

   dissolve 73.4 mg cGMP, sodium salt, in 2 ml water.

The actual concentrations of the nucleotide solutions are determined and adjusted accordingly by measurement of the absorbance at 260 nm, using an absorption coefficient of 1.18 l × $mmol^{-1}$ × mm^{-1}. The pH value of these solutions is adjusted to 7.0 for the determination of A_{260} and for storage.

Other solutions

7. Zinc acetate (125 mmol/l):

   dissolve 27.44 g $(CH_3COO)_2Zn \cdot 2\ H_2O$ in 1 l water.

8. Sodium carbonate (125 mmol/l):

   dissolve 35.77 g $Na_2CO_3 \cdot 10\ H_2O$ in 1 l water.

9. Tris-HCl buffer (0.1 mol/l, pH 7.5):

   dissolve 12.11 g Tris in 0.9 l water, adjust the pH value to 7.5 by HCl, 1 mol/l, and fill up to 1 l with water.

10. Perchloric acid (0.2 mol/l):

    dilute 21.9 ml $HClO_4$ (60%, w/v) to 1 l with water.

11. Sodium acetate (0.25 mol/l):

    dissolve 340.2 g $CH_3COONa \cdot 3\ H_2O$ in 10 l water.

Working solutions (prepared fresh for every assay)

12. Standard component mixture:

    thaw the required frozen stock solutions and mix the following components to give a final volume sufficient for 100 tubes (or for the expected tube number with an excess volume for 3 to 5 tubes); add solid substances of creatine-P-$Na_2 \cdot 4\ H_2O$ and creatine kinase, lyophilized, ca. 380 U/mg (37°C).

Triethanolamine-HCl buffer (1)	500 µl
$MgCl_2$ solution (2)	500 µl
IBMX solution (4)	1000 µl

cGMP solution (6)	100 μl
GTP solution (5)	100 μl
Creatine-P-$Na_2 \cdot 4\,H_2O$	16.4 mg
Creatine kinase	4 mg (1.52 kU, 37°C)
[α-^{32}P]GTP	$2-5 \times 10^7$ cpm
Water to a final volume of	5000 μl.

13. Mixture for determination of product recovery:

prepare a mixture as required for 10 tubes, consisting of the same components as given above except for [α-^{32}P]GTP being replaced by [^{32}P]cGMP (2000 – 10000 cpm/tube). Alternatively, tritiated cGMP can be used for the determination of product recovery; in contrast to ^{32}P, the use of ^{3}H requires the use of an organic scintillation cocktail for counting.

A heavy metal chelator (e.g. EDTA or EGTA, ethylene glycol-bis-(β-aminoethyl-ether)N,N,N′,N′-tetraacetic acid) can be included but may affect guanylate cyclase activity by a redox mechanism. If a thiol, e.g. dithiothreitol, is required for guanylate cyclase protection and for stimulation by some NO-containing drugs, the thiol should be added with the enzyme preparation and the stimulatory drug, respectively. $MgCl_2$ can be replaced by $MnCl_2$ if high basal guanylate cyclase activity is desired, e.g., in studies localizing the enzyme activity; however, some regulations of guanylate cyclase activity may not be seen in the presence of manganese ions.

Stability of solutions: store all working solutions at 0 – 4°C during use to avoid microbial growth, decomposition of creatine-P and GTP and auto-activation of crude soluble guanylate cyclase preparations. The working solutions can only be used for one series of assays.

Procedure

Preparation of specimen: purification of soluble and particulate guanylate cyclases from various sources has been described [3 – 6, 18]. In addition, guanylate cyclase activity can be determined in high-speed centrifugation supernatants and membrane fractions after tissue disintegration.

Sample solution a. Guanylate cyclase solution for determination of enzyme activity: soluble guanylate cyclase is very sensitive to oxidation. Therefore, a thiol, e.g. glutathione, dithiotreitol or 2-mercaptoethanol, should be included in the buffer. Mix the solution containing the enzyme (purified enzyme, supernatant or particulate suspension) with thiol, protecting protein and buffer as required.

Sample solution b. Guanylate cyclase solution for determination of product recovery and blank values: this solution should contain the same components as given above except for the guanylate cyclase preparation being replaced by denatured enzyme; this is obtained by heating an aliquot of the enzyme solution for 5 min at 100°C.

Stability of enzyme preparations: a thiol compound should be added to the buffer to avoid oxidation of the enzyme. Freezing of crude soluble enzyme preparations is possible. Purified soluble guanylate cyclase can be stored at −70°C with addition of glycerol to a final concentration of 50% (v/v) for several months.

Assay conditions

Constant time assays should be carried out in a total volume of 100 µl in conical, 1.5 ml plastic tubes (*Eppendorf* type). Prepare the following samples (preferably in triplicates):

(1) samples containing [α-^{32}P]GTP solution (12) and guanylate cyclase (sample solution a), for the determination of [^{32}P]cGMP formed during the incubation period under the various assay conditions tested,
(2) samples containing [α-^{32}P]GTP solution (12) and denatured guanylate cyclase (sample solution b), for the determination of the blank value of the [α-^{32}P]GTP used and of the amount of [^{32}P]cGMP formed non-enzymatically,
(3) samples containing ^{32}P- or ^{3}H-labelled cGMP solution (13) instead of [α-^{32}P]GTP solution (12), for the determination of the product recovery,
(4) for the determination of the amount of [α-^{32}P]GTP added per assay tube, pipette 50 µl of the solution (12) into scintillation vials,
(5) for the determination of the amount of labelled cGMP added per tube, pipette 50 µl of solution (13) into scintillation vials.

Time course assays: for a given number (n) of time points, pipette the (n + 1)-fold amounts each of

(1) the standard component mixture (12) (except for the substrate),
(2) variable additions as required,
(3) the guanylate cyclase (sample solution a),
(4) start the reaction by the addition of [α-^{32}P]GTP in a small volume (15 s intervals between tubes),
(5) stop the reaction at the desired time points by transferring a 100 µl aliquot into a 1.5 ml conical plastic tube and immediately adding 400 µl of zinc acetate solution (7),
(6) add 500 µl sodium carbonate solution (8).

If pre-incubation of guanylate cyclase is desired, e.g. in studies on the auto-activation of the soluble enzyme form, the reaction can alternatively be started by addition of the substrate mixture. If more than 2 variable additions are required, another 10 or 20 µl of free volume can be gained by reducing the volume of standard component mixture (12) and/or enzyme solution.

Measurement

Pipette successively into 1.5 ml tubes (on ice):		concentration in assay	
standard component mixture (12)	50 µl	TEA buffer $MgCl_2$ IBMX cGMP GTP creatine-P CK	50 mmol/l 5 mmol/l 1 mmol/l 1 mmol/l 0.1 mmol/l $2-5 \cdot 10^5$ cpm 5 mmol/l 152 U/l
variable addition A (e.g. a drug) variable addition B (e.g. a thiol)	10 µl 10 µl	as required as required	
mix thoroughly, transfer vials at 15 s intervals to water-bath (30 or 37 °C). Allow to pre-incubate for thermal equilibration (2 – 5 min), add			
sample solution a exactly on time (15 s intervals)	30 ml	volume fraction	0.3
mix thoroughly, incubate for exactly the desired period of time, stop by adding			
zinc acetate solution (7) sodium carbonate solution (8)	400 µl 500 µl		50 mmol/l 62 mmol/l

Product isolation: labelled cGMP formed by the guanylate cyclase is separated from the remaining substrate and from side products such as 5′-nucleotides, inorganic phosphate and pyrophosphate by zinc carbonate co-precipitation of these side products; this step is followed by chromatography on a short alumina column.

Centrifuge the samples for 5 – 10 min at about 10000 *g*. If only a low speed centrifuge (1000 – 2000 *g*) is available, freeze the samples prior to centrifugation, in order to form a coarser precipitate. After centrifugation, transfer 800 µl of the supernatant fluid to a neutral alumina column. After draining of the sample, apply 5 ml Tris-HCl buffer (solution 9), to be followed (after draining) by another 5 ml of this buffer. Collect the entire eluate in a liquid scintillation vial (20 ml capacity). ANDA (7-amino-1,3-naphthalene disulphonate in HCl, 0.1 mol/l, to give an acid pH value) can be added to a final concentration of 0.01 – 0.1% (w/v), in order to increase the counting efficiency in liquid scintillation counters with old photomultiplier tubes. Determine the amount of ^{32}P present in the eluate by measuring the *Čerenkov* radiation (settings of the counter as for ^3H).

Lower blank values will result if the product isolation procedure is performed on acid alumina columns. Prior to the application of the sample (zinc carbonate supernatant fraction), add 2 ml of perchloric acid (solution 10) to the alumina column. While the acid is still draining, add the sample. Wash the column with 8 ml of water. Elute cGMP with 10 ml of sodium acetate (solution 11). Collect this eluate and count as described above.

Calculation: the catalytic activity related to the mass of protein (specific activity) is expressed as the formation of 1 pmol cGMP per min and mg protein. The amount of cGMP formed is calculated from the fraction of [α-^{32}P]GTP converted to cGMP, considering the specific activity of GTP and the product recovery. The catalytic activity content is

$$z_c/m_c = \frac{C_p - C_{bl}}{R \times X \times m \times t} \quad \text{pmol} \times \text{min}^{-1} \times \text{mg}^{-1} \qquad (\text{mU/g})$$

where

C_p counting rate of the individual sample measured, cpm
C_{bl} counting rate of samples with denatured guanylate cyclase, cpm
R recovered fraction of labelled cGMP added to separate samples, determined from the ratio of isolated to added labelled cGMP in separate samples
X specific radioactivity of total GTP added per tube, cpm/pmol
m protein added per tube, mg
t incubation time, min.

Determination of blank values C_{bl} is required in every assay, as their magnitude varies with the batch and age of [α-^{32}P]GTP used. In addition, non-enzymatic formation of cGMP can occur [14] and may be affected by components of the incubation mixture. The determination of the blank values should be performed with separate samples containing all assay components and denatured, i.e. heated, guanylate cyclase. Blanks should be 0.01 – 0.1% with labelled GTP prepared as described [12, 13] and with commercial substrate. Product recovery (R) is determined in separate samples which contain a known amount of ^{32}P- or ^{3}H-labelled cGMP (about 2000 – 10000 cpm/tube) instead of [α-^{32}P]GTP. Due to aliquoting and partial absorption of cGMP by zinc carbonate, the product recovery is generally about 60%.

Validation of Method

Precision, accuracy, detection limits and sensitivity: the standard deviation of guanylate cyclase determinations performed in triplicates is generally within 10%. On the other hand, the accuracy of these determinations, which has not been determined, is poor as it is affected by a number of factors including oxygen. The detection limit has also not been determined. It is possible to increase the sensitivity of the present method significantly by reducing the assay volume, increasing the specific activity of the substrate and other measures.

Sources of error: spontaneous activation of soluble guanylate cyclase has been mentioned above. In addition, guanylate cyclase stimulation or inhibition are possible with all kinds of factors and conditions affecting the redox state of the enzyme.

Specificity: the specificity of the assay depends largely on the labelled substrate used. Soluble guanylate cyclase will not only convert GTP but also dGTP and ITP to the respective cyclic nucleotides; in the presence of an NO-containing stimulatory drug and Mn^{2+}, the enzyme will even convert ATP to cAMP [17].

Reference ranges: specific activities of soluble and particulate guanylate cyclase forms vary from tissue to tissue [1 – 3]. Specific activities of about 0.1 and 10 U have been reported for soluble guanylate cyclase purified from bovine lung when measured under basal and stimulated conditions, respectively [18].

References

[1] *N. D. Goldberg, M. K. Haddox,* Cyclic GMP Metabolism and Involvement in Biological Regulation, Annu. Rev. Biochem. *46*, 823 – 896 (1977).

[2] *C. K. Mittal, F. Murad,* Guanylate Cyclase: Regulation of Cyclic GMP Metabolism, in: *J. A. Nathanson, J. W. Kebabian* (eds.), Handbook of Experimental Pharmacology *58/I,* Cyclic Nucleotides I, Springer, Berlin, Heidelberg, New York 1982, pp. 225 – 260.

[3] *D. L. Garbers, E. W. Radany,* Characteristics of the Soluble and Particulate Forms of Guanylate Cyclase, Adv. Cyclic. Nucleotide Res. *14*, 241 – 254 (1981).

[4] *D. L. Garbers,* Sea Urchin Sperm Guanylate Cyclase – Purification and Loss of Cooperativity, J. Biol. Chem. *251*, 4071 – 4077 (1976).

[5] *R. Gerzer, E. Böhme, F. Hofmann, G. Schultz,* Soluble Guanylate Cyclase Purified from Bovine Lung contains Heme and Copper, FEBS Lett. *132*, 71 – 74 (1981).

[6] *E. H. Ohlstein, K. S. Wood, L. J. Ignarro,* Purification and Properties of Heme-deficient Hepatic Soluble Guanylate Cyclase-effects of Heme and other Factors on Enzyme Activation by NO, NO-heme, and Protopohyrin-IX, Arch. Biochem. Biophys. *218*, 187 – 198 (1982).

[7] *S. Klumpp, J. E. Schultz,* Characterization of a Ca^{2+}-dependent Guanylate Cyclase in the Excitable Ciliary Membrane from Paramecium, Eur. J. Biochem. *124*, 317 – 324 (1982).

[8] *L. J. Ignarro, H. Lippton, J.C. Edwards, W. H. Baricos, A. L. Hyman, P. J. Kadowitz, C. A. Gruetter,* Mechanism of Vascular Smooth Muscle Relaxation by Organic Nitrates, Nitrites, Nitroprusside and Nitric Oxide – Evidence for the Involvement of S-nitrosothiols as Active Intermediates, J. Pharmacol. Exp. Ther. *218*, 739 – 749 (1981).

[9] *G. Schultz,* General Principles of Assays of Adenylate Cyclase and Guanylate Cyclase Activity, in: *S. P. Colowick, N. O. Kaplan* (eds.), Methods in Enzymology, Vol. *XXXVIII,* Academic Press, New York 1974, pp. 115 – 125.

[10] *D. L. Garbers, F. Murad,* Guanylate Cyclase Assay Methods, Adv. Cyclic. Nucleotide Res. *10*, 57 – 67 (1979).

[11] *G. Schultz, K. H. Jakobs,* Adenylate Cyclase, in: *H. U. Bergmeyer* (ed.), Methods of Enzymatic Analysis, 3rd edit., Vol. *IV*, Verlag Chemie, Weinheim 1984, pp. 369 – 378.

[12] *R. A. Johnson, T. F. Walseth,* The Enzymatic Preparation of (α-^{32}P)ATP, (α-^{32}P)GTP, (^{32}P)cAMP, and (^{32}P)cGMP, and their Use in the Assay of Adenylate and Guanylate Cyclases and Cyclic Nucleotide Phosphodiesterases, Adv. Cyclic Nucleotide Res. *10*, 135 – 167 (1979).

[13] *T. F. Walseth, R. A. Johnson,* The Enzymatic Preparation of (α-^{32}P)Nucleoside Triphosphates, Cyclic (^{32}P)cAMP, and (^{32}P)cGMP, Biochim. Biophys. Acta *562*, 11 – 31 (1979).

[14] *K. H. Jakobs, E. Böhme, G. Schultz,* Determination of Cyclic GMP in Biological Material, in: *J. E. Dumont, B. L. Brown, N. J. Marshall* (eds.), Eukaryotic Cell Function and Growth. Regulation by Intracellular Cyclic Nucleotides, Plenum, New York, London 1976, pp. 295 – 311.

[15] *H. Kimura, F. Murad,* Nonenzymatic Formation of Guanosine 3′,5′-monophosphate from Guanosine Triphosphate, J. Biol. Chem. *249*, 329 – 331 (1974).

[16] *G. Schultz, E. Böhme, J. G. Hardman,* Separation and Purification of Cyclic Nucleotides by Ion-exchange Resin Column Chromatography, in: *S. P. Colowick, N. O. Kaplan* (eds.), Methods in Enzymology, Vol. *XXXVIII,* Academic Press, New York 1974, pp. 9 – 20.

[17] *C. K. Mittal, F. Murad,* Formation of Adenosine 3′,5′-monophosphate by Preparations of Guanylate Cyclase from Rat Liver and Other Tissues, J. Biol. Chem. *252*, 3136 – 3140 (1977).

[18] *R. Gerzer, F. Hofmann, G. Schultz,* Purification of a Soluble, Sodium Nitroprusside Stimulated Guanylate Cyclase from Bovine Lung, Eur. J. Biochem. *116*, 479 – 486 (1981).

Appendix

1 Symbols, Quantities, Units and Constants

Units and symbols in ⟨ ⟩ should not be used.

m	Metre, m	g	Gram, g
mm	Millimetre, 10^{-3} m	mg	Milligram, 10^{-3} g
µm	Micrometre, 10^{-6} m	µg	Microgram, 10^{-6} g
nm	Nanometre, 10^{-9} m	ng	Nanogram, 10^{-9} g
		pg	Picogram, 10^{-12} g
l	Litre, 10^{-3} m^3	fg	Femtogram, 10^{-15} g
ml	Millilitre, 10^{-3} l	ag	Attogram, 10^{-18} g
µl	Microlitre, 10^{-6} l		
h	Hour	kg	kilogram, 10^3 g
min	Minute	mg	Megagram, 10^6 g
s	Second		

t	Time, s, (h, min)
Δt	Interval between measurements, s (h, min)
t	Temperature, °C
T	Temperature, K
K	Kelvin
V	Volume (usually volume of assay mixture), ml, l
v	Volume (usually volume of sample in assay mixture), ml, l
φ	Volume fraction of sample in assay mixture
m_s	Mass of substance, g
c	Substance concentration, mol/l
ρ	Mass concentration, g/l
n_c/m_s	Substance content, mol/kg
w	Mass fraction
⟨ppm	Parts per million⟩
⟨ppb	Parts per billion⟩
%	Percentage
% (v/v)	Percentage, volume related to volume
% (v/w)	Percentage, volume related to weight
% (w/v)	Percentage, weight related to volume
% (w/w)	Percentage, weight related to weight
M_r	Molecular weight, relative molecular mass

v	Rate of reaction, $mol \times l^{-1} \times s^{-1}$ ($\mu mol \times ml^{-1} \times min^{-1}$)
V	Maximum rate of reaction, $mol \times l^{-1} \times s^{-1}$
$\dot{\xi}$	Rate of conversion, $mol \times s^{-1}$ ($\mu mol \times min^{-1}$)
ν_i	Stoichiometric coefficient
kat	Katal, $mol \times s^{-1}$
U	International unit (for enzymes), $\mu mol \times min^{-1}$
mU	International milliunit, 10^{-3} U
kU	International kilounit, 10^3 U
MU	International megaunit, 10^6 U
Inh.U	Inhibitor unit, $\mu mol \times min^{-1}$
z	Catalytic activity, U, kat
b	Catalytic activity concentration, U/l, kat/l
z_c/m_s	Catalytic activity content (specific catalytic activity), U/g, kat/kg
ε	Linear molar absorption coefficient, $l \times mol^{-1} \times mm^{-1}$
d	Light path, mm
A	Absorbance
F	Fluorescence intensity
I	Light intensity
T	Transmission
Bq	Becquerel, s^{-1}
⟨Ci	Curie, s^{-1}⟩
cpm	Counts per minute, min^{-1}
dpm	Disintegrations per min, min^{-1}
C	Counting rate (cpm) of the radioactive product
η	Counting efficiency
X	Specific radioactivity, $Bq \times mol^{-1}$ ⟨$Ci \times mol^{-1}$⟩
Y	Number of counts measured
Z	Decay rate, disintegrations per minute, dpm
k	Reaction constant
K	Equilibrium constant
K'	Apparent equilibrium constant
K_m	Michaelis constant, mol/l
K_I	Inhibitor constant, mol/l
pH	Negative logarithm of the hydrogen ion concentration
pK	Negative logarithm of the dissociation constant, $-\log K$
$[\alpha]_D^{20}$	Specific rotation (sodium D-line at 20 °C)
sp.gr.	Specific gravity at 20 °C relative to water at 4 °C
g	Acceleration due to gravity; $9.81\ m \times s^{-2}$
rpm	Revolutions per minute, min^{-1}

J	Joule, $m^2 \times kg \times s^{-2}$
R	Gas constant, 8.312 $J \times mol^{-1} \times K^{-1}$
h	*Planck's* constant
ν	Frequence of emitted light
$\bar{x}$	Mean value
s, SD	Standard deviation
RSD	Relative standard deviation
⟨CV	Coefficient of variation⟩

2 Abbreviations for Chemical and Biochemical Compounds

It is unavoidable with the numerous abbreviations in use that one abbreviation occasionally is used for different compounds. In such cases the correct meaning can be obtained from the next. Only the unequivocal abbreviations are used in the book without further explanation.

A	Adenosine
ABTS®	2,2′-Azino-di-(3-ethylbenzthiazoline)-sulphonate
Acetoacetyl-CoA	Acetoacetyl-coenzyme A
Acetyl-CoA	Acetyl-coenzyme A
AChE	Acetylcholinesterase
AcP	Acid phosphatase
ADA	Adenosine deaminase
ADH	Alcohol dehydrogenase
ADP	Adenosine 5′-diphosphate
ADPglucose	Adenosine 5′-diphosphoglucose
ALAT (ALT, AlaAT)	Alanine aminotransferase
Alcohol-OD	Alcohol oxidase
ALD	Aldolase
AK	Acetate kinase
AK	Adenylate kinase
Ammediol	2-Amino-2-methyl-propane-1,3-diol
AMP	Adenosine 5′-monophosphate
A-2-MP	Adenosine 2′-monophosphate
A-3-MP	Adenosine 3′-monophosphate
A-3,(2)-MP	Adenosine 3′(2′)-monophosphate
A-3 : 5-MP (cAMP)	Adenosine 3′ : 5′-monophosphate, cyclic
AOD	Amino acid oxidase
AP	Alkaline phosphatase
Ap_5A	Diadenosine 5′-pentaphosphate
APAD	Acetylpyridine-adenine dinucleotide
APADH	Acetylpyridine-adenine dinucleotide, reduced
ASAT (AST; AspAT)	Aspartate aminotransferase
Ascorbate-OD	Ascorbate oxidase
ATP	Adenosine 5′-triphosphate
ATPase	Adenosine 5′-triphosphatase
BAEE	Benzoyl-L-arginine ethyl ester
BAPNA	N-Benzoyl-arginine-4-nitroanilide
Benzoyl-CoA	Benzoyl-coenzyme A
BES	Bis(2-hydroxyethylamino)ethanesulphonic acid
Bicine	N,N-Bis(2-hydroxyethyl)glycine

BMTD	6-Benzamido-4-methoxy-3-toluidinediazonium chloride
BSA	Bovine serum albumin
C	Cytidine
CDP	Cytidine 5′-diphosphate
CDPglucose	Cytidine 5′-diphosphoglucose
CE	Cholesterol esterase
Cellosolve	Ethylene glycol monomethyl ether
CHA	Cyclohexylammonium
ChE	Cholinesterase
ChOD	Cholesterol oxidase
CK	Creatine kinase
CL	Citrate lyase
CMP	Cytidine 5′-monophosphate
C-2-MP	Cytidine 2′-monophosphate
C-2:3-MP	Cytidine 2′:3′-monophosphate, cyclic
C-3-MP	Cytidine 3′-monophosphate
C-3,(2)-MP	Cytidine 3′(2′)-monophosphate
CoA	Coenzyme A
COX	Cytochrome c oxidase
CS	Citrate synthase
CTP	Cytidine 5′-triphosphate
Cyt-c	Cytochrome c
d	deoxy (prefix)
DAP	Dihydroxyacetone phosphate
DEA	Diethanolamine
DEAE	Diethylaminoethyl
DFP	Diisopropylfluorophosphate
DM-POPOP	2,2′-(1,4-Phenylene)bis(4-methyl-5-phenyloxazole)
DMSO	Dimethylsulphoxide
DNase	Deoxyribonuclease
DNP	Dinitrophenylhydrazine
DTNB	5,5′-Dithiobis(2-nitrobenzoic acid)
DTT	Dithiothreitol
EDTA	Ethylenediamine tetraacetate
EGTA	Ethyleneglycol-bis-(β-aminoethyl ether) N,N′-tetraacetic acid
FAD	Flavin-adenine dinucleotide
FDH	Formate dehydrogenate
FDNB	1-Fluoro-2,4-dinitrobenzene
FH_2	Dihydrofolate
FH_4	Tetrahydrofolate

FiGlu	N-Formimino-L-glutamate
FMN	Flavin mononucleotide
F-1-P	Fructose 1-phosphate
F-1,6-P_2	Fructose 1,6-bisphosphate
F-2,6-P_2	Fructose 2,6-bisphosphate
F-6-P	Fructose 6-phosphate
F-6-PK	Fructose-6-phosphate kinase
GABAse	Mixture of GAB-AT and SS-DH
GAB-AT	γ-Aminobutyrate aminotransferase
Gal-DH	Galactose dehydrogenase
Gal-OD	Galactose oxydase
GAP	Glyceraldehyde phosphate
GAPDH	D-Glyceraldehyde-3-phosphate dehydrogenase
GDH	L-Glycerol-3-phosphate dehydrogenase (glycerol-1-phosphate dehydrogenase; α-glycerophosphate dehydrogenase)
GDP	Guanosine 5′-diphosphate
GlDH	L-Glutamate dehydrogenase
GMP	Guanosine 5′-monophosphate
G-2-MP	Guanosine 2′-monophosphate
G-3-MP	Guanosine 3′-monophosphate
G-3,(2)-MP	Guanosine 3′(2′)-monophosphate
G-3:5-MP (cGMP)	Guanosine 3:5-monophosphate, cyclic
GMPK	GMP kinase
GOD	Glucose oxidase
GOT	Glutamate-oxaloacetate transaminase/Aspartate aminotransferase
G-1-P	Glucose 1-phosphate
G-1,6-P_2	Glucose 1,6-bisphosphate
G-6-P	Glucose 6-phosphate
G6P-DH	Glucose-6-phosphate dehydrogenase
GPT	Glutamate-pyruvate transaminase/Alanine aminotransferase
GR	Glutathione reductase
GSH	Glutathione, reduced
GSSG	Glutathione, oxidized
γ-GT	γ-Glutamyltransferase
GTP	Guanosine 5′-triphosphate
Hb	Haemoglobin
Hepes	N-2-Hydroxyethyl piperazine-N′-2-ethanesulphonic acid
HK	Hexokinase
HMG-CoA	3-Hydroxy-3-methylglutaryl-coenzyme A
I	Inosine
IDP	Inosine 5′-diphosphate

Ig	Immunoglobulin
IMP	Inosine 5′-monophosphate
INT	2-(4-Iodophenyl)-3-(4-nitrophenyl)-5-phenyltetrazolium-chloride
ITP	Inosine 5′-triphosphate
LDH (L-LDH)	L-Lactate dehydrogenase
D-LDH	D-Lactate dehydrogenase
MDH	L-Malate dehydrogenase
MES	2-(N-Morpholino)ethanesulphonic acid
4-Met-um	4-Methylumbelliferon (4-Methylumbelliferyl)
MK	Myokinase/adenylate kinase
Mops	3-(N-Morpholino)propanesulphonic acid
4-NA	4-Nitroaniline
NAC	N-Acetyl cysteine
NAD	Nicotinamide-adenine dinucleotide
NADH	Nicotinamide-adenine dinucleotide, reduced
NADP	Nicotinamide-adenine dinucleotide phosphate
NADPH	Nicotinamide-adenine dinucleotide phosphate, reduced
NBT	Nitro-BT-tetrazolium salt; 2,2′-bis(4-nitrophenyl-5,5′-diphenyl-3,3′-(-dimethoxy-4,4′-diphenylene)-ditetrazolium chloride
NBTH	N-Methyl-2-benzothiazolone hydrazone
NDPK	Nucleoside diphosphate kinase
NMN	Nicotinamide mononucleotide
NMPK	Nucleoside monophosphate kinase
NP	Nucleoside phosphorylase
4-NP	4-Nitrophenol (4-Nitrophenyl)
4-NP-G_1	4-Nitrophenyl α-D-glucopyranoside
4-NP-G_2	4-Nitrophenyl [α-D-glucopyranosyl-(1 → 4)-α-D-glucopyranoside]/4-Nitrophenyl α-D-maltoside
4-NP-G_3	4-Nitrophenyl {di[α-D-glucopyranosyl-(1 → 4)]-α-D-glucopyranoside}/4-Nitrophenyl α-D-maltotrioside
4-NP-G_4	4-Nitrophenyl {tri[α-D-glucopyranosyl-(1 → 4)]-α-D-glucopyranoside}/4-Nitrophenyl α-D-maltotetraoside
4-NP-G_5	4-Nitrophenyl {tetra[α-D-glucopyranosyl-(1 → 4)]-α-D-glucopyranoside}/4-Nitrophenyl α-D-maltopentaoside
4-NP-G_6	4-Nitrophenyl {penta[α-D-glucopyranosyl-(1 → 4)]-α-D-glucopyranoside}/4-Nitrophenyl α-D-maltohexaoside
4-NP-G_7	4-Nitrophenyl {hexa[α-D-glucopyranosyl-(1 → 4)]-α-D-glucopyranoside}/4-Nitrophenyl α-D-maltoheptaoside
4-NPgal	4-Nitrophenyl galactoside
4-NPgluc	4-Nitrophenyl glucoside

4-NPman	4-Nitrophenyl mannoside
4-NPP	4-Nitrophenyl phosphate
NTA	Nitriloacetate
OMP	Orotidine 5′-monophosphate
PALP	Pyridoxal 5′-phosphate
PAMP	Pyridoxamine 5′-phosphate
PDE	Phosphodiesterase
PEP	Phosphoenolpyruvate
6-PGDH	6-Phosphogluconate dehydrogenase
PGK	3-Phosphoglycerate kinase
PGluM	Phosphoglucomutase
PGM	Phosphoglycerate mutase
P_i	Inorganic phosphate
Pipes	Piperazine-N,N′-bis(2-ethanesulphonic acid)
PK	Pyruvate kinase
PL-A	Phospholipase A
PL-C	Phospholipase C
PL-D	Phospholipase D
PMS	Phenazine methosulphate
POD	Peroxidase
POPOP	2,2′-(1,4-Phenylene)bis[5-phenyloxazole]
PPase	Pyrophosphatase, inorganic
PP_i	Inorganic pyrophosphate
PPO	2,5-Diphenyloxazole
PRPP	5-Phospho-α-D-ribose 1-diphosphate
PTA	Phosphotransacetylase
PyDC	Pyruvate decarboxylase
RNA	Ribonucleic acid
rRNA	Ribosomal ribonucleic acid
tRNA	Transfer ribonucleic acid
RNase	Ribonuclease
SOD	Superoxide dismutase
SS-DH	Succinate-semialdehyde dehydrogenase
ST	Sialyltransferase
SUPHEPA	N-Succinyl-L-phenylalanine-4-nitroanilide
T	Ribosylthymidine
dT	Thymidine
TA	Transaldolase
TAT (TyrAT)	Tyrosine aminotransferase
dTDP	Thymidine 5′-diphosphate

dTDPglucose	Thymidine 5′-diphosphoglucose
TCA	Trichloroacetic acid
TEA	Triethanolamine
THF	Tetrahydrofolic acid
TIM	Triosephosphate isomerase
TPP	Thiamine pyrophosphate
Tricine	N,N,N-Tris(hydroxymethyl)methylglycine
Tris	Tris(hydroxymethyl)aminomethane
dTTP	Thymidine 5′-triphosphate
U	Uridine
UDP	Uridine 5′-diphosphate
UDPG (UDPglucose)	Uridine 5′-diphosphoglucose
UDPAG	Uridine 5′-diphospho-N-acetylglucosamine
UDPGA	Uridine 5′-diphosphoglucuronate
UDPGal	Uridine 5′-diphosphogalactose
UDPG-DH	UDPglucose dehydrogenase
UDPG-PP	UDPglucose pyrophosphorylase
UMP	Uridine 5′-monophosphate
U-2-MP	Uridine 2′-monophosphate
U-2:3-MP	Uridine 2′:3′-monophosphate, cyclic
U-3-MP	Uridine 3′-monophosphate
U-3,(2)-MP	Uridine 3′(2′)-monophosphate
UTP	Uridine 5′-triphosphate
X	Xanthosine
XOD	Xanthine oxidase

3 Formulae

Experimental data must be converted to concentrations or contents; only then one can speak of a result (cf. Vol. II, chapter 3.1). According to international recommendations [1, 2] the following symbols and units are equivalent:

Metabolites		*Enzymes*	
Substance concentration c	mol/l	Catalytic activity concentration b	kat/l; U/l
	mmol/l		mkat/l
Mass concentration ρ	g/l		μkat/l
	mg/l		nkat/l
Substance content n_c/m_s	mol/kg	Catalytic activity content z_c/m_s	kat/kg
	mmol/g		(U/g, U/mg)
Mass fraction w	1		mkat/g
	0.001		μkat/g

The use of the basic kind of quantities of the SI (cf. Vol. I, pp. 10, 22) involves some changes in the symbols, quantities and units formerly used. For example, the base unit of length is the metre (m) and thousands or thousandths of it. The path length of a photometer cuvette is in general 10 mm (not 1 cm). Accordingly, the unit of the absorption coefficient also changes ($l \times mol^{-1} \times mm^{-1}$, instead of $l \times mol^{-1} \times cm^{-1}$).

For example, for NADH at 339 nm $\varepsilon = 6.3 \times 10^2$ $l \times mol^{-1} \times mm^{-1}$ instead of 6.3×10^3 $l \times mol^{-1} \times cm^{-1}$. For practical purposes it is proposed to use the path length $d = 10$ mm and the concentration c in mmol/l (μmol/ml). Then for NADH at 339 nm $\varepsilon \times d = 6.3$ $l \times mmol^{-1}$ applies.

The following symbols, kind of quantities and units are customary in this context. Throughout this book we use v, V for reaction rates and v, V for volumes (deviating from international recommendations). We also write T for transmission of light and T for thermodynamic temperature.

Symbols in radiometry are still not completely coherent with those in other laboratory sciences, also for various commonly used terms apparently no official symbols exist (cf. [3]).

General

c	substance concentration, mol/l
ρ	mass concentration, g/l
n_c/m_s	substance content, mol/kg
w	mass fraction, 1

MW	mass of one millimole or mole of substrate, mg/mmol, g/mol
V	assay volume, l
v	volume of sample used in assay, l
φ	volume fraction of sample in assay (incubation) mixture, v/V
t	time, s (min, h)
z	catalytic activity, kat, U
b	catalytic activity concentration, kat/l, U/l
z_c/m_s	catalytic activity content (specific catalytic activity), kat/kg, U/g
ν_i	stoichiometric coefficient

Photometry, fluorimetry, luminometry

A	absorbance
ε	linear millimolar absorption coefficient, $l \times mmol^{-1} \times mm^{-1}$
I	light intensity
F	fluorescence intensity
d	light path, mm

For experimental data obtained by fluorimetric methods, F and ΔF are used instead of A and ΔA. In luminometry I and I_0 are used.

Radiometry

Y	number of counts measured
Z	decay rate, disintegrations per minute, dpm
C_b	background counting rate, cpm
C_p	sample (probe) counting rate, cpm
C_n	net counting rate, cpm
η	counting efficiency, cpm/dpm
X	specific radioactivity, Bq/mol (Ci/mol)
v_r	volume taken for scintillation counting, l
cpm	counts per minute
dpm	disintegrations per minute
2.22×10^{12}	factor for conversion of decay rate Ci to dpm (1 Ci = 2.22×10^{12} dpm)
3.7×10^{10}	factor for conversion of Ci to Bq (1 Ci = 3.7×10^{10} Bq)
2.7×10^{-11}	factor for conversion of Bq to Ci (1 Bq = 2.7×10^{-11} Ci)

Metabolites

Photometry, fluorimetry, luminometry

From *Lambert-Beer*'s law (cf. Vol. I, p. 283) follows

$$\text{(a)} \qquad c = \frac{\log I_0/I}{\varepsilon \times d} = \frac{A}{\varepsilon \times d} \quad \text{mmol/l}.$$

For chemical reactions this gives

$$c_1 - c_2 = \frac{A_1 - A_2}{\varepsilon \times d}; \qquad \Delta c = \frac{\Delta A}{\varepsilon \times d} \quad \text{mmol/l}$$

and for complete conversion ($c_2 = 0$)

(a$_1$) $$c = \frac{\Delta A}{\varepsilon \times d} \quad \text{mmol/l} \qquad \text{(in the cuvette)}.$$

For the determination of the concentration of the analyte in the sample the ratio of assay volume : sample volume (V : v) or the volume fraction of sample in the assay (v/V) = φ), respectively, must be considered:

(b) $$c = \frac{\Delta A \times \mathrm{V}}{\varepsilon \times d \times \mathrm{v}} = \frac{\Delta A}{\varepsilon \times d \times \varphi} \quad \text{mmol/l} \qquad \text{(in the sample)}$$

(b$_1$) $$\rho = \frac{\Delta A \times \mathrm{V} \times \mathrm{MW}}{\varepsilon \times d \times \mathrm{v}} = \frac{\Delta A \times \mathrm{MW}}{\varepsilon \times d \times \varphi} \quad \text{mg/l} \qquad \text{(in the sample)}.$$

The substance content of the sample n_c/m_s (i.e. of the analyte in the material under investigation) is

(c) $$n_c/m_s = \frac{\Delta A \times \mathrm{V}}{\varepsilon \times d \times \mathrm{v} \times \rho_{\text{sample}}} = \frac{\Delta A}{\varepsilon \times d \times \varphi \times \rho_{\text{sample}}} \quad \text{mmol/g}.$$

The mass fraction of the analyte in the sample (g/g) is

(c$_1$) $$w = \frac{\Delta A \times \mathrm{V} \times \mathrm{MW}}{\varepsilon \times d \times \mathrm{v} \times \rho_{\text{sample}}} = \frac{\Delta A \times \mathrm{MW}}{\varepsilon \times d \times \varphi \times \rho_{\text{sample}}}.$$

If two or more moles of light-absorbing reaction products (e.g. NADH) are formed or consumed per mole substrate that reacts, the corresponding stoichiometric coefficient ν_i (cf. Vol. I, p. 10) also appears in the denominator of eqns. (b) and (c). If several moles of one substrate go to form the unit of substance on which the measurement is based, the corresponding factor appears in the numerator.

The result for the sample in relation to a standard is

(d) $$Y = \frac{c_{\text{sample (measured)}}}{c_{\text{standard (measured)}}} \times c_{\text{standard (weighed out)}} \quad \text{mmol/l}$$

or

(d$_1$) $$c = \frac{A_{\text{sample (measured)}}}{A_{\text{standard (measured)}}} \times c_{\text{standard (weighed out)}} \quad \text{mmol/l}.$$

From the concentration of the substance in the sample solution (e.g. tissue extract), the content of the substance in the material under investigation is calculated by relating to its mass concentration in the standard solution:

$$\text{(e)} \qquad n_c/m_s = \frac{c_{\text{sample (measured)}}}{\rho_{\text{standard (weighed out)}}} \qquad \frac{\text{mmol/l}}{\text{mg/l}} = \text{mmol/mg}\,,$$

$$\text{(f)} \qquad w = \frac{\rho_{\text{sample (measured)}}}{\rho_{\text{standard (weighed out)}}} \qquad \frac{\text{mg/l}}{\text{mg/l}} = 1\,.$$

Example

Determination of fructose 1,6-bisphosphate (molecular weight: 340) in rat liver. Enzymatic analysis with aldolase/triosephosphate isomerase/glycerophosphate dehydrogenase. Two moles of NADH are transformed per mole of fructose-1,6-P_2. To prepare an "extract", 1 g of fresh liver was homogenized in 7.25 ml of $HClO_4$. With a value of 75% (w/w) for the fluid content of the liver, the volume of the extract is 7.25 + 0.75 = 8.00 ml. To neutralize and remove perchlorate, 0.2 ml K_2CO_3 solution was added to 6 ml of the extract. The volume of the perchlorate-free extract is thus 6.2 ml. The dilution factor for the extract is 6.2/6.0 = 1.033, and that for the tissue is 8 × 6.2/6.0 = 8.267. The experimental data must be multiplied by these values to express the results per 1 ml of acid extract or per 1 g of tissue.

The measured change in absorbance at 339 nm ($\varepsilon = 0.63\ \text{l} \times \text{mmol}^{-1} \times \text{mm}^{-1}$) was $\Delta A = 0.120$; the volume of the assay solution was 3×10^{-3} l, the volume of sample was 1.5×10^{-3} l, and the light path was 10 mm. The concentration in the perchlorate-free sample used for the assay was:

according to eqn. (b)

$$c = \frac{0.120 \times 3 \times 10^{-3}}{0.63 \times 10 \times 1.5 \times 10^{-3} \times 2} = 0.0190\ \text{mmol/l}$$

or according to eqn. (c)

$$\rho = \frac{0.120 \times 3 \times 10^{-3} \times 340}{0.63 \times 10 \times 1.5 \times 10^{-3} \times 2} = 6.46\ \text{mg/l}\,.$$

Multiplication by the dilution factor gives the concentration in the acid extract:

$$c = 0.0190 \times 1.033 = 0.0196\ \text{mmol/l}$$

$$\rho = 6.46 \times 1.033 = 6.67\ \text{mg/l}\,.$$

The substance content of fructose-1,6-P_2 in the tissue is

$$n_c/m_s = 0.0190 \times 8.267 = 0.157 \text{ mmol/kg}$$

or its mass fraction (g/g)

$$w = 6.46 \times 10^{-6} \times 8.267 = 5.4 \times 10^{-5}.$$

Radiometry

For calculation of the analyte concentration and content from the net counting rate C_n of sample and blank, the following relationships are valid.

Substance concentration

(g) $$c = \frac{\Delta C_n}{\eta \times 2.22 \times 10^{12} \times X \times \varphi \times v_r} \text{ mol/l (in the sample)}$$

(units: C_n in cpm; X in Ci/mol or Bq/mol; $\varphi = v/V$ in l/l; v_r in l).

Mass concentration

(g₁) $$\rho = \frac{\Delta C_n \times MW}{\eta \times 2.22 \times 10^{12} \times X \times \varphi \times v_r} \text{ g/l (in the sample)}$$

(units as for eqn. (g); MW in g/mol).

Substance content

(h) $$n_c/m_s = \frac{\Delta C_n}{\eta \times 2.22 \times 10^{12} \times X \times \varphi \times v_r \times \rho_{sample}} \text{ mol/g (in the sample)}$$

(units as for eqn. (g); ρ_{sample} in g/l).

Mass fraction

(h_1) $$w = \frac{\Delta C_n \times MW}{\eta \times 2.22 \times 10^{12} \times X \times \varphi \times v_r \times \rho_{sample}}$$

(units as for eqn. (g_1); ρ_{sample} in g/l).

Enzymes

Photometry, fluorimetry, luminometry

For measurement of the catalytic activity z of enzymes the rate of the catalyzed substrate conversion per time unit is used, µmol/min or mol/s.

Concerning the stoichiometric coefficient ν_i cf. p. 403.

According to eqn. (a)

catalytic activity (conversion in mol* per time unit) is

(i) $$z = \frac{\Delta c \times V}{\Delta t} = \frac{\Delta A \times V}{1000 \times \varepsilon \times d \times \Delta t} \quad \text{mol/s (kat)}$$

(units: ε in $l \times mmol^{-1} \times mm^{-1}$; V of the assay volume in l; d in mm; t in s)

(i_1) $$z = \frac{\Delta c \times V}{\Delta t} = \frac{\Delta A \times V \times 1000}{\varepsilon \times d \times \Delta t} \quad \mu\text{mol/min (U)}$$

(units as for eqn. (i); t in min).

The catalytic activity concentration in the sample is

(k) $$b = \frac{\Delta A \times V}{1000 \times \varepsilon \times d \times \Delta t \times v} = \frac{\Delta A}{1000 \times \varepsilon \times d \times \Delta t \times \varphi} \quad \text{mol} \times \text{s}^{-1} \times \text{l}^{-1} \text{(kat/l))}$$

(units as for eqn. (i); v in l; φ = v/V in l/l).

(k_1) $$b = \frac{\Delta A \times V \times 1000}{\varepsilon \times d \times \Delta t \times v} = \frac{\Delta A \times 1000}{\varepsilon \times d \times \Delta t \times \varphi} \quad \mu\text{mol} \times \text{min}^{-1} \times \text{l}^{-1} \text{(U/l)}$$

(units as for eqn. (i_1); v in l; φ = v/V in l/l).

The catalytic activity related to the mass of protein, catalytic activity content z_c/m_s (specific activity) is

(l) $$z_c/m_s = \frac{\Delta A \times V}{1000 \times \varepsilon \times d \times \Delta t \times v \times \rho_{\text{protein}}} = \frac{\Delta A}{1000 \times \varepsilon \times d \times \Delta t \times \varphi \times \rho_{\text{sample}}} \quad \text{kat/g}$$

(units as for eqn. (k); ρ_{protein} in g/l)

(l_1) $$z_c/m_s = \frac{\Delta A \times V \times 1000}{\varepsilon \times d \times \Delta t \times v \times \rho_{\text{protein}}} = \frac{\Delta A \times 1000}{\varepsilon \times d \times \Delta t \times \varphi \times \rho_{\text{protein}}} \quad \text{U/g}$$

(units as for eqn. (k_1); ρ_{protein} in g/l).

* $c \times V$ (mol/l) × l = mol.

Example

Determination of the (specific) catalytic activity of an enzyme. The measurements were made at 339 nm; $\Delta A/\Delta t = 0.063/60$ s in a 3 ml assay mixture ($V = 3 \times 10^{-3}$ l). The volume of the sample was 2×10^{-4} l. The sample diluted over thousandfold for measurement contained 10 g of enzyme protein per litre.

In the assay mixture the catalytic activity is according to eqn. (i) or (i_1), respectively

$$z = \frac{0.063 \times 3 \times 10^{-3}}{10^3 \times 0.63 \times 10 \times 60} = 5 \times 10^{-10}\,\text{kat} = 0.5\,\text{nkat}$$

or

$$z = \frac{0.063 \times 3 \times 10^{-3} \times 10^3}{0.63 \times 10 \times 1} = 0.03\ \text{U}.$$

The catalytic activity concentration in the sample solution according to eqn. (k) or (k_1), respectively, is

$$b = \frac{0.063 \times 3 \times 10^{-3}}{10^3 \times 0.63 \times 10 \times 60 \times 2 \times 10^{-4}} = 2500\,\text{nkat/l}$$

or

$$b = \frac{0.063 \times 3 \times 10^{-3} \times 10^3}{0.63 \times 10 \times 1 \times 2 \times 10^{-4}} = 150\,\text{U/l}\,.$$

Related to the mass of protein (dilution factor 1000), the catalytic activity content is

$$z_c/m_s = \frac{2500 \times 1000}{10} \; \frac{\text{nkat/l}}{\text{g/l}} = 250\,\mu\text{kat/g}$$

or

$$z_c/m_s = \frac{150 \times 1000}{10} \; \frac{\text{U/l}}{\text{g/l}} = 15\,\text{U/g}.$$

The result for the sample in relation to an enzyme standard solution is

$$\text{(m)} \qquad b = \frac{b_{\text{sample (measured)}}}{b_{\text{standard (measured)}}} \times b_{\text{standard (indicated)}} \quad \text{nkat/l (U/l)}$$

or simply

$$(\text{m}_1) \qquad b = \frac{A_{\text{sample (measured)}}}{A_{\text{standard (measured)}}} \times b_{\text{standard (indicated)}} \quad \text{nkat/l (U/l)}$$

Radiometry

According to eqns. (k) and (k_1), respectively, and using eqns. (g) and (h)

catalytic activity is

(n) $$z = \frac{\Delta C_n \times V}{\eta \times 2.22 \times 10^{12} \times 10^{-9} \times X \times v_r \times \Delta t} \quad \text{nmol/s (nkat)}$$

(units: C_n in cpm; X in Ci/mol or Bq/mol; v_r in l; t in s),

(n_1) $$z = \frac{\Delta C_n \times V}{\eta \times 2.22 \times 10^{12} \times 10^{-6} \times X \times v_r \times \Delta t} \quad \mu\text{mol/min (U)}$$

(units as for eqn. (n); t in min).

The catalytic activity concentration in the sample is

(o) $$b = \frac{\Delta C_n}{\eta \times 2.22 \times 10^{12} \times 10^{-9} \times X \times \varphi \times v_r \times \Delta t} \quad \text{nkat/l}$$

(units as in eqns. (n), (n_1); φ = v/V in l/l),

(o_1) $$b = \frac{\Delta C_n}{\eta \times 2.22 \times 10^{12} \times 10^{-6} \times X \times \varphi \times v_r \times \Delta t} \quad \text{U/l}$$

(units as in eqn. (o); t in min).

The catalytic activity related to the mass of protein, catalytic activity content z_c/m_s (specific activity), is

(p) $$z_c/m_s = \frac{\Delta C_n}{\eta \times 2.22 \times 10^{12} \times 10^{-9} \times X \times \varphi \times v_r \times \Delta t \times \rho_{protein}} \quad \text{nkat/g}$$

(units as for eqn. (o); $\rho_{protein}$ in g/l),

(p_1) $$z_c/m_s = \frac{\Delta C_n}{\eta \times 2.22 \times 10^{12} \times 10^{-6} \times X \times \varphi \times v_r \times \Delta t \times \rho_{protein}} \quad \text{U/g}$$

(units as for eqn. (o_1); $\rho_{protein}$ in g/l).

Statistics

For parameters and sample statistics, for mathematical evaluation of experiments (mean value, standard deviation, imprecision, inaccuracy) and for statistical evaluation of results, especially for regression analysis, cf. Vol. II, chapter 3.2.

Refering the Experimental Results to Biological Material

In the analysis of organ extracts, blood, serum, etc., the dilution resulting from deproteinization of samples must also be taken into account along with the fluid content of the sample. The following values give reasonable accuracy: blood 80% (w/w), tissue (liver, kidney, muscle, heart) 75% (w/w). Tissue samples are weighed out; a sp. gr. of 1.06 is used for the conversion of blood volumes into mass.

Apart from the volume (in the case of serum, plasma, blood, urine, etc.), other reference quantities that may be used for biological material are the fresh weight, dry weight, total nitrogen, protein content, protein nitrogen; cell count, e.g. erythrocyte count; haemoglobin content, cytochrome c content, and dry weight of the cell-free sample solution.

Examples

Determination of the dilution factor for blood.

The specific gravity of blood is taken as 1.06, and its fluid content is taken as 80% (w/w). 2 ml of blood are deproteinized with 3 ml of perchloric acid and centrifuged, and 2.5 ml of the supernatant are neutralized with 1 ml K_2CO_3 solution. The blood is therefore diluted by the following factor:

$$F = \frac{2 \times 1.06 \times 0.8 + 3}{2} \times \frac{2.5 + 1}{2.5} = \frac{1.696 + 3}{2} \times \frac{3.5}{2.5} = 3.29\,.$$

The experimental result obtained with the neutralized blood extract must be multiplied by this factor to obtain the content of the metabolite in the blood.

Calculation of the dilution factor for multi-stage assay mixtures.
In the determination of maltose in biological fluids, the various reaction steps have different pH-optima. The assay begins with an incubation: 0.5 ml sample with 0.5 ml acetate buffer and 0.02 ml α-glucosidase. After inactivation of the enzyme, the second step is the determination of the glucose in 0.2 ml of the incubation solution (V = 3.42 ml). The dilution factor is thus:

$$F = \frac{0.5 + 0.5 + 0.02}{0.5} \times \frac{3.42}{0.2} = 2.04 \times 17.1 = 34.88\,.$$

The experimental result, after division by 2 (1 maltose ≙ 2 glucose), must be multiplied by this factor to obtain the maltose content in 1 ml sample.

To calculate the metabolite content of the cells of a tissue, the metabolite content of the blood in this tissue must be taken into account.

The mass fraction of blood in the tissue is determined according to *Bücher et al.* [4] from absorbance measurements at 578, 560, and 540 nm. Assuming that the proportion of oxyhaemoglobin (HbO_2) in the circulating blood and the tissue is approximately the same, it follows that the fraction of blood w in the tissue is

$$w = \frac{\Delta A_{HbO_2} \times F_1 \times d_1}{\Delta A'_{HbO_2} \times F_2 \times d_2} \quad (g/g)$$

where

ΔA_{HbO_2} is absorbance difference for tissue extract
$\Delta A'_{HbO_2}$ is absorbance difference for blood dilution
F_1 and F_2 are dilution factors
d_1 and d_2 are light paths of the cuvettes

ΔA_{HbO_2} and $\Delta A'_{HbO_2}$ are calculated [4] from the absorbance measurements at 578, 560, and 540 nm, according to the formula:

$$\Delta A_{HbO_2} \quad \text{or} \quad \Delta A'_{HbO_2} = (A_{578} - A_{560}) + [(A_{540} - A_{578}) \times 0.47].$$

If the metabolite concentration in the tissue sample is to be referred to the true volume of cellular fluid in order to give the physiological concentration, the result (in mmol/l of tissue extract) is multiplied by the following factor (*G. Michal,* unpublished):

$$F = \frac{V_{\text{after neutralization}}}{V_{\text{before neutralization}}} \times \frac{\dfrac{m_{\text{tissue}}}{\text{spec. gr.}} \times \varphi + V_d}{\dfrac{m_{\text{tissue}}}{\text{spec. gr.}} \times \varphi}$$

φ is the volume fraction of liquid in the tissue volume; m_{tissue} is tissue wet weight; specific gravity relates to the sample. The latter quantity may be taken as unity in most cases. V_d is volume of reagent solution (e.g. $HClO_4$) used for deproteinization.

A similar calculation for erythrocytes has been published by *Bürgi* [5] (cf. Vol. II, chapter 1.1.1.3, p. 13).

References

[1] International Union of Pure and Applied Chemistry (IUPAC) and International Federation of Clinical Chemistry (IFCC): Approved Recommendation (1978), Quantities and Units in Clinical Chemistry, Clin. Chim. Acta *96*, 157 F – 183 F (1979).

[2] Expert Panel on Nomenclature and Principles of Quality Control in Clinical Chemistry; Committee on Standards (IFCC): Approved Recommendation (1978) on Quality Control in Clinical Chemistry, Part 1: General Principles and Terminology, J. Clin. Chem. Clin. Biochem. *18*, 69 – 77 (1980).

[3] International Commission on Radiation Units and Measurements: ICRU Report 33 "Radiation Quantities and Units", 1980, 7910 Woodmont Ave., Washington D.C. 20014, USA.

[4] *H. J. Hohorst, F. H. Kreutz, Th. Bücher,* Über Metabolitgehalte und Metabolit-Konzentrationen in der Leber der Ratte, Biochem. Z. *332*, 18 – 46 (1950).

[5] *W. Bürgi,* The Volume Displacement Effect in Quantitative Analysis of Red Blood Cell Constituents, J. Clin. Chem. Clin. Biochem. *7*, 458 – 460 (1969).

4 Absorption Coefficients of NAD(P)H

The absorption curves and absorption coefficients depend on the temperature, pH, and the ionic strength of the solution. The best-investigated cases [1 – 4] are NADH and NADPH. At λ = 334 nm the temperature dependence of ε is approximately zero. The value of ε falls with rising temperature at wavelengths $\lambda > 334$ nm (including the maximum of the absorption curve) and increases at $\lambda < 334$ nm; the absorption maximum is shifted accordingly.

Table 1. Molar decadic absorption coefficients ($l \times mol^{-1} \times mm^{-1}$) for β-NADH and β-NADPH (measured in triethanolamine/HCl buffer, 0.1 mol/l; pH = 7.6) [1].

	°C	Hg 334 nm	339 nm	340 nm	339.85 nm^{+}	Hg 365 nm
β-NADH	25	6.176×10^2	no measurement	$6.317 \times 10^{2++}$	6.292×10^2	3.441×10^2
		6.182×10^2*	cf. Table 2	–	6.298×10^2*	3.444×10^2*
β-NADH	30	6.187×10^2*	–	–	–	3.427×10^2*
β-NADPH	25	6.178×10^2*	–	–	–	3.532×10^2*
β-NADPH	30	6.186×10^2*	–	–	–	3.515×10^2*

$^{+}$ Checking the photometer revealed 339.85 nm to be correct, instead of 340 nm.
$^{++}$ Acc. to [2] in Tris buffer, 0.1 mol/l; pH 7.8.
* Values were not corrected for beam convergence or intrinsic absorbance of the oxidized coenzyme.

The above mentioned investigations [1 – 4] on NAD(P)H show:

- ε is different for NADH and NADPH,
- ε is temperature-dependent,
- ε depends on the pH and the ionic strength of the solution to be measured,
- ε cannot be determined sufficiently accurately at 37 °C because of the instability of the coenzyme,
- the absorption maximum of NAD(P)H is not located at exactly 340 nm; 339 nm can be taken as a first approximation,
- the absorption maximum is temperature-dependent,
- the differences in the value of ε due to the factors mentioned above are smallest at Hg 334 nm.

All of the influences mentioned above lead to deviations of the value of ε, not exceeding 0.5% at Hg 334 nm. Consequently, it is best, since practically independent of the conditions of measurement, to perform measurements at this wavelength. Moreover, the values of ε are identical here for both coenzymes.

However, the values of ε at 25 °C and 30 °C at 340 nm (or 339 nm) and Hg 365 nm are also sufficiently close for practical purposes (< 1% error at 340 (339) nm; about 2% at Hg 365 nm) and are independent of the other conditions of measurement. For measurement at Hg 365 nm, however, one must distinguish between the values of ε for NADH and NADPH.

In a routine laboratory it would not be practical to use the exact values of the absorption coefficients given above. It would then be necessary, under certain circumstances, to determine the exact value of ε for each experiment. The figures recommended for practical purposes in the routine laboratory vary, according to temperature and other measurement conditions, by less than 1 to 2%, which is within the limits of error attainable for routine enzymatic analyses.

Molar decadic absorption coefficients ($l \times mol^{-1} \times mm^{-1}$) for NADH and NADPH at temperatures of 25 °C and 30 °C [3] are for practical use:

	Hg 334 nm (334.15 nm)	*340 nm (339 nm)*	*Hg 365 nm (365.3 nm)*
NADH	6.18×10^2	6.3×10^2	3.4×10^2
NADPH	6.18×10^2	6.3×10^2	3.5×10^2

References

[1] *J. Ziegenhorn, M. Senn, T. Bücher,* Molar Absorptivities of β-NADH and β-NADPH, Clin. Chem. *22*, 151 – 160 (1976).

[2] *R. B. McComb, L. W. Bond, R. W. Burnett, R. C. Keech, G. N. Bowers, jr.,* Determination of the Molar Absorptivity of NADH, Clin. Chem. *22*, 141 – 150 (1976).

[3] *H. U. Bergmeyer,* Neue Werte für die molaren Extinktions-Koeffizienten von NADH und NADPH zum Gebrauch im Routine-Laboratorium, J. Clin. Chem. Clin. Biochem. *13*, 507 – 508 (1975).

[4] *Th. Bücher, G. Lüsch, H. Krell* in: *G. Anido, E. J. van Kampen, S. B. Rosalki, M. Rubin* (eds.), Temperature Dependence of Difference-Absorption Coefficients of NADH Minus NAD^+ and NADPH Minus $NADP^+$ in the Near Ultraviolet, Quality Control in Clinical Chemistry, Walter de Gruyter, Berlin, New York 1975, pp. 301 – 310.

5 Numbering and Classification of Enzymes

Extract of Enzyme Nomenclature*

1. Oxidoreductases

1.1 Acting on the CH-OH group of donors

1.1.1 With NAD^+ or $NADP^+$ as acceptor
1.1.2 With a cytochrome as acceptor
1.1.3 With oxygen as acceptor
1.1.99 With other acceptors

1.2 Acting on the aldehyde or oxo group of donors

1.2.1 With NAD^+ or $NADP^+$ as acceptor
1.2.2 With a cytochrome as acceptor
1.2.3 With oxygen as acceptor
1.2.4 With a disulphide compound as acceptor
1.2.7 With an iron-sulphur protein as acceptor
1.2.99 With other acceptors

1.3 Acting on the CH-CH group of donors

1.3.1 With NAD^+ or $NADP^+$ as acceptor
1.3.2 With a cytochrome as acceptor
1.3.3 With oxygen as acceptor
1.3.7 With an iron-sulphur protein as acceptor
1.3.99 With other acceptors

1.4 Acting on the CH-NH_2 group of donors

1.4.1 With NAD^+ or $NADP^+$ as acceptor
1.4.2 With a cytochrome as acceptor
1.4.3 With oxygen as acceptor
1.4.4 With a disulphide compound as acceptor
1.4.7 With an iron-sulphur protein as acceptor
1.4.99 With other acceptors

1.5 Acting on the CH-NH group of donors

1.5.1 With NAD^+ or $NADP^+$ as acceptor
1.5.3 With oxygen as acceptor
1.5.99 With other acceptors

---

* Enzyme Nomenclature, Recommendations (1978) of the Nomenclature Committee of the International Union of Biochemistry, Published for the International Union of Biochemistry by Academic Press, Inc., New York, United Kingdom Edition Academic Press, Inc., London (Ltd.) (1979), pp. 19–26.

1.6 Acting on NADH or NADPH

1.6.1 With NAD^+ or $NADP^+$ as acceptor
1.6.2 With a cytochrome as acceptor
1.6.4 With a disulphide compound as acceptor
1.6.5 With a quinone or related compound as acceptor
1.6.6 With a nitrogenous group as acceptor
1.6.7 With an iron-sulphur protein as acceptor
1.6.99 With other acceptors

1.7 Acting on other nitrogenous compounds as donors

1.7.2 With a cytochrome as acceptor
1.7.3 With oxygen as acceptor
1.7.7 With an iron-sulphur protein as acceptor
1.7.99 With other acceptors

1.8 Acting on a sulphur group of donors

1.8.1 With NAD^+ or $NADP^+$ as acceptor
1.8.2 With a cytochrome as acceptor
1.8.3 With oxygen as acceptor
1.8.4 With a disulphide compound as acceptor
1.8.5 With a quinone or related compound as acceptor
1.8.7 With an iron-sulphur protein as acceptor
1.8.99 With other acceptors

1.9 Acting on a haem group of donors

1.9.3 With oxygen as acceptor
1.9.6 With a nitrogenous group as acceptor
1.9.99 With other acceptors

1.10 Acting on diphenols and related substances as donors

1.10.1 With NAD^+ or $NADP^+$ as acceptor
1.10.2 With a cytochrome as acceptor
1.10.3 With oxygen as acceptor

1.11 Acting on hydrogen peroxide as acceptor

1.12 Acting on hydrogen as donor

1.12.1 With NAD^+ or $NADP^+$ as acceptor
1.12.2 With a cytochrome as acceptor
1.12.7 With an iron-sulphur protein as acceptor

1.13 Acting on single donors with incorporation of molecular oxygen (oxygenases)

1.13.11 With incorporation of two atoms of oxygen
1.13.12 With incorporation of one atom of oxygen (internal monooxygenases or internal mixed function oxidases)
1.13.99 Miscellaneous (requires further characterization)

1.14 Acting on paired donors with incorporation of molecular oxygen

1.14.11 With 2-oxoglutarate as one donor, and incorporation of one atom each of oxygen into both donors
1.14.12 With NADH or NADPH as one donor, and incorporation of two atoms of oxygen into one donor
1.14.13 With NADH or NADPH as one donor, and incorporation of one atom of oxygen
1.14.14 With reduced flavin or flavoprotein as one donor, and incorporation of one atom of oxygen
1.14.15 With a reduced iron-sulphur protein as one donor, and incorporation of one atom of oxygen
1.14.16 With reduced pteridine as one donor, and incorporation of one atom of oxygen
1.14.17 With ascorbate as one donor, and incorporation of one atom of oxygen
1.14.18 With another compound as one donor, and incorporation of one atom of oxygen
1.14.99 Miscellaneous (requires further characterization)

1.15 Acting on superoxide radicals as acceptor

1.16 Oxidizing metal ions

1.16.3 With oxygen as acceptor

1.17 Acting on -CH_2 groups

1.17.1 With NAD^+ or $NADP^+$ as acceptor
1.17.4 With a disulphide compound as acceptor

1.97 Other oxidoreductases

2. Transferases

2.1 Transferring one-carbon groups

2.1.1 Methyltransferases
2.1.2 Hydroxymethyl-, formyl- and related transferases
2.1.3 Carboxyl- and carbamoyltransferases
2.1.4 Amidinotransferases

2.2 Transferring aldehyde or ketonic residues

2.3 Acyltransferases

2.3.1 Acyltransferases
2.3.2 Aminoacyltransferases

2.4 Glycosyltransferases

2.4.1 Hexosyltransferases
2.4.2 Pentosyltransferases
2.4.99 Transferring other glycosyl groups

2.5 Transferring alkyl or aryl groups, other than methyl groups

2.6 Transferring nitrogenous groups

2.6.1 Aminotransferases
2.6.3 Oximinotransferases

2.7 Transferring phosphorus-containing groups

2.7.1 Phosphotransferases with an alcohol group as acceptor
2.7.2 Phosphotransferases with a carboxyl group as acceptor
2.7.3 Phosphotransferases with a nitrogenous group as acceptor
2.7.4 Phosphotransferases with a phosphate group as acceptor
2.7.5 Phosphotransferases with regeneration of donors (apparently catalysing intramolecular transfers)
2.7.6 Diphosphotransferases
2.7.7 Nucleotidyltransferases
2.7.8 Transferases for other substituted phosphate groups
2.7.9 Phosphotransferases with paired acceptors

2.8 Transferring sulphur-containing groups

2.8.1 Sulphurtransferases
2.8.2 Sulphotransferases
2.8.3 CoA-transferases

3. Hydrolases

3.1 Acting on ester bonds

3.1.1 Carboxylic ester hydrolases
3.1.2 Thiolester hydrolases
3.1.3 Phosphoric monoester hydrolases
3.1.4 Phosphoric diester hydrolases

3.1.5 Triphosphoric monoester hydrolases
3.1.6 Sulphuric ester hydrolases
3.1.7 Diphosphoric monoester hydrolases
3.1.11 Exodeoxyribonucleases producing 5′-phosphomonoesters
3.1.13 Exoribonucleases producing 5′-phosphomonoesters
3.1.14 Exoribonucleases producing other than 5′-phosphomonoesters
3.1.15 Exonucleases active with either ribo- or deoxyribonucleic acids and producing 5′-phosphomonoesters
3.1.16 Exonucleases active with either ribo- or deoxyribonucleic acids and producing other than 5′-phosphomonoesters
3.1.21 Endodeoxyribonucleases producing 5′-phosphomonoesters
3.1.22 Endodeoxyribonucleases producing other than 5′-phosphomonoesters
3.1.23 Site-specific endodeoxyribonucleases: cleavage is sequence-specific
3.1.24 Site-specific endodeoxyribonucleases: cleavage is not sequence-specific
3.1.25 Site-specific endodeoxyribonucleases: specific for altered bases
3.1.26 Endoribonucleases producing 5′-phosphomonoesters
3.1.27 Endoribonucleases producing other than 5′-phosphomonoesters
3.1.30 Endonucleases active with either ribo- or deoxyribonucleic acids and producing 5′-phosphomonoesters
3.1.31 Endonucleases active with either ribo- or deoxyribonucleic acids and producing other than 5′-phosphomonoesters

3.2 Acting on glycosyl compounds

3.2.1 Hydrolysing *O*-glycosyl compounds
3.2.2 Hydrolysing *N*-glycosyl compounds
3.2.3 Hydrolysing *S*-glycosyl compounds

3.3 Acting on ether bonds

3.3.1 Thiolether hydrolases
3.3.2 Ether hydrolases

3.4 Acting on peptide bonds (peptide hydrolases)

3.4.11 α-Aminoacylpeptide hydrolases
3.4.12 Peptidylamino-acid hydrolases or acylamino-acid hydrolases
3.4.13 Dipeptide hydrolases
3.4.14 Dipeptidylpeptide hydrolases
3.4.15 Peptidyldipeptide hydrolases
3.4.16 Serine carboxypeptidases
3.4.17 Metallo-carboxypeptidases
3.4.21 Serine proteinases
3.4.22 Thiol proteinases
3.4.23 Carboxyl (acid) proteinases
3.4.24 Metalloproteinases
3.4.99 Proteinases of unknown catalytic machanism

3.5 Actin on carbon-nitrogen bonds, other than peptide bonds

3.5.1 In linear amides
3.5.2 In cyclic amides
3.5.3 In linear amidines
3.5.4 In cyclic amidines
3.5.5 In nitriles
3.5.99 In other compounds

3.6 Acting on acid anhydrides

3.6.1 In phosphoryl-containing anhydrides
3.6.2 In sulphonyl-containing anhydrides

3.7 Acting on carbon-carbon bonds

3.7.1 In ketonic substances

3.8 Acting on halide bonds

3.8.1 In C-halide compounds
3.8.2 In P-halide compounds

3.9 Acting on phosphorus-nitrogen bonds

3.10 Acting on sulphur-nitrogen bonds

3.11 Acting on carbon-phosphorus bonds

4. Lyases

4.1 Carbon-carbon lyases

4.1.1 Carboxyl-lyases
4.1.2 Aldehyde-lyases
4.1.3 Oxo-acid-lyases
4.1.99 Other carbon-carbon lyases

4.2 Carbon-oxygen lyases

4.2.1 Hydro-lyases
4.2.2 Acting on polysaccharides
4.2.99 Other carbon-oxygen lyases

4.3 Carbon-nitrogen lyases

4.3.1 Ammonia-lyases
4.3.2 Amidine-lyases

4.4 Carbon-sulphur lyases

4.5 Carbon-halide lyases

4.6 Phosphorus-oxygen lyases

4.99 Other lyases

5. Isomerases

5.1 Racemases and epimerases

5.1.1 Acting on amino acids and derivatives
5.1.2 Acting on hydroxy acids and derivatives
5.1.3 Acting on carbohydrates and derivatives
5.1.99 Acting on other compounds

5.2 Cis-trans isomerases

5.3 Intramolecular oxidoreductases

5.3.1 Interconverting aldolases and ketoses
5.3.2 Interconverting keto- and enol-groups
5.3.3 Transposing C=C bonds
5.3.4 Transposing S–S bonds
5.3.99 Other intramolecular oxidoreductases

5.4 Intramolecular transferases

5.4.1 Transferring acyl groups
5.4.2 Transferring phosphoryl groups
5.4.3 Transferring amino groups
5.4.99 Transferring other groups

5.5 Intramolecular lyases

5.99 Other isomerases

6. Ligases (synthetases)

6.1 Forming carbon-oxygen bonds

6.1.1 Ligases forming aminoacyl-tRNA and related compounds

6.2 Forming carbon-sulphur bonds

6.2.1 Acid-thiol ligases

6.3 Forming carbon-nitrogen bonds

6.3.1 Acid-ammonia (or amine) ligases (amide synthetases)
6.3.2 Acid-amino-acid ligases (peptide synthetases)
6.3.3 Cyclo-ligases
6.3.4 Other carbon-nitrogen ligases
6.3.5 Carbon-nitrogen ligases with glutamine as amido-*N*-donor

6.4 Forming carbon-carbon bonds

6.5 Forming phosphate ester bonds

Index